Natural Resource Management and Policy
Volume 62

There is a growing awareness to the role that natural resources, such as water, land, forests and environmental amenities, play in our lives. There are many competing uses for natural resources, and society is challenged to manage them for improving social well-being. Furthermore, there may be dire consequences to natural resources mismanagement. Renewable resources, such as water, land and the environment are linked, and decisions made with regard to one may affect the others. Policy and management of natural resources now require interdisciplinary approaches including natural and social sciences to correctly address our society preferences.This series provides a collection of works containing most recent findings on economics, management and policy of renewable biological resources, such as water, land, crop protection, sustainable agriculture, technology, and environmental health. It incorporates modern thinking and techniques of economics and management. Books in this series will incorporate knowledge and models of natural phenomena with economics and managerial decision frameworks to assess alternative options for managing natural resources and environment.

João Carlos Correia Leitão • Wael Khairy
Dina Pereira • Jelena Zascerinska
Pedro Dinis Gaspar
Editors

Sustainable Water Management and Irrigation Systems

Tackling Climate Change

Springer

Editors
João Carlos Correia Leitão
Department of Management and Economics
University of Beira Interior
Covilhã, Portugal

Wael Khairy
Faculty of Engineering
Heliopolis University
Cairo, Egypt

Dina Pereira
CEG-IST
University of Lisbon
Lisboa, Portugal

Jelena Zascerinska
University of Wismar
Wismar, Mecklenburg-Vorpommern, Germany

Pedro Dinis Gaspar
Dept of Electromechanical Engg
University of Beira Interior
COVILHÃ, Portugal

ISSN 0929-127X ISSN 2511-8560 (electronic)
Natural Resource Management and Policy
ISBN 978-3-031-93586-2 ISBN 978-3-031-93587-9 (eBook)
https://doi.org/10.1007/978-3-031-93587-9

This Springer imprint is published by the registered company Springer Nature Switzerland AG.
The registered company address is: Gewerbestrasse 11, 6330 Cham, Switzerland

Acknowledgments

Within the project's scope, the editors' team acknowledges the following support: MEDiterranean irrigation and Water supply for smallholder farmers by providing Efficient, low-cost and nature-based Technologies and practices (MED-WET). Project ID: 1646, reference PRIMA/0008/2020 (https://doi.org/10.54499/PRIMA/0008/2020).

The first editor also acknowledges the support of The Portuguese Foundation for Science and Technology (Grants and NECE—UIDB/04,630/2020).

Additional details about the MED-WET project can be found here: https://www.medwet-eu.com/.

Contents

Technological and Climate Change for Sustainable Water Management . 1
João Carlos Correia Leitão, Wael Khairy, Dina Pereira, Jelena Zascerinska, and Pedro Dinis Gaspar

Emergent Challenges and Technologies 11

Sustainable Water Management Based on Micro-Irrigation Ecosystems: Research Framework and Policy Action Agenda 13
João Carlos Correia Leitão, Dina Pereira, and Ângelo Novo

Water Security in Central Asia: Towards Sustainable Development and Resilience 47
Iskandar Abdullaev and Shavkat Rakhmatullaev

Challenges and Trends in Water System Management 71
Hermilio Vilarinho and Ana S. Camanho

Integrated Aquaculture-Agriculture System: A Tool for Sustainable and Optimum Water Management 91
Magdy A. Salah El Deen and Wael Khairy

Integrated Solar Power and Rainwater Harvesting System 117
Yuvraj Rathore, Tapash Nayak, Ramesh N, Shalabh Kumar Mishra, and Nawnit Kumar

Case Studies 127

Exploring Sustainable Practices of Micro-irrigation in Farming: A Case Study of SLECI in Morocco 129
Jan Siering, Bassou Bouazzama, Sabri Ahmed, Nadya Wahid, Hicham El Yousfi, Jelena Zascerinska, and Regina Krause

Impact of Climate Change on Sustainable Water Management: A Study on Urban Bangladesh 167
A. K. M. Mahmudul Haque, Imran Hossain, Md. Sohel Rana, Md. Asfaq Salehin, and Md. Shamim Hossen

Sustainable Water Management Practices Using Micro-irrigation Technology: A Case Study in Portugal and Future Research Directions . 187
Artur Frazão, Pedro Dinis Gaspar, Cristina Fael, Isabel Falorca, Dina Pereira, and João Carlos Correia Leitão

The Impact of the Irrigation Project on the Health Status of Local Communities: An Analysis of the Teesta Barrage Irrigation Project in Bangladesh 231
Md. Abu Shahen and Md. Nurul Momen

Inventories of Existing Financial Support Mechanisms and Identifying Financing Tools Promoting Irrigation Efficiency for Smallholder Farmers in the Mediterranean Region 251
Nadya Wahid, Joseph Piscopo, Reno Micallef, Wael Khairy, Mariana Salvado, and Regina Krause

Analysis of the Performance and Productivity of "Regadio" Irrigation Systems and Water Management in the Cova da Beira Region, Portugal . 271
Dina Pereira and João Carlos Correia Leitão

Capital Investment Planning of Infrastructure Assets in the Water Sector . . 303
Pedro Aguiar Costa, Jaime Gabriel Silva, Ligia B. Ramos, Nuno Marques de Almeida, and Ana S. Camanho

Index 333

About the Editors

João Carlos Correia Leitão is an Associate Professor with habilitation in Technological Change and Entrepreneurship, and the Director of the UBIExecutive, Business School at the University of Beira Interior (UBI), Portugal. He further is a research fellow at the NECE Research Center for Business Sciences at UBI, and an Associate Researcher at the Instituto de Ciências Sociais (ICS), University of Lisbon, Portugal. Leitão is an invited scholar at the Università Degli Studi di Bergamo, Italy. He is a prolific author and a series editor of the Springer book series Studies in Entrepreneurship, Structural Change, and Industrial Dynamics. Leitão is also an editorial board member of several international journals.

Pedro Dinis Gaspar is an Associate Professor with habilitation in Mechanical Engineering, and leads the Department of Electromechanical Engineering at the University of Beira Interior, Portugal. He is a researcher at the Centre for Mechanics and Aerospace Science and Technologies (CMAST), coordinator of Laboratory of Innovation and Technologies for Sustainability (LITecS), and collaborates with the collaborative laboratory Food4Sustainability. His work focuses on energy efficiency, sustainability, and automation/robotics in agriculture, contributing to numerous international and national R&D projects. He has authored over 180 publications in books, book chapters, international journals, and conference proceedings and co-authored six patents. He has edited several international journals and books.

Wael Khairy is a Senior Academician and Researcher Engineer. He has a blend of academic, managerial, strategic, technical, communication, and operational experiences developed over a career spanning about 30 years. He has conducted extensive academic and research work focused on water science, such as surface water hydrology, river engineering, water resources planning, management, and development, and watershed management using numerical modeling, earth observation, and statistical hydrology. Dr. Khairy contributed positively and constructively to community development and environmental sustainability. He has published various technical papers and book chapters in prestigious and renowned journals in the areas of integrated watershed modeling, drainage water reuse, wastewater treatment technologies, monitoring and evaluation, water quality assessment, pollution control, climate change, and environmental management.

Dina Pereira is the Executive Director of UBImedical, Incubator of University of Beira Interior (UBI), Portugal, and Researcher at CEGIST—Centro de Estudos de Gestão | CEGIST—Centre for Management Studies, Instituto Superior Técnico—Universidade de Lisboa, in the fields of innovation and entrepreneurship. She is COO and Co-Founder of NeuroSoV, a startup that developed a drug candidate to slow down the progression of Parkinson disease and is further working on other innovative therapies for neurodegenerative diseases and dementias.

Jelena Zascerinska is a Scientific Assistant and experienced Project Manager at Hochschule Wismar, Germany. Also, Dr. Zascerinska is a leading researcher at the Centre for Education and Innovation Research, Latvia. She is an expert of EU programs (Horizon 2020, Horizon Europe, COST, EUREKA, Eurostars, etc.) and National Research Agencies (Latvia, Poland, Slovakia). She has authored and co-authored about 300 scientific publications. She is active in peer reviewing in several research journals. Dr. Zascerinska is Latvia Top 100 Scientists and Latvia Top 3 Scientists in Education.

Technological and Climate Change for Sustainable Water Management

João Carlos Correia Leitão, Wael Khairy, Dina Pereira, Jelena Zascerinska, and Pedro Dinis Gaspar

Abstract This chapter presents the main challenges arising from climate change, calling for the rapid introduction and spread of low-cost, nature-based technological innovations, in order to mitigate the effects of that change, contributing to greater water efficiency and savings, in the context of irrigation and micro-irrigation systems. It presents the main missions and trends in the new water economy and explains the sequencing in structuring this work, organized in two main parts, namely, Part I. Sustainable Water Management: State of the Art, Challenges, and Landscapes (5 contributions) and Part II. Smart Irrigation Technologies and Climate Challenges: Benchmarks and Case Studies (8 contributions).

Keywords Agriculture · Climate change · Sustainable development · Water economy · Technology

Climate change and technology are greatly transforming how we manage water resources, making sustainable management of water one of the greatest challenges of the twenty-first century. This work is pioneering in exploring how forms of sustainable water management and new nature-based technology can mitigate the

---

J. C. Correia Leitão (✉)
Department of Management and Economics, University of Beira Interior, Covilhã, Portugal
e-mail: jleitao@ubi.pt

W. Khairy
Faculty of Engineering, Heliopolis University, Cairo, Egypt

D. Pereira
CEG-IST, University of Lisbon, Lisboa, Portugal

J. Zascerinska
University of Wismar, Wismar, Mecklenburg-Vorpommern, Germany

P. D. Gaspar
Dept of Electromechanical Engg, University of Beira Interior, COVILHÃ, Portugal

J. C. Correia Leitão et al. (eds.), *Sustainable Water Management and Irrigation Systems*, Natural Resource Management and Policy 62,
https://doi.org/10.1007/978-3-031-93587-9_1

effects of climate change, to face up better to the increasing shortage of water and ensure its availability for future generations, in the context of a global water crisis.

Climate change is having significant impacts on precipitation patterns, the availability of drinking water, and the frequency of extreme weather events. These impacts mean unprecedented challenges for sustainable water management and for the planet's sustainable balance.

Significant alterations in precipitation patterns are observed worldwide. Some regions are experiencing prolonged periods of drought, while others face more intense and frequent rain. The latter cannot be taken advantage of due to degradation of the soil and the lack of technological hardware and software, which would allow storage and eco-efficient reuse of a scarce natural resource. These changes have a direct effect on both the availability of freshwater and the capacity to replenish aquifers.

Global warming has accelerated the melting of glaciers everywhere. Although this can temporarily increase the flow of water in some regions, in the long term it represents a significant threat to the water supply in many areas that depend on the seasonal thaw.

Also connected to global warming, an increase in both the frequency and intensity of extreme weather events has been observed, for example, severe droughts and flooding. These events have devastating impacts in relation to water resources, affecting the quantity and quality of water available, which explains the emergence of the abovementioned global water crisis.

According to Mazucatto and Mazzucato and Zaqout (2024), the global water crisis requires a new approach to water governance. The authors advocate two critical aspects to address, namely:

1. Several regions have too much, too little, or too dirty water, thus affecting billions of people worldwide.
2. There is a need for a holistic and global response, for three main reasons:
    (a) The hydrological cycle is globally interconnected, and human activities are destabilizing it.
    (b) Water cuts across all economic sectors and Sustainable Development Goals (SDG), requiring an economy-wide approach.
    (c) Addressing the crisis requires collective action, as no single country or region can solve it alone.

The same authors criticize the current economic approaches to water governance, which are based on conceptualizing water as an economic good (public, private, club, or common-pool resource). In addition, they recommend starting to approach water through its three main functions: public health service, production resource, and ecosystem service.

Mazucatto and Mazzucato and Zaqout (2024) also emphasize the need for a more ambitious role for the state in water governance, moving beyond the market-fixing approach suggested by traditional economic theory and suggesting a new paradigm for conceptualizing and governing water resources to address the global water crisis effectively.

Following the GCEW's vision (2024), to rethink how we value water and manage the water cycle as a global common good, we require a new water economics. Its fundamental idea lies in understanding the relationship between social justice, environmental sustainability, and economic efficiency.

To alleviate the water shortage, the GCEW proposes the exploration of five mission areas, namely:

- Mission 1: Start a new food systems revolution.
- Mission 2: Preserve and repair natural habitats that are essential to safeguarding green water.
- Mission 3: Create a water economy that is circular.
- Mission 4: Facilitate a highly AI-enabled, clean-energy era with much reduced water intensity.
- Mission 5: By 2030, make sure that no child dies from drinking contaminated water.

Given the urgent challenges and missions presented above, this work intends to contribute to this debate, arguing first of all that technological innovations should have a crucial role not only in the search for sustainable solutions to water management but also in providing low-cost, usable tools. These can both increase efficiency and mitigate the impacts of climate change, through forecasting and anticipation, giving a central, participatory role to the users of technology in general and water technology in particular, with effective state regulation.

One of the principal innovations in sustainable water management, worthy of special note in the applications presented in this work, is the monitoring, in real time, of all activities in the water cycle. Applying advanced sensors and wireless sensor networks, providing alternative accesses to broadband Internet, and eliminating undesirable situations of lack of coverage, above all in areas of low population density, allow uninterrupted monitoring of the quality and flow of water. This technology allows the rapid detection of pollution, identifies leaks, and forecasts extreme weather events that can affect water resources.

The priority technology for sustainable water management covers a wide range of innovative solutions aiming to optimize the use, treatment, and preservation of water resources. This technology is transforming how we manage water, making it more efficient, sustainable, and resilient in the face of climate and environmental challenges. Below, we deal summarily with the areas of applying that technology considered as priorities for sustainable water management, namely: (1) monitoring and advanced analysis, (2) water treatment and reuse, (3) management and intelligent distribution of water, (4) nature-based solutions, and (5) advanced disinfection of water.

First of all, concerning monitoring and advanced analysis, including forecasting, simulation, and creating scenarios, it is worth highlighting the advances registered regarding the spread of the Internet of Things (IoT) and smart sensors. These devices allow monitoring, in real time, of water quality, consumption, and detection of leaks in the whole water infrastructure, facilitating rapid responses to problems and optimizing water distribution.

In addition, artificial intelligence (AI) and machine learning are used to forecast the demand for water, optimize treatment, and identify consumption patterns. This technology can significantly improve the operational efficiency of water management systems.

Secondly, regarding the treatment and reuse of water, special mention should be made of energy-efficient treatment systems. These systems have advanced considerably, with the twin aim of reducing energy consumption and minimizing waste production. Furthermore, desalinization technology has followed an evolutionary path, becoming more affordable, efficient, and a viable, more widely used solution for regions with a shortage of fresh water.

Added to the list of priority technology are nanotechnology and advanced membranes, with the use of graphene membranes in reverse osmosis representing a significant innovation. These membranes provide more efficient and less costly filtering in terms of energy, increasing the viability of desalination and the treatment of salt water.

Thirdly, concerning management and intelligent distribution of water, blockchain technology and decentralized management systems are developing technology that cannot be ignored.

At present, blockchain technology is used to track the use of water and ensure transparency in its management. In turn, decentralized management systems allow more efficient distribution of water resources.

Another fundamental aspect to go deeper into concerns renewable energy in production and distribution, as the integration of renewable energy, such as solar, wind, and geothermal, in systems for treating and pumping water can mean a substantial reduction in the agricultural sector's carbon footprint.

Fourthly, for nature-based solutions, the vision must be more wide-ranging and come to include, for example, projects to recuperate urban rivers, as in major cities there are obvious problems in revitalizing rivers or streams. This requires new investment that contributes to improving water quality, creating pleasant spaces for public enjoyment, and raising the quality of life for the resident population and visitors.

Another exploratory path that should be intensified concerns rainwater solutions. This implies capturing and storing it, in private housing but also in public buildings, parks, sports facilities, old people's homes, hotels, firms, and agriculture, making an effective contribution to conserving water resources.

Fifthly, concerning technology for advanced water disinfection, the growing use of ozone, and ultraviolet irradiation systems is highlighted. This has been used to destroy pathogens, without adding chemical products, thereby avoiding the formation of harmful sub-products with unquestionable implications for public health and people's quality of life.

Agriculture is one of the production activities that consumes most water, with the urgent need to join up intelligent irrigation systems that use ground humidity sensors and meteorological data in real time, to help farmer to optimize water use, reducing waste and increasing efficiency. But also the testing, prototyping, and provision of low-cost, micro-irrigation technology, as occurred with the SLECI technology developed and patented in Germany, which was tested successfully, in the scope of the PRIMA MED-WET[1] Project, in countries with different soils, such as Portugal, Morocco, and Malta, in different productions including cherries, peaches, olives, oranges, tangerines, lemons, and grapes, giving water savings above 86% (Pereira et al., 2023).

To face effectively the challenges imposed by climate change, it is also necessary to adopt a holistic, integrated approach to water management. This implies adopting integrated management practices for water resources, to promote, in coordination, the development and management of water, land, and related resources. This ecosystemic approach aims to maximize economic and social well-being, without jeopardizing the sustainability of vital ecosystems.

From an economic perspective of an innovation cycle, one path of exploration to ensure the sustainability of these vital ecosystems lies in the introduction, adoption, and subsequent spread of innovations linked to the concept of the circular economy applied to water management, promoting the reuse and recycling of water in various sectors, whenever possible through nature-based approaches. This type of approach not only reduces the pressure on natural and water resources but also minimizes waste and carbon dioxide emissions.

Pioneering initiatives linked to forming wetland solutions, as it was possible to prototype and test in the Egyptian desert, also in the context of the PRIMA MED-WET Project referred to above, i.e., nature-based solutions, including the recovery of wetlands and the creation of urban green infrastructure, are gaining prominence as effective ways to improve water management and reuse wastewater. These solutions not only help to regulate the water flow but also improve the water quality and increase the resilience to climate change, using ancestral knowledge combined with new technology, including that connected to renewable energy production.

In regions where the shortage of freshwater is a chronic problem, desalinization is emerging as a viable solution. New, more efficient, and sustainable desalinization techniques, in terms of energy, are making this process more affordable and environmentally friendly. This work describes the pioneering experience on the islands of Malta and Gozo, which are now advancing towards prototyping an eco-efficient solar house, following on from the PRIMA MED-WET Project, making it possible to use desalinized water to develop hydroponics, using that same water with compounds, meaning production capacity can increase in places where agriculture is difficult.

It should be highlighted that besides technological and management solutions, education and greater awareness play a fundamental role in promoting sustainable water management.

Here, it is essential to design and implement environmental education programs focused on the importance of water conservation and sustainable practices when using water, in order to promote long-term changes in behavior.

---

[1] For more information about the project: *Improving MEDiterranean irrigation and Water supply for smallholder farmers by providing Efficient, low-cost and nature-based Technologies and practices* (MED- WET) Project ID: 1646, reference PRIMA/0008/2020 (https://doi.org/10.54499/PRIMA/0008/2020) consult: https://www.medwet-eu.com/

Besides environmental education initiatives, both formal and informal, public awareness campaigns can help to inform people about the challenges related to water, as well as encouraging conservation and reuse. Such campaigns can include simple hints to save water every day, such as closing the tap when brushing your teeth, reusing rainwater to water plants, or installing devices in houses to collect and use rainwater.

In conclusion, climate change raises unprecedented challenges for sustainable water management. Nevertheless, technological innovations and integrated management approaches provide promising opportunities to face these challenges. The combination of advanced monitoring schemes, water conservation technology, integrated management approaches, and education and awareness initiatives can help to ensure the availability and quality of water for future generations.

The technology described is at the forefront of sustainable water management, providing innovative solutions to global water challenges. The integration of that technology in intelligent management systems, combining the IoT, AI, and data analysis, has set in motion a real revolution regarding how we monitor, treat, and distribute water. By adopting these advanced solutions, states, governments, industries, and communities not only improve operational efficiency and reduce costs but also demonstrate a clear commitment to meeting the sustainable development goals (SDG) and wide-ranging action of social responsibility and innovation, namely regarding SDG3—Good Health and Well-being, to ensure healthy lives and promote well-being for all ages; SDG6—Clean Water and Sanitation, to ensure the availability and sustainable management of water and sanitation for all; and SDG 12—Responsible Production and Consumption, to ensure sustainable consumption and production patterns.

Despite the above, concerted efforts will be required by the different levels of the government, industry, universities, and community, including nongovernmental organizations and citizens. The latter should be the focus in changing towards pro-efficiency behavior regarding the natural resource of water, following a user-centric approach to innovation. Only through collective action and a change of paradigm in how we value and manage water will we be able to ensure a safe, fair, and sustainable future for all nations and individuals, irrespective of their geographical location and level of income.

Concerning structure, this work begins with this chapter devoted to the umbrella topic of Technological and Climate Change for Sustainable Water Management, by João Leitão, Wael Khairy, Dina Pereira, Jelena Zascerinska, and Pedro Dinis Gaspar. It contextualizes the central theme of the work, i.e., Sustainable Water Management, in relation to present and future challenges, technological change, and the intelligent struggle against the effects of climate change, aiming to contribute to fulfilling the Sustainable Development Goals (SDG). It also provides a guide to the structure of this volume, which is developed in two main parts.

Part I: Sustainable Water Management: State of the Art, Challenges, and Landscapes includes five scientific contributions, describing present and future scenarios, as a starting point for the set of applications and benchmarks presented afterwards.

In the second chapter, João Leitão, Dina Pereira, and Ângelo Novo carry out a systematic review of the literature on technological and climate change for sustainable water management, aiming, firstly, to determine how climate and technological changes affect sustainable water management and, secondly, to map the effects of climate change on the sustainability of irrigation systems. Applying the PRISMA research protocol, the authors take a sample of 202 articles published on Clarivate's Web of Science, from which a bibliometric analysis is produced. This serves as the basis for presenting a research framework and a policy action agenda. The main results indicate that sustainable management of water resources implies efficient use of water by adopting agricultural micro-irrigation ecosystems. Furthermore, in ecosystems with great demands for agricultural irrigation, water savings to face up to a future water shortage is not enough to ensure the future sustainability of water resources, so it is essential to improve sustainable management of storage, distribution, and micro-irrigation in order to prevent extreme scenarios of stress on water resources.

In the third chapter, Iskandar Abdullaev and Shavkat Rakhmatullaev address the relevant issue of water security in relation to Central Asia, focusing on sustainable development and resilience. The authors describe the state of the art concerning water security, positioning return flows, groundwater, and dam safety as important adaptation and mitigation strategies. They also advocate that energy efficiency is a crucial component of resilience, at the operational level of water management. A fruitful discussion ensues, pointing out effective ways to manage shared irrigation infrastructure on transboundary watercourses, which is essential for peace and prosperity, in the context of climate change and adaptive management of Central Asian water resources.

In the fourth chapter, Hermilio Vilarinho and Ana Camanho address the challenges and most recent trends in water system management. The authors highlight the importance of residual water treatment and potable water supplies for environmental sustainability and public health, and they refer to the increased efforts to enhance water supply management services. By adopting a comprehensive perspective on the complex issues of managing water sustainability and the need to enhance management techniques, they also emphasize the role that water management plays in fostering resilience and satisfying the demands of present and future generations.

In the fifth chapter, Magdy El Deen and Wael Khairy approach the integrated aquaculture-agriculture system (IAAS), as an efficient way of using water to increase productivity and reduce the risks associated with water shortages. This type of approach is especially important in the context of developing countries, when the agricultural community's ability to engage in intensive aquaculture is constrained. The authors present and discuss several case studies, exploring aquaponics as a single-loop recirculating system that combines the elements of hydroponics with a recirculating aquaculture system. With the ability to reduce the chemical and fertilizer inputs needed in conventional food production, aquaponics presents a viable method for growing vegetables and rearing fish while using less water and land.

In the sixth chapter, Yuvraj Rathore, Tapash Nayak, Amjed Dar, Shalabh Mishra, and Nawnit Kumar offer a unique perspective on the connection between rainwater collection systems and integrated solar electricity. In addition to analyzing and assessing the device's performance and implications, the authors propose an autonomous tracking solar plate that may also be utilized for water harvesting.

Part II: Smart Irrigation Technologies and Climate Challenges: Benchmarks and Case Studies presents eight contributions that besides providing various international benchmarks open the door to exploring different case studies covering smart water management systems, low-cost micro-irrigation technology, and irrigation systems in different geographical locations, namely, in Africa, Europe, and Asia and in the specific context of the Mediterranean Basin, as studied in the scope of the PRIMA: MED-WET Project.

In Chap. 7, Jan Siering, Bassou Bouazzama, Sabri Ahmed, Nadya Wahid, Hicham El Yousfi, Jelena Zascerinska, and Regina Krause examine environmentally friendly methods of using SLECI (self-regulating, low energy, clay-based irrigation) micro-irrigation technology. It is possible to draw the conclusion that SLECI is a sustainable method of micro-irrigation in agriculture based on analysis of its application in Morocco. SLECI produced favorable outcomes when seen from the social and environmental aspects of sustainability. Compared to traditional drip irrigation systems, the economic expenses of using SLECI in Morocco seemed to be higher.

In Chap. 8, Imran Hossain and A. Mahmudul Haque say that in addition to implementing technological improvements, there is a need to improve risk resilience against climate change and the management of water resources and water efficiency. Through an interdisciplinary lens, the authors also emphasize the need for coordinated efforts to lessen the consequences of climate change on Bangladesh's water resources. Additional justifications are offered for funding R&D, stakeholder involvement, and integrated strategies in order to advance sustainable water management and ensure universal access to clean water.

In Chap. 9, by Artur Frazão, Pedro Dinis Gaspar, Cristina Fael, Isabel Falorca, Dina Pereira, and João Leitão, contrasting the Self-Regulating Low Energy Clay tube Irrigation Systems (SLECIS) technology with conventional Drip Irrigation Systems (DIS) in Mediterranean climates, specifically in Portuguese cherry and peach orchards, the authors assess the sustainability and economic feasibility of a novel micro-irrigation system. The empirical evidence demonstrates that SLECIS outperforms DIS in terms of water conservation rates due to its adaptive response to soil moisture levels. The authors conclude that SLECIS might be a calculated step in the direction of sustainable agriculture, providing a workable answer to the increasing demands of water scarcity and the expanding need for agricultural products.

In Chap. 10, Md. Abu Shahen and Md. Nurul Momen provide a useful case study on the Teesta Barrage Irrigation Project (TBIP), which aimed to ensure farmers' access to irrigation water by building a canal from the Teesta barrage at Dalia Point in Bangladesh's Lalmonirhat district. The case study reveals there have been some noteworthy advances in health status, including the use of private vehicles for patient transport, improved primary healthcare for the majority of farmers, modern diagnostic and Caesarean services, and notable improvements in living and sanitary conditions.

In Chap. 11, Nadya Wahid, Joseph Piscopo, Reno Micallef, Wael Khairy, Mariana Salvado, and Regina Krause examine the current financial framework for implementing innovative, effective irrigation systems that have been researched in a few Mediterranean nations, including Morocco, Portugal, and Malta. Given the particular funding programs and the demand funding matrix of each country studied, the authors offer a number of financing mechanism solutions aimed at implementing a financial planning roadmap.

In Chap. 12, Dina Pereira and João Leitão analyze the effectiveness and output of water management strategies and "Regadio" irrigation systems in Portugal's Cova da Beira Region. The "Regadio" system is necessary to support agricultural activities and address the region's water scarcity problems. The primary findings indicate that the irrigation system's performance is positively correlated with the size of the units of exploration. The region's economic returns have a significant and favorable impact on the system's performance, according to the panel estimation. However, quantile regression does not show this effect, while the adverse effects of the environment are more pronounced.

Finally, in Chap. 13, the issue of capital investment planning of infrastructure assets in the water industry is addressed by Pedro Costa, Jaime Silva, Ligia Ramos, Nuno Marques de Almeida, and Ana Camanho. The Infrastructure Value Index (IVI), one of the indicators that is being evaluated in the Portuguese water industry, is calculated using a more practical method suggested by the authors. A revised IVI formulation is suggested, and an empirical case study using water infrastructure run by the Portuguese company Águas do Douro e Paiva (AdDP) is used to illustrate it.

References

GCEW. (2024). *The economics of water: Valuing the hydrological cycle as a global common good.* Global Commission on the Economics of Water (GCEE) OECD Environment Directorate Climate, Biodiversity and Water Division. https://economicsofwater.watercommission.org/report/economics-of-water.pdf

Mazzucato, M., & Zaqout, M. (2024). *Rethinking the global water challenge through a common good lens.* UCL Institute for Innovation and Public Purpose. Working Paper Series (IIPP WP2024–06). https://www.ucl.ac.uk/bartlett/public-purpose/publications/2024/mar/rethinking-global-water-challenge-through-common-good-lens

Pereira, D., Leitao, J. C. C., Gaspar, P. D., Fael, C., Falorca, I., Khairy, W., Wahid, N., El Yousfi, H., Bouazzama, B., Siering, J., Hansmann, H., Zascerinska, J., Camilleri, S., Busuttil, F., Borg, M., Mizzi, J., Micallef, R., & Cutajar, J. (2023). Exploring irrigation and water supply technologies for smallholder farmers in the mediterranean region. *Sustainability, 15*, 6875. https://doi.org/10.3390/su15086875

Emergent Challenges and Technologies

Sustainable Water Management Based on Micro-Irrigation Ecosystems: Research Framework and Policy Action Agenda

João Carlos Correia Leitão, Dina Pereira, and Ângelo Novo

Abstract The effects of climate change raise new challenges for the design and management of water ecosystems oriented towards sustainable irrigation, showing the need to promote innovative micro-irrigation ecosystems. This systematic literature review aims firstly to determine how climate change affects the sustainability of water management systems. Secondly, it will map the effects of climate change on the sustainability of irrigation systems. Applying the PRISMA research protocol, a sample of 202 articles published on Clarivate's Web of Science was selected. From this, a bibliometric analysis was produced, serving as the basis for presenting a research framework and policy action agenda. With the results, and given the intensifying effects of climate change, the study contributes to better understanding of how sustainable management of water resources implies efficient use of water through the adoption of agricultural micro-irrigation ecosystems. In addition, in ecosystems with a great demand for agricultural irrigation, water saving with a view to future water shortage is not enough to ensure the future sustainability of water resources. It is therefore essential to improve the sustainable management of storage, distribution, and micro-irrigation, in order to prevent extreme scenarios of water resource stress.

J. C. Correia Leitão (✉)
Department of Management and Economics, University of Beira Interior, Covilhã, Portugal
e-mail: jleitao@ubi.pt

D. Pereira
CEG-IST, University of Lisbon, Lisboa, Portugal
e-mail: dina@ubi.pt

Â. Novo
Faculty of Social and Human Sciences, Department of Management and Economics, University of Beira Interior, Covilhã, Portugal
e-mail: angelo.novo@ubi.pt

J. C. Correia Leitão et al. (eds.), *Sustainable Water Management and Irrigation Systems*, Natural Resource Management and Policy 62,
https://doi.org/10.1007/978-3-031-93587-9_2

Keywords Climate change · Irrigation · Micro-irrigation · Sustainability · Water management

1 Introduction

The increased human pressure on freshwater resources worldwide accompanies other anthropogenic changes in the global system, from climate to changes in ecosystems, leading to scientific indications that humans have entered a new geological age known as the Anthropocene (Rockström et al., 2014).

Water systems can become more reliable by improving the supply or reducing the demand for water. Increased water storage capacity is a fundamental strategy to create a buffer zone between seasonal precipitation, melting snow, and the demand for water during the agricultural growing season (Zhao & Boll, 2022).

Recent hydrological modeling and observations of the Earth have located and quantified alarming rates of underground water disappearing throughout the world. This depletion is due mainly to indiscriminate extraction of water for irrigation purposes (Dalin et al., 2017).

The availability of water can limit the capacity to ensure food safety for a steadily growing population. Extracting the available water to provide food will limit the amount of water available to preserve the ecosystems (Archer et al., 2010a, 2010b).

Agriculture is a sector that uses a lot of water, especially in arid areas prone to drought that depend on irrigation to sustain agricultural production. In recent years, different types of pressure arising from climate change, population growth, and intensive agricultural irrigation have exacerbated the excessive pumping of underground water and caused the fall of water tables, which greatly affects the hydrologic cycle (Wang et al., 2022).

Irrigated agriculture is the biggest stakeholder in the demand for water, but domestic and industrial demand for water has increased significantly, due to population growth, increased urbanization and industrialization, and also today's higher standard of living. Water needs to produce food and energy have also recorded substantial increases (Laghari et al., 2011).

Intensive and extensive agriculture with high water consumption, for example, cotton, combined with bad management of water resources, has resulted in a significant increase in the demand for water, often addressed by overexploitation of underground water resources. Measures should therefore be taken to manage water demand, aiming to reduce the demand for it in agriculture. New policy measures on the demand for water could mean reducing the pumping of underground water to sustainable levels, helping to restore depleted underground water resources (Loukas et al., 2007).

Irrigation consumes more water than any other human activity, with the challenges of water sustainability and food safety being closely linked. Improving water productivity by reducing the amount of water necessary for agricultural production is a fundamental strategy to achieve food safety and water sustainability in a world with growing demands in these areas (Brauman et al., 2013).

Worldwide, irrigation accounts for around 70% of all water taken from the surface and underground supplies and 90% in areas with extensive irrigated agriculture. Therefore, improving the productivity of irrigation water means reducing the loss of water in the agricultural production system, keeping crop production at a certain level. In the last decade, great progress has been achieved, both by applying a single technology and through a combination of various types (Kang et al., 2017). In the future, water shortage in agriculture is expected to intensify due to continued economic development, rapid population growth, and the intensified effects of global climate change (Kang et al., 2017).

Between 33% and 40% of food produced globally originates in irrigated land, which increases the importance of obtaining precise, timely information about global irrigation. This is necessary to manage increasingly scarce water resources and for improved food security, given crop failures, climate change, and extreme events such as drought, torrential rain, tornados, hurricanes, flooding, and heatwaves (Salmon et al., 2015).

The deficit irrigation system has the potential to reduce by 50% the amount of water used for irrigation purposes. This great potential to reduce underground water extraction adopting water-conserving irrigation techniques requires policies and regulations that can help farmers to change to more sustainable water management practices (Wang et al., 2022).

Irrigation is the key to maintaining high crop production and to improving efficiency in the use of water. Regionally, increased efficiency in water use is an important objective in agricultural water management, but this has not yet been achieved due to the complex influences of multiple factors, such as soil variations, hydrologic variability, and agronomic handling, on crop growth and efficient use of water (Fang et al., 2010).

The last thousand years have been a period of great swings in temperature and combinations of extreme weather events. The past provides the means to test the models we need to forecast, plan, and anticipate the future, in relation to key ecological processes, changing socioenvironmental conditions, and the intensity of impacts. Human society responds to climate change in multiple ways, including collapse or failure, migration, and creative invention through discovery. For example, extreme drought brought about both social breakdown and ingenious management of water through irrigation (Costanza et al., 2007).

Climate change is a large-scale environmental risk, challenging environmental health and the sustainability of global development. Irrigation with wastewater can make a great contribution to reducing water needs, recycling nutrients, improving the health of the soil, and reducing the amount of pollution ending up in water courses (Hanjra et al., 2012).

To cope better with the effects of climate change, water managers can implement locally adapted strategies, incorporating conventional measures to deal with the effects of water stress (De Silva et al., 2007).

In general, this study aims to carry out a systematic review of the literature (SLR) on sustainable water management systems. Specifically, the aim is to draw up an integrative research framework and guidelines for policy action, as well as proposing strategies to be followed to increase the sustainability of water systems and mitigate the effects of climate change and water shortage.

This SLR addresses two research questions (RQ), namely: (1) How does climate change affect the sustainability of water management systems? and (2) What are the effects of climate change on the sustainability of irrigation systems? To this end, a process of systematizing publications in Clarivate's Web of Science was followed. This included adopting the PRISMA protocol, together with criteria for including and excluding publications. Based on the concepts associated with the research questions formulated, namely, "water management," "irrigation," and "sustainability," after various attempts, a combination of terms was established that would lead to results considered appropriate to fulfil the goals of this SLR.

Therefore, the aim is first to determine how climate change affects the sustainability of water management systems. Secondly, it intends to map the effects of climate change on the sustainability of irrigation systems.

It studies the management of water resource systems oriented towards sustainable irrigation, aiming to contribute to better systematization of available knowledge on the different practices used in facing up to climate change and its associated effects.

The process of implementing the methodological design of this SLR was based on five steps proposed by Higgins and Green (2008), namely: (1) clear formulation of the RQs for the SLR, by establishing criteria for including and excluding studies; (2) including in the research all important documents or those that could have some impact on the conclusions of the review; (3) carrying out a review and selecting studies, with consequent data extraction; (4) analyzing the methodological quality of studies; and (5) presenting the results obtained.

2 Setting the Scenario

Agriculture is greatly influenced by the weather and climate. Although farmers are often found to be very flexible in relation to climate change and variations from year to year, there is still a high degree of adaptation to local climate through farming practices arising from personal experience. Therefore, climate change is expected to have impacts on farming that can affect its sustainability but also provide opportunities for improvement (Gornall et al., 2010).

For Olesen and Bindi (2002), crop growth depends greatly on climatic conditions on the surface of the soil (solar radiation, temperature, precipitation, physical characteristics of the soil, etc.). Considering that artificial management of the crop/water can only be adjusted to production in the open air, irrigation is the key factor that can control the crop's growth, development, and production (Gornall et al., 2010).

It is also forecast that global warming will lead to agriculture migrating northwards, as these areas will be more suitable and have a shorter growing period for certain crops (e.g., cereals) (Olesen & Bindi, 2002).

Assessment of the effects of climate change on natural resources, especially the water resources available, is essential for objective analysis of the problems identified, in order to formalize policies that can lessen the effects of climate change on the water available for agriculture (Golfam & Ashofteh, 2022).

Climate change explains between 30% and 50% of the variation in crop production (Fujihara et al., 2008). This is due mainly to the shortage of water, which is directly or indirectly linked to periods of drought (Gado et al., 2017; Ezenne et al., 2019). Climate change, water, and farming interact greatly (Vaghefi et al., 2017). That is why a change in precipitation can have a substantial effect on agricultural production and food safety (Mostafa et al., 2021). Therefore, studying the possible effects of climate change on crop production and productivity is essential to ensure the sustainability of the agricultural sector (Araya et al., 2020; Paul et al., 2020).

Climate change is a global challenge for the sustainable production of various crops due to creating a stressful environment for plant growth (Farooq & Gheewala, 2020). It is fundamentally important to understand how crop irrigation is affected for efficient management of water resources (Woznicki et al., 2015).

Since the agricultural sector consumes most water, besides the importance of ensuring professional management of water resources, it is essential to forecast periods of drought, in order to create water reservoirs (Latif et al., 2021), especially in arid and semiarid regions (Apaydin et al., 2020).

With an increasing population and climate change, it is more and more difficult to ensure the supply of quality freshwater, as it is necessary to satisfy a growing demand for food worldwide, especially in developing countries (Falkenmark, 2013; Wang et al., 2016).

According to the forecasts of the Intergovernmental Panel on Climate Change (IPCC), in 2050, around 2 thousand million people will not have access to a suitable water supply, which represents an increase of approximately twice the current population. Moreover, in 2080, it is forecast that half the world's population will face a scenario of a shortage of drinking water, due to the rising temperature and changes in rainfall (Morrison et al., 2009).

The impacts of climate change on the availability of water for irrigation is one of the biggest concerns regarding the sustainability of agricultural activity (Elliott et al., 2014).

The growing demand for agricultural products is putting unprecedented pressure on water resources (Dalin et al., 2017; D'Odorico et al., 2019). This market pressure, combined with the serious water shortage in many regions of the world, represents a challenge to ensure simultaneously food safety and sustainable management of water resources (Mekonnen & Hoekstra, 2016), considering the future growth in population (Mancosu et al., 2015).

Farming is an economic activity that needs to use a lot of water, especially in dry areas prone to drought that depend on irrigation to sustain production. In recent years, excessive extraction from underground, aggravated by climate change, a growing population, and intensive irrigation, has led to a fall in water levels and influenced the hydrologic cycle. It is essential to understand the changes in hydrologic processes in order to achieve water sustainability (Wang et al., 2022).

Adaptive management has been suggested as a viable strategy for better coordination of surface and underground water resources, in order to cope better with the effects of prolonged periods of drought. Increased water storage should be a priority through increasing networks including reservoirs, dams, natural bodies of water, and aquifers as platforms of available water (Zhao & Boll, 2022).

Decision-making processes related to natural resource management and the spread of eco-efficient technology are gaining more attention from the local to the international level (Megdal et al., 2017).

Here, the case of biochar should be highlighted. This technology could help to achieve sustainable water management in agriculture, due to its ability to increase the water retention capacity of the soil. Despite this opportunity, a limited number of studies prevent quantification of the impacts of biochar in storing water in the ground or characterizing the impacts of biochar additions in the composition of vegetal water (Lyon et al., 2022).

Therefore, irrigation and water management can be used to alleviate the greatest global challenges of resource management. Water management practices could increase food production and sustainability (Finley & Seiber, 2014).

Achieving sustainable use of water resources, considering the increased demand for food and biofuels and the possible impacts arising from climate change, will require optimization of irrigation systems, no easy task given unpredictable rainfall and the multiple soil-plant-atmosphere interactions (Vico & Porporato, 2010).

The most modern principles applicable to irrigation systems promote more adjusted irrigation that implemented through the intermediary of micro-irrigation technology (e.g., dripping or maintenance irrigation, micro-sprinkling, or subsurface emitters (Vico & Porporato, 2010), or water-withholding irrigation (Pereira et al., 2023a), which add water continuously to avoid water stress, minimizing losses through drainage and infiltration. Adoption of this technology can also lead to reduced leaching of soluble chemical products, for example, pesticides and fertilizers (Böhlke, 2002).

Irrigation efficiency is defined as the relation between the amount of water absorbed by a crop and the amount of water applied in irrigation. That efficiency in water use is influenced by irrigation techniques and methods. Those techniques can be placed in three major categories: (1) surface (75% of farmland), (2) sprinkling irrigation (20% of farmland), and (3) micro-irrigation (5% of farmland) (Mateos, 2016). For Al-Agele et al. (2021), the efficiency of applying these general categories varies greatly. Micro-irrigation is generally the most efficient (80–91%) and surface irrigation the least efficient (50–70%), while the efficiency of sprinkling varies between 54% and 80%. However, the most efficient techniques are the least used ones.

Irrigation based on the demand for water implies irrigation when the soil humidity or the crop's humidity status reaches a predetermined level of stress. Each watering restores the soil humidity to a fixed level. Modern micro-irrigation involves very frequent, but shallow, applications of water, to stop the humidity of the soil around the root falling below the ideal levels. Demand-based irrigation strategies are generally more efficient than programmed irrigation applications. Programmed irrigation applications provide a fixed amount of water, irrespective of humidity of the plant and soil (Vico & Porporato, 2011).

All those irrigation strategies based on the demand for water can be framed in a "generalized irrigation scheme" where the key parameters are the point of intervention and the target level (Vico & Porporato, 2011).

3 Methodological Design

This study carries out an SLR about management of water resource ecosystems oriented towards sustainable irrigation, in order to contribute to better systematization of current knowledge about the different practices used to combat the effects of climate change and especially about the different forms of water resource management used in adverse or extreme periods.

To implement the methodological design of this SLR, the five phases proposed by Higgins and Green (2008) are followed, using the procedures applied by Leitão et al. (2023) and Pereira et al. (2023b), in developing application of the PRISMA protocol.

3.1 Research Questions and Systematization of Studies

The SLR aims to answer the following research questions (RQ):

RQ1: How does climate change affect the sustainability of the water management system?

QI2: What are the effects of climate change on the sustainability of irrigation systems?

A process of systematizing studies was followed, with criteria for inclusion and exclusion of publications, based on searches made in Clarivate's Web of Science, as shown in Table 1 below.

The key concept of this SLR is water management which, despite covering a vast area, is mainly set in environmental science, water resources, economics, agronomy, and sustainable green science technology.

Based on the concepts associated with the research questions formulated, namely, "water management," "irrigation," and "sustainability," after various attempts, a combination of terms that would lead to appropriate results was established.

Therefore, in the fields of "Title," "Abstract," and "Keywords," the combination of terms used was "water management," "irrigation," and "sustainability," adding as complementary search filters "articles" written in "English" for all the thematic areas available.

The search was made in December 2022, resulting in a total of 495 scientific articles available on Clarivate's Web of Science.

Table 1 Criteria for including and excluding publications

Inclusion criteria	Exclusion criteria
Published in the period of reference: 1999–January 2023	Books, book chapters, gray literature such as nonacademic reports and research, work documents and reports or conference minutes
Presence on the database: Clarivate's web of science (ISI)	Classified as bibliographic review, systematic review, information review (information about IC), bibliometric study, general review, point of view, scientometric analysis, meta-analysis, among others
Presence of the combination of terms "water management," "irrigation," and "sustainability" related to the Boolean operator "and," in the fields of "title," "abstract," and "keywords"	Articles with anticipated access
Peer-reviewed scientific articles published in English	Peer-reviewed scientific articles published in other languages

Source: Own elaboration

3.2 Selection and Review of Documents

Analysis of the title, abstract, and keywords of each article in the total sample of 495 articles, with the twin aim of identifying and ensuring a link with the issue studied, resulted in a total of 202 articles considered appropriate to carry out this SLR, using the PRISMA research protocol presented in Fig. 1 and following the methodological recommendations proposed by Pereira et al. (2023b).

3.3 Descriptive Analysis

Figure 2 shows a gradual increase in the number of articles published annually on water management, as well as the number of citations. It is noted that in the period studied—1999–January 2023—the greatest number of articles published (n = 39) was reached in 2021, while the maximum number of citations in a single year was achieved in 2022, with approximately 1100 citations. The number of documents reached a total of 199 articles, produced by 1019 authors and coauthors (Graph 1).

Table 2 below presents the articles with most citations among the 202 selected to carry out this SLR, showing the title, authors and coauthors, journals and year of publication, and citations.

Table 3 presents the journals publishing the greatest number of articles, of the 202 selected for this SLR, as well as the total number of articles per journal.

Table 4 shows the principle authors in the sample of 202 articles selected for this SLR, together with the corresponding number of articles.

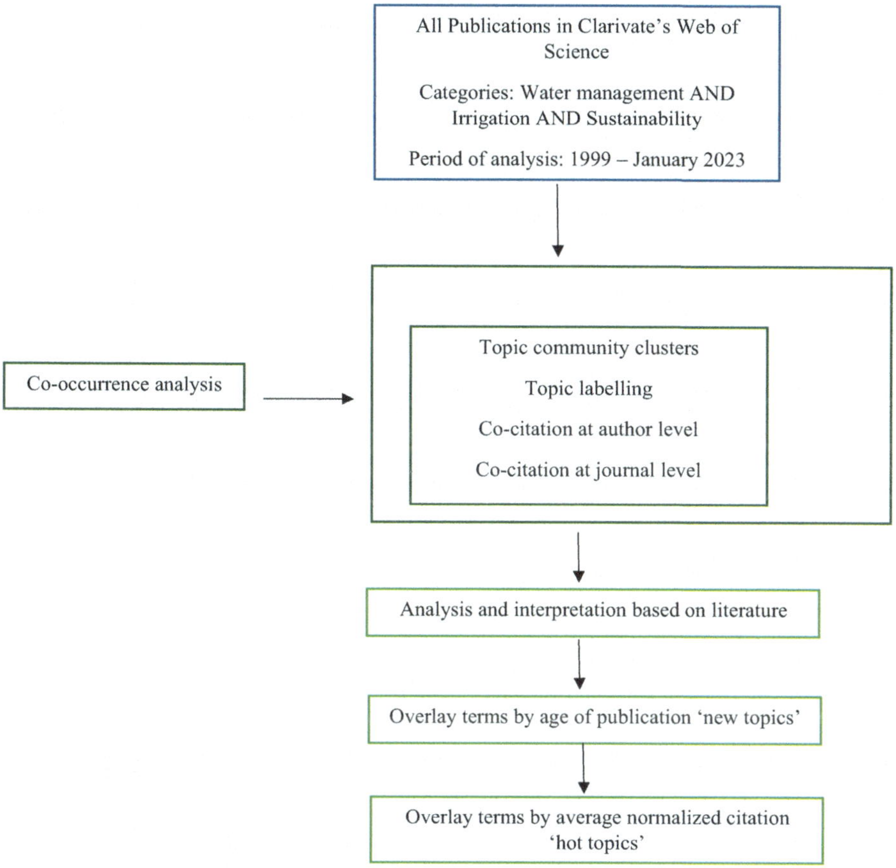

Fig. 1 Application of the PRISMA protocol. (Source: Elaborated from Pereira et al. (2023))

The "citnet" program was used to identify the main articles included in this SLR, first mapping four clusters represented by different colors, as observed in Fig. 2 below.

Regarding co-citation relations, four clusters were identified. Over time, the first cluster (purple) originated in the work by Vörösmarty et al. (2000), followed by the studies by Allen et al. (2007), Rodell et al. (2009), and Sauer et al. (2010), the guiding theme being the sustainability of water resources in relation to the effects of climate change focusing on periods of drought. Evolution of the topic, over time, can be described as follows: "World water resources: Vulnerability to climate change and demographic growth," "Integrated systems of irrigation in crop and stockbreeding land in dry climates," "Satellite estimates of exhausted underground water supplies in India," and "Agriculture and resource availability in a changing world: The role of irrigation."

The second cluster (blue) originated in two more representative studies. The first concerns the contribution of Hu et al. (2005), interconnecting with the second by

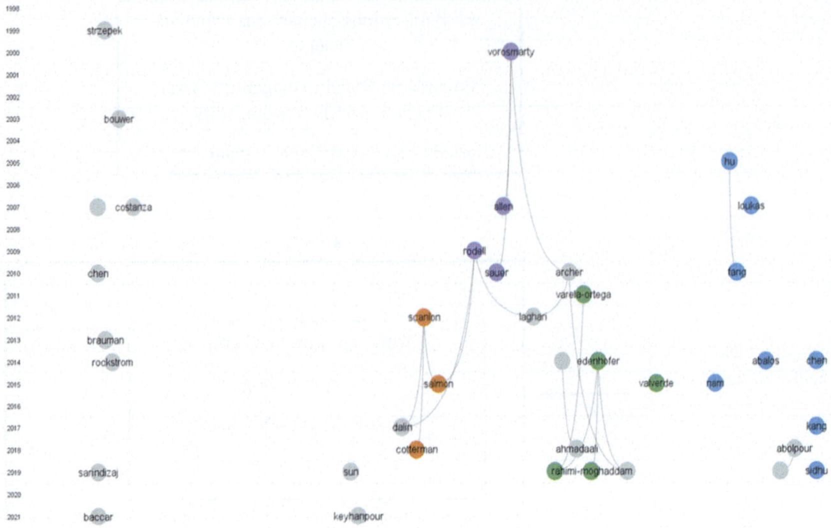

Fig. 2 Most cited articles according to citnet

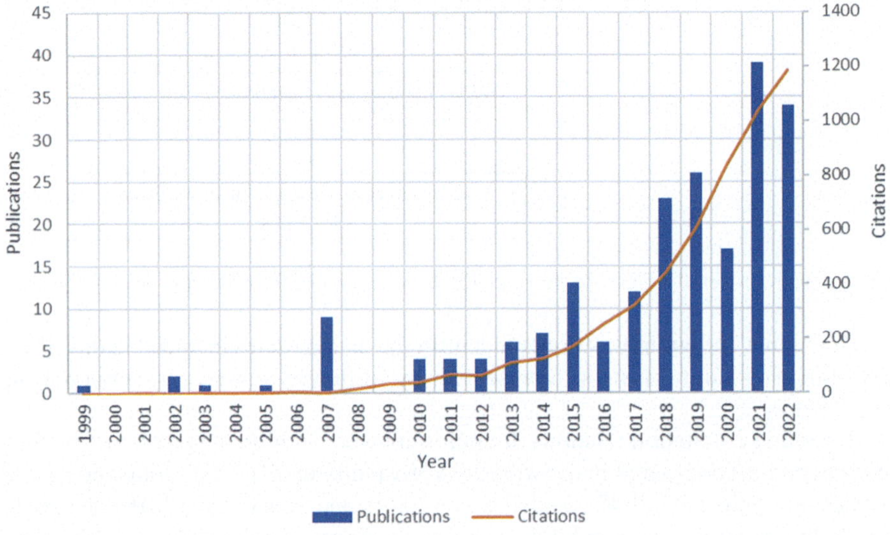

Graph 1 *Number of articles and citations per year.* (Source: Own elaboration from data available on Clarivate's Web of Science)

Fang et al. (2010). A set of related work was also identified, including Loukas et al. (2007), Chen et al. (2014), Abalos et al. (2014), Nam et al. (2015), Kang et al., 2017, and Sidhu et al. (2019), dealing with the efficiency of water use in irrigation processes for agriculture. As for the topic's development over time, initially Hu et al. (2005) evaluated the use of underground water for wheat in the region of Luancheng

Table 2 Publications with the greatest number of citations, according to Web of Science in December 2022

N°	Title	Author(s),	Journal and year	Citations
1	Groundwater depletion embedded in international food trade	Dalin, C; Wada, Y; Kastner, T; Puma, MJ	Nature (2017)	389
2	Improving agricultural water productivity to ensure food security in China under changing environment: From research to practice	Kang, SZ; Hao, XM; Du, TS; Tong, L; Su, XL; Lu, HN; Li, XL; Huo, ZL; Li, SE; Ding, RS	Agricultural Water Management (2017)	317
3	Sustainability or to collapse: What can we learn from integrating the history of humans and the rest of nature?	Costanza, R; Steffen, W; Crumley, C; Querido, J; Hibbard, K; Leemans, R; Redman, C; Schimel, D; Graumlich, L	Ambio (2007)	211
4	Wastewater irrigation and environmental health: Implications for water governance and public policy	Hanjra, MA; Carr, G; Zhang, FH; Jackson, TM; Blackwell, J	International Journal of Hygiene and Environmental Health (2012)	170
5	Water resources and water use efficiency in the North China Plain: Current status and agronomic management options	Fang, QX; Verde, TR; Yu, Q; Wang, TD; Ahuja, LR; Mãe, L	Agricultural Water Management (2010)	168
6	Predicting the impacts of climate change - A case study of paddy irrigation water requirements in Sri Lanka	De Silva, CS; Weatherhead, EK; Knox, JW; Rodriguez-Diaz, JA	Agricultural Water Management (2007)	164
7	Global rain-fed, irrigated, and paddy croplands: A new high resolution map derived from remote sensing, crop inventories and climate data	Salmão, JM; Friedl, MA; Frolking, S; Wisser, D; Douglas, EM	International Journal of Applied Earth Observation and Geoinformation (2015)	153
8	Improvements in crop water productivity increase water sustainability and food security-a global analysis	Rockstrom, J; Falkenmark, M; Allan, T; Folke, C; Gordon, L; Jagerskog, A; Kummu, M; Lannerstad, M; Meybeck, M; Molden, D; Postel, S.; Savenije, HHG; Svedin, U; Turton, A; Varis, O	Ecohydrology (2014)	142

(continued)

Table 2 (continued)

N°	Title	Author(s),	Journal and year	Citations
9	The unfolding water drama in the Anthropocene: towards a resilience-based perspective on water for global sustainability	Brauman, KA; Siebert, S; Foley, JA	Environmental Research Letters (2013)	142
10	Sustainability of water resources management in the Indus Basin under changing climatic and socio economic conditions	Archer, DR; Forsythe, N; Fowler, HJ; Xá, SM	Hydrology and Earth System Sciences (2010)	133

Source: Own elaboration from data available in Clarivate's Web of Science

Table 3 Journals with the greatest number of articles

N°	Journal	Number of articles
1	Environmental sciences	84
2	Water resources	77
3	Agronomy	50
4	Green sustainable science technology	24
5	Environmental studies	18
6	Geosciences multidisciplinary	17
7	Engineering environmental	16
8	Engineering civil	12
9	Meteorology atmospheric sciences	11
10	Agriculture multidisciplinary	10

Source: Own elaboration from Clarivate's Web of Science

Table 4 Principle authors and number of articles

N°	Author	Number of articles
1	Bergez, Jacques-Eric	3
2	Hyndman, David W.	3
3	Jat, Mangi Lal	3
4	Kendall, Anthony D.	3
5	Lorite, Ignacio J.	3
6	Ma, Lin	3
7	Ward, frank A.	3
8	Abalos, Diego	2
9	Ahmad, Shakil	2
10	Baccar, Mariem	2

Source: Own elaboration from data available on Clarivate's Web of Science

Xian and the potential implications for water conservation in the northwest of the North China Plain. This interconnects with the study by Fang et al. (2010), on the "water resources and efficiency of water use in the North China Plain: Current situation and options for agronomic management." The remaining studies in this cluster are "A modelling system to evaluate strategies for water resource management in Thessaly, Greece"; "Energy assessment of the contribution of irrigation water and its use in three agricultural systems in China"; "Management of watering frequency and fertilization with nitrogen to mitigate emissions of greenhouse and nitrous oxide from irrigated crops"; "Assessment of the vulnerability of irrigation in relation to the risk of water supply to agriculture for adaptive management of climate change in South Korea"; "Improving the productivity of water in agriculture to ensure food safety in China in a changing environment: From research to practice"; and "Sub-surface drip fertigation with conservation agriculture in a rice-wheat system: Advancing efficiency in the use of water and nitrogen."

The third cluster (green) contains a unifying study, the contribution by Varela-Ortega et al. (2011), followed by the work of Edenhofer et al. (2014), Valverde et al. (2015), Ward et al. (2019), and Rahimi-Moghaddam et al. (2019), who analyze the effects of climate change on farming activities, as well as the alternatives to protect water resources coming under pressure. The first study in the table for this cluster is by Varela-Ortega et al. (2011): "Balance between conserving underground water and means of rural subsistence in the context of water and climate uncertainties: An integrated framework of hydro-economic modelling," which links to three other studies, "Climate Change 2014 Mitigation of Climate Change (Work Group III) Contribution to the Fifth Evaluation Report of the Intergovernmental Panel on Climate Change Preface"; "The economics of plans to protect aquifers subject to climatic water stress: New perspectives from hydro-economic models"; and "Optimal genotype × environment × management as a strategy to increase the productivity of maize and efficiency of water use in environments with water limitations and rising temperatures." There is also one publication without any link to the previous ones, belonging to Valverde et al. (2015), which studies the impacts of climate change on irrigated agriculture in the Guadiana Basin (Portugal).

The fourth cluster (orange) spreads out from the study by Scanlon et al. (2012), to include the subsequent work by Salmon et al. (2015) and Cotterman et al. (2018), making contributions to study of the sustainability of underground water for the purpose of irrigating farmland. The first article by Scanlon et al. (2012) studies the exhaustion of underground water and the sustainability of irrigation in the high plains and central valley of the USA, being followed by "Global rain-fed, irrigated, and paddy croplands: dry land, irrigated land, and paddy fields: A new high-resolution map derived from remote detection, crop inventories, and climate data" and "Exhaustion of underground water and climate change: future perspectives of vegetable production in the Central Plain Aquifer."

3.4 Bibliometric Analysis: Co-Authorship, Co-Occurrence, and Citation Networks

VOSviewer software was used to perform a bibliometric analysis, based on co-authorship, co-occurrence, and citation networks. Figure 3 maps the authors of articles with 5 or more citations, noting that the total number of authors in these circumstances is 688, despite choosing to present the authors who besides having 5 citations are also related with each other.

Figure 4 shows the institutions presenting at least 1 article among the 202 selected for this SLR, with a minimum of 5 citations. The institutions presenting the greatest number of publications are "Chinese acad sci" with 12 publications, "ars" with 6 publications, and "China agr univ," "univ politecn madrid," "univ calif davis," "texas a&m univ," "hoai univ.," and "univ toulose" with 5 publications each. The organizations with the greatest number of citations are respectively "int inst appl syst anal" with 507 citations, "nasa" with 416 citations, "columbia univ" with 411

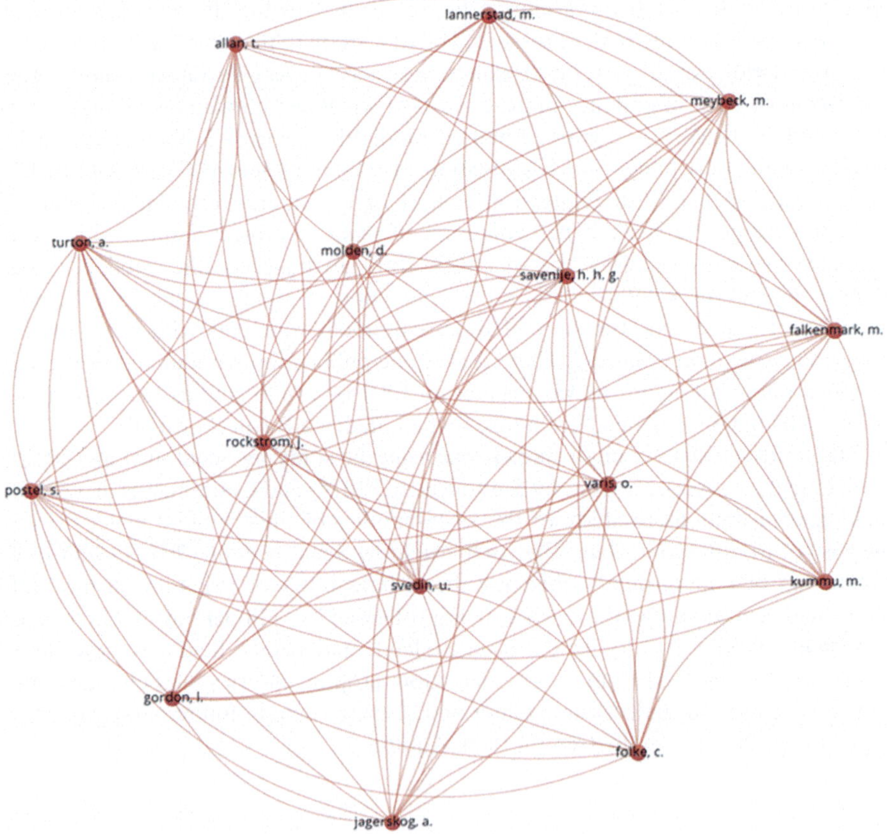

Fig. 3 Co-authorship of authors

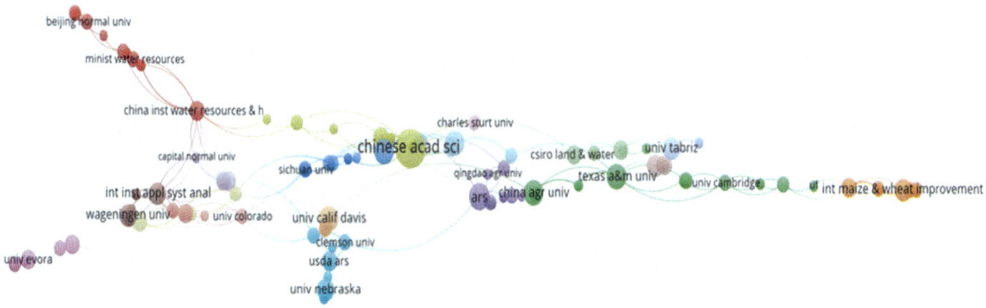

Fig. 4 Co-authorship of institutions

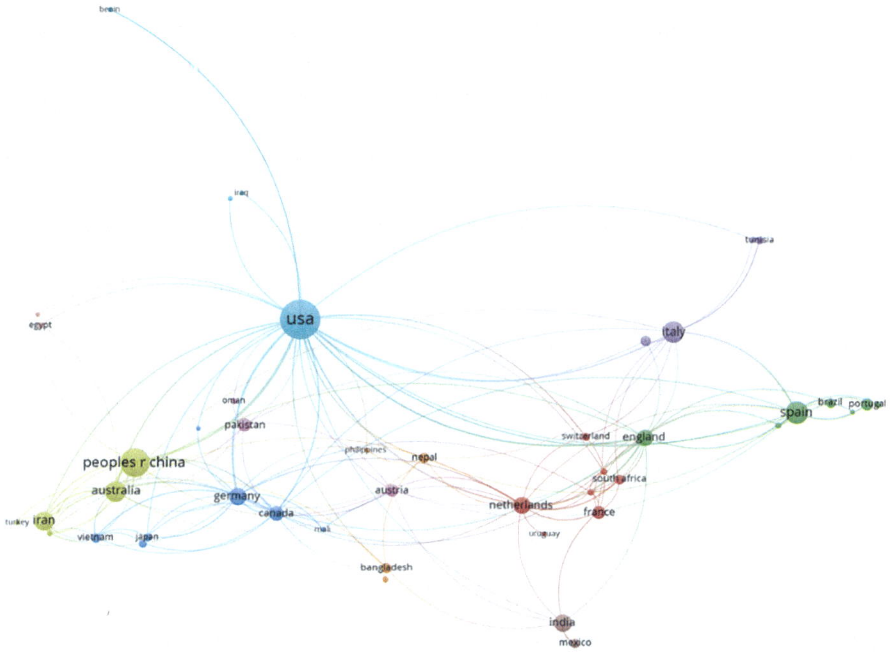

Fig. 5 Co-authorship by country

citations, "chinese acad sci" with 391 citations, and "ucl," "univ utrecht," and "alpen adria univ klagenfurt" with 388 citations.

Similarly, a new map of co-authorship networks was elaborated, considering once again at least one article and five citations, but now according to country. The countries with the greatest number of documents are the USA, with 64 documents, the People's Republic of China with 35, Spain with 21, and Italy with 20. Concerning citations, the USA remains in the first place with 2109, followed by the People's Republic of China with 1280, England with 1157, and the Netherlands with 1016, as seen in Fig. 5.

Fig. 6 Citations per document

Figure 6 shows the citations per document, the analysis performed with Vosviewer considering documents with five or more citations. However, of 135 documents satisfying this requirement, only 7 are linked to each other, highlighting the relative importance of studies (Varela-Ortega et al., 2011), speaking about the balance in conserving underground water and means of rural subsistence faced with water and climate uncertainties: an integrated hydro-economic modeling structure, dealing with the impacts of climate change on irrigated farming in the Guadiana Basin in Portugal (Valverde et al., 2015).

An analysis of the citation network per source was carried out next, applying a criterion of cumulative observance of at least five articles per source, and each article having five or more citations, obtaining a total of seven sources, despite finding that only three are linked to each other, namely, agricultural water management, water resources research, and science of the total environment, as shown in Fig. 7.

Tests were also performed for citations of authors, countries, and institutions, using the following criteria: at least 2 articles (per author/country and institution, respectively) with 4 or more citations (cf. as can be observed in Figs. 10, 11, and 12). Following application of these criteria, 30 authors satisfied the requirements of the model, but only 4 showed links with each other, namely, David W. Hyndman, Anthony D. Kendall, Jillian M. Deines, and Bruno Basso (see Fig. 8).

As observed in Fig. 9, 34 countries meet the requirements of having at least 2 articles and presenting 4 or more citations in each. However, only 21 present interconnections, highlighting the dominance of authors and co-authors located in the USA, China, Australia, Spain, and Italy. The most cited countries are the USA with 2109, the People's Republic of China with 1280, England with 1157, and the Netherlands with 1016, similarly to what was found in Fig. 5.

Figure 10 presents the institutions which, besides meeting the requirements, are interlinked, forming a total of 45 institutions. It is noted, however, that the total number of institutions with 2 or more articles, presenting 4 or more citations, is 87. The organizations with the greatest number of citations are "int inst appl syst anal" with 507 citations, "nasa" with 416, "columbia univ" with 411, "chinese acad sci" with 391, and "ucl" and "univ utrecht" with 388.

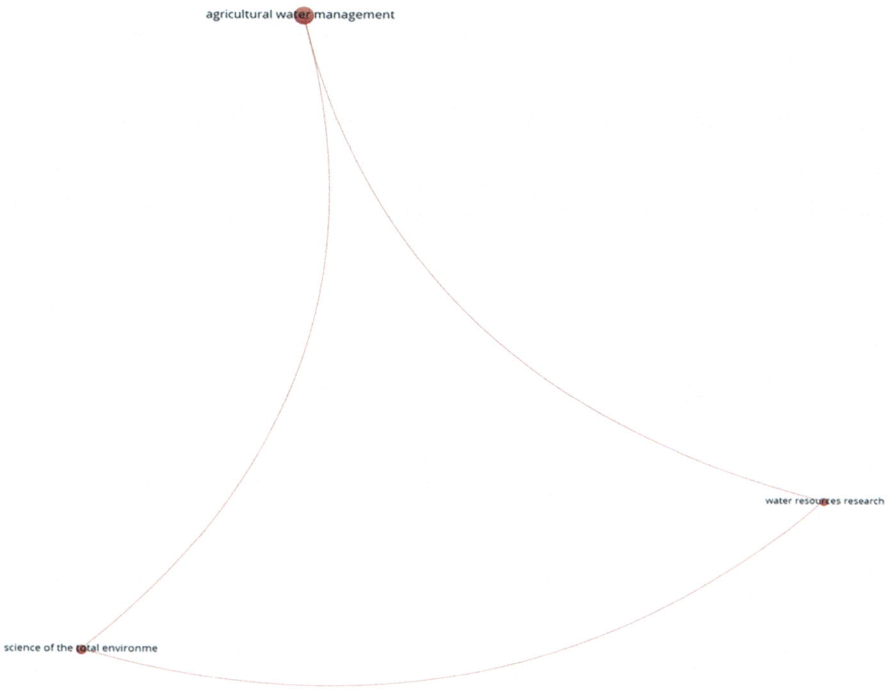

Fig. 7 Citations per source

Fig. 8 Citations per author

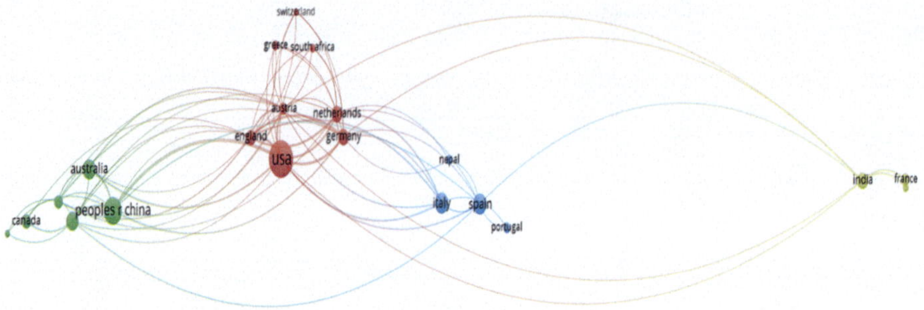

Fig. 9 Citations per country

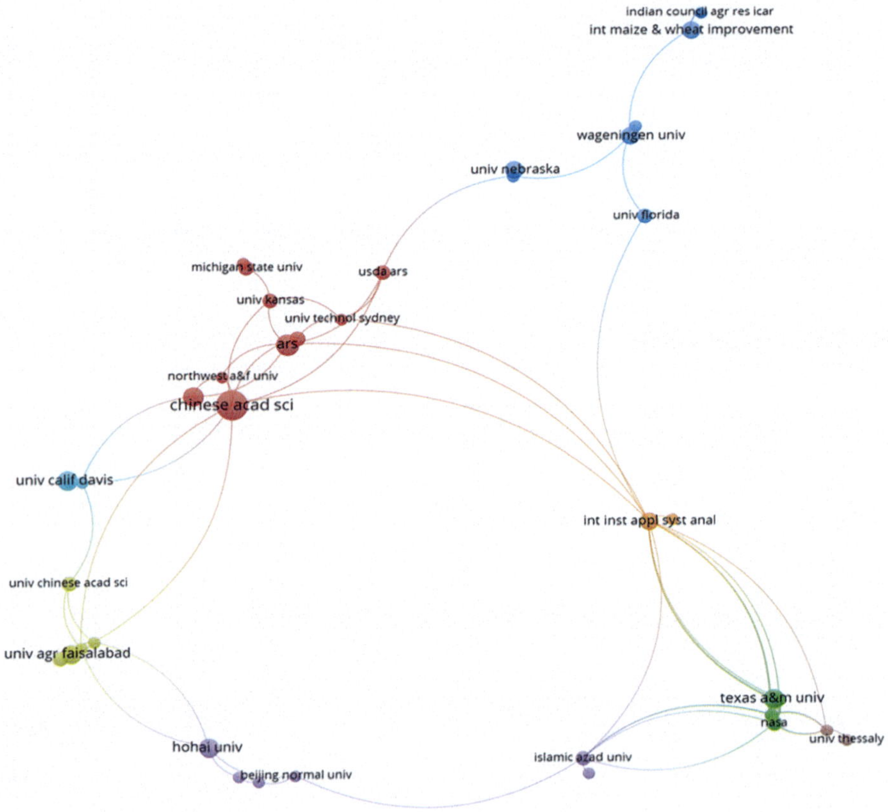

Fig. 10 Citations per institution

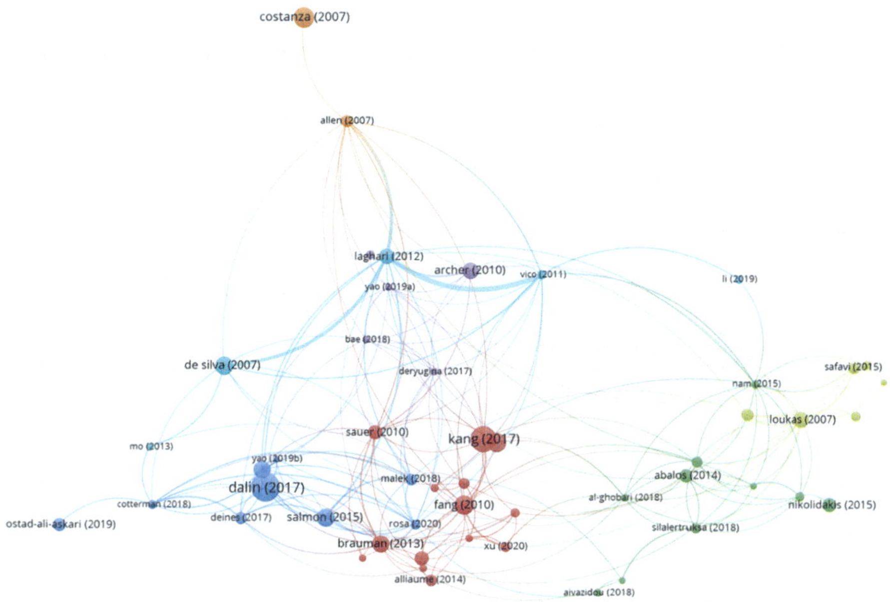

Fig. 11 Bibliographic coupling of documents

3.5 Coupling Analysis: Bibliographic and Co-Citations

Continuing with the SLR, a bibliographic and co-citation coupling analysis was carried out. Concerning the bibliographic coupling, following De Brito and Leitão (2021), a minimum requirement of 25 citations was considered, which gave a total of 53 documents, despite only 51 presenting interconnections (cf. Fig. 11).

Next, the bibliographic coupling between the different sources is presented, considering as cumulative requirements: a minimum of 2 articles per source and a minimum of 4 citations, which allowed the inclusion of all sources (journals), as all were interconnected, highlighting the polarizing nature of Agricultural Water Management (see Fig. 12).

The next three tests concern the bibliographic coupling of authors, organizations, and countries, considering as the cumulative requirement: two articles, with four or more citations. For the bibliographic coupling of authors, 30 authors were identified, but only 29 had connections with each other (see Fig. 13). As for the bibliographic coupling of organizations, all the 87 meeting the criteria presented interconnections (see Fig. 14). The same is found for the bibliographic coupling of countries, as the 34 identified all presented interconnections (cf. Fig. 15).

Tests applied to co-citation networks were also performed. The first dealt with the co-citations of references cited for a minimum number of 5 citations, allowing 58 studies of reference to be mapped, as can be seen in Fig. 16.

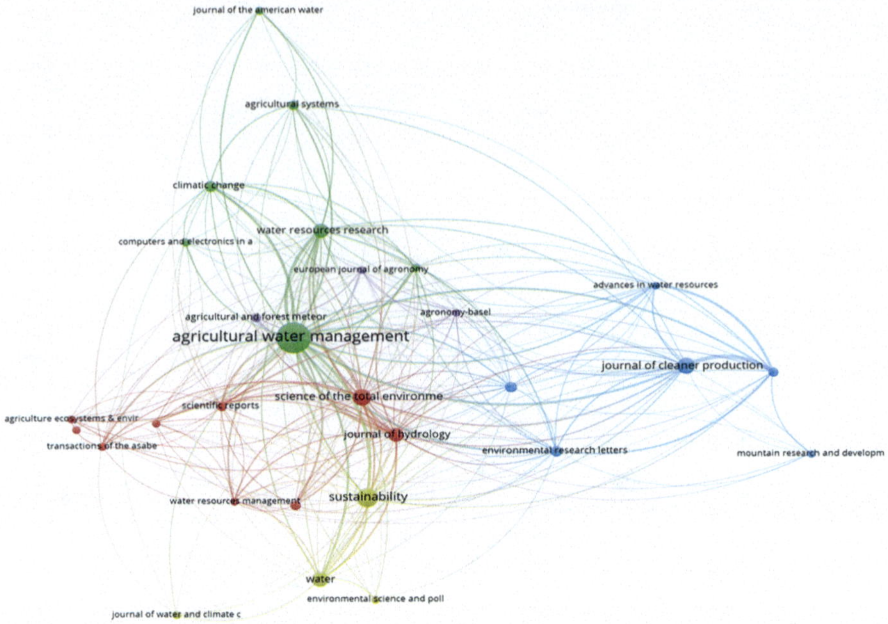

Fig. 12 Bibliographich coupling per source

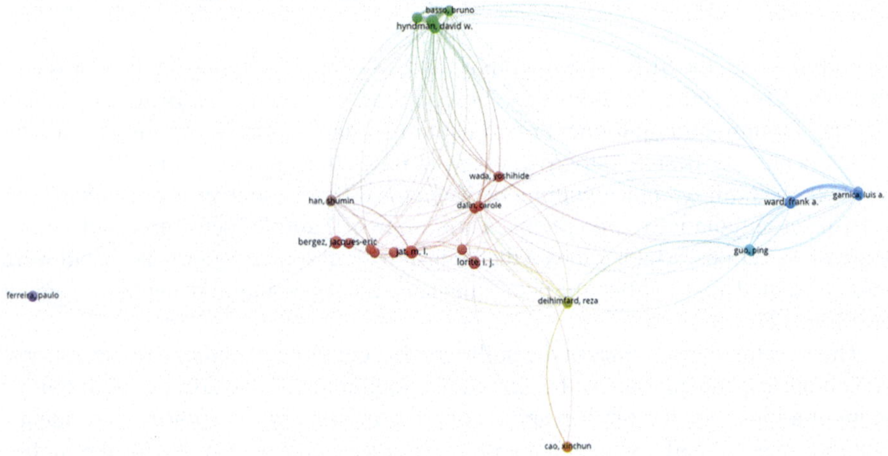

Fig. 13 Bibliographic coupling of authors

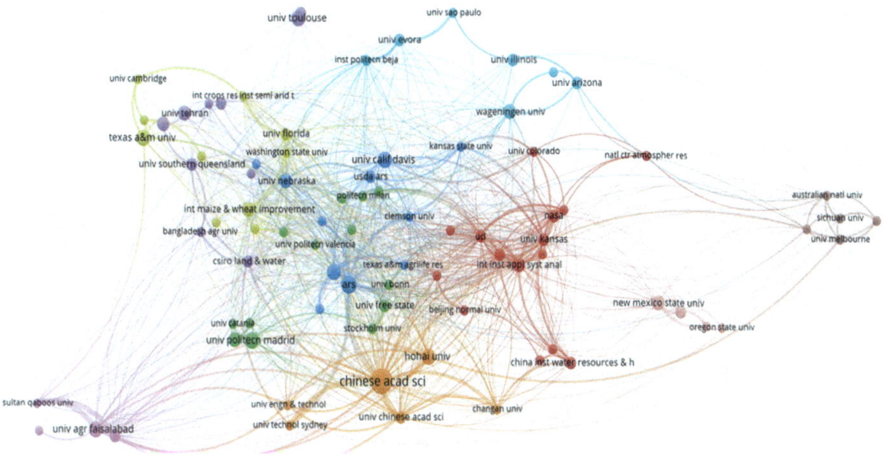

Fig. 14 Bibliographic coupling of organisations

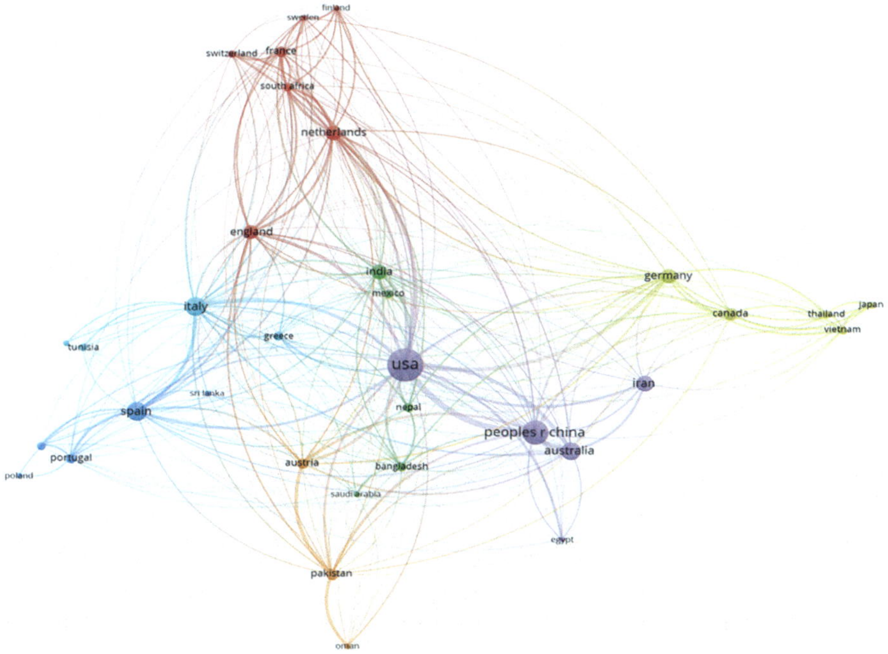

Fig. 15 Bibliographic coupling of countries

Next, the co-citations per source cited were studied, for a minimum number of 25 citations per sources, as shown in Fig. 17.

Co-citations per cited authors were also analyzed, for a minimum of 10 citations per author, giving a total of 66 authors, with 65 being interconnected (see Fig. 18).

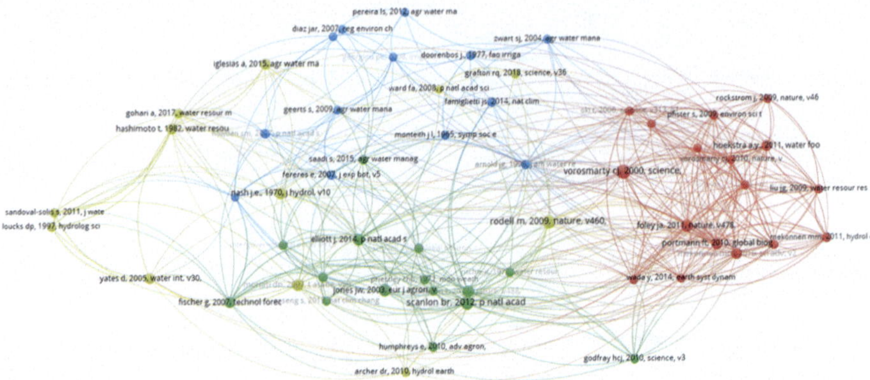

Fig. 16 Co-citations per cited reference

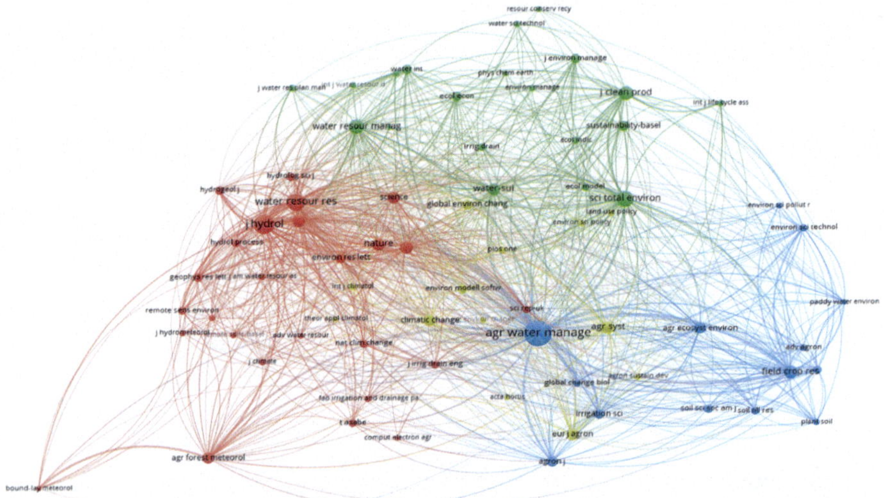

Fig. 17 Co-citations per cited source

In this SLR, 5 clusters were identified (cf. Fig. 19 below), the first, in red, being climate change and represented by words such as "CO_2," "change impacts," "precipitation," temperature," etc. The second cluster, in green, represents the effects of climate change, with words such as "Carbon footprint," "footprint," greenhouse-gas emissions," etc. The third cluster, in blue, represents water cultures, with words such as "irrigation," "deficit irrigation," "water-use," etc. The fourth cluster, in yellow, portrays precision agriculture and water efficiency, being represented by keywords such as "agriculture," "water-management," "water-use efficiency," "food security," etc. The fifth cluster, in purple, represents sustainability, containing keywords such as "optimization," ecosystem services," "management," "resources," "demand," etc.

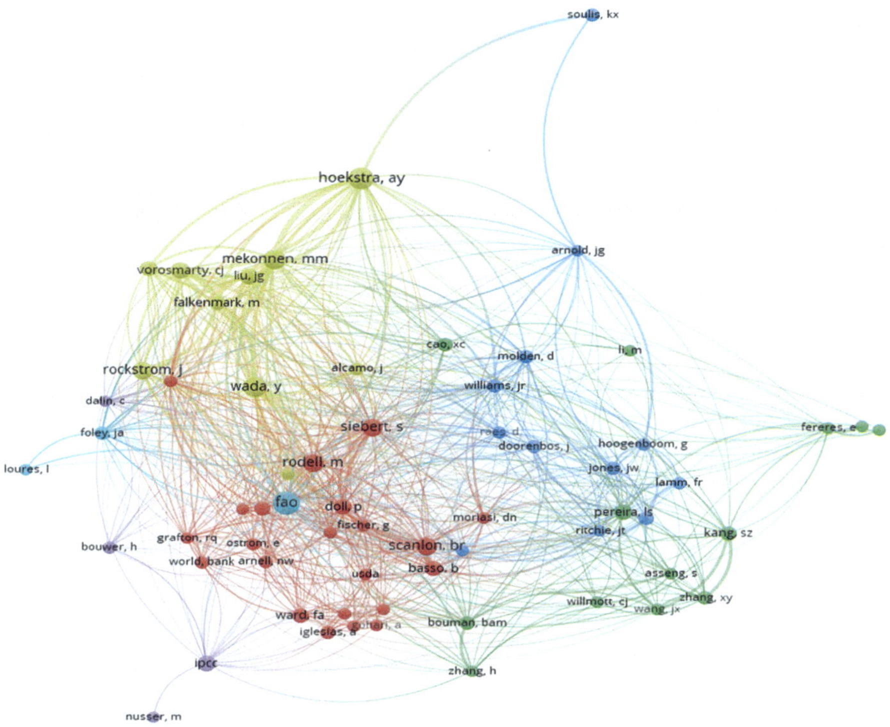

Fig. 18 Co-citations per cited authors

Both Figs. 19, above, and 20, below, show networks of representations of dominant words/expressions with greatest relevance in the title and abstracts of the articles in the SLR. However, Fig. 6 only shows the words/expressions appearing in at least five different articles, highlighting climate change, irrigation, and model (Fig. 20).

Figure 21 identifies the networks of the most frequent and relevant keywords, only considering those used in a minimum of five articles in this SLR, highlighting the relative importance of climate change, irrigation, and sustainability.

3.6 Cluster Analysis and Research Framework Proposal

Based on the analysis of keyword co-occurrence (cf. Fig. 19 above), from which five clusters were identified, each cluster is first characterized. Then a research framework is proposed, to situate in an integral way the advance of knowledge on sustainable water management ecosystems and set the bases for a new research agenda.

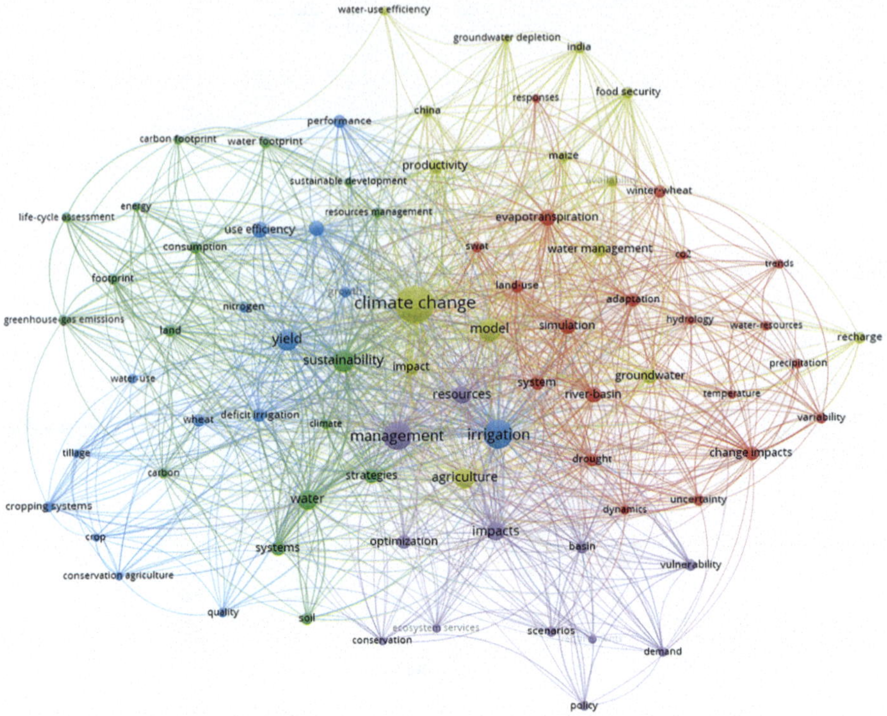

Fig. 19 Co-occurrence of key-words

3.6.1 Cluster 1: Climate Change

Due to climate change, hotter, drier climates have been recorded (Hordofa et al., 2022). These climate change scenarios allow anticipation of the occurrence of more frequent and severe droughts (Zhao & Boll, 2022), with agricultural production in arid and semiarid regions being particularly vulnerable to these changes (Bonetti et al., 2022). Climate change is a matter of concern for hydrologists, climatologists, environmentalists, and agronomists, as it is expected to have an increasingly serious effect on water resources, infrastructure, agriculture, biodiversity, and functioning of the ecosystem and human health (Guan et al., 2021; Teklay et al., 2022). Climate change is also expected to be particularly severe in areas where socioecological systems are greatly dependent on rain and/or water flows (Shrestha & Nepal, 2016).

3.6.2 Cluster 2: Effects of Climate Change

Climate change has increased extreme climatic events such as droughts, heatwaves, and the seriousness of plagues and weeds, which in turn requires a number of changes in farming practices (Liu et al., 2021). Climate change scenarios suggest intensified pressure on water resources, especially in regions that are already

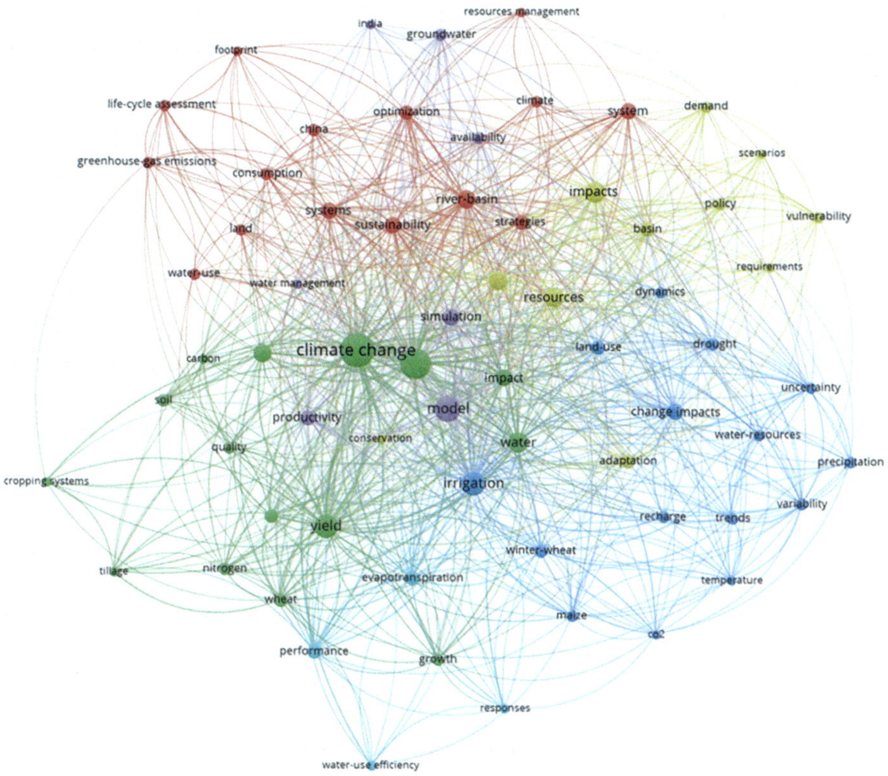

Fig. 20 Co-occurrence of key-words plus

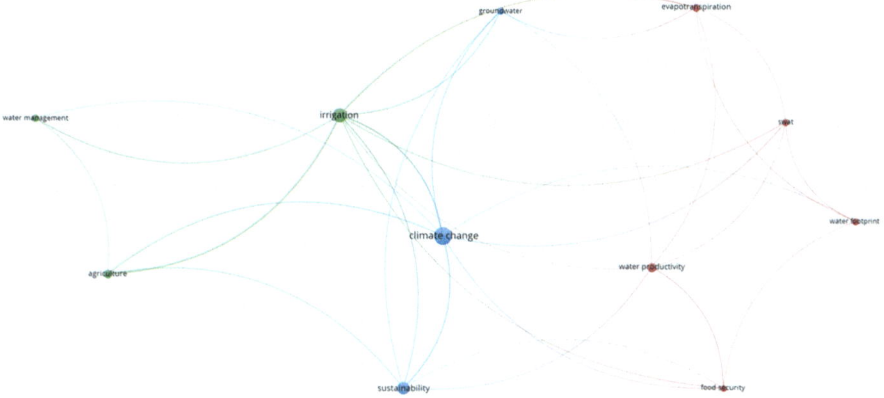

Fig. 21 Co-occurrence of authors' key-words

vulnerable (Bonetti et al., 2022). In dry areas, the warmer and drier climate in the future is expected to cause a general reduction in wheat production in winter, in the productivity of water culture, and in the productivity of irrigation water (Hordofa et al., 2022). A fall in the average temperature will probably delay plant growth and reduce the rate of chlorophyll in the leaves, contributing to lower productivity (Risal et al., 2022).

3.6.3 Cluster 3: Water Cultures

Irrigation accounts for 70% of all use of water, and between 33% and 40% of global food production originates in irrigated land (Salmon et al., 2015). Modernizing irrigation infrastructure has been considered in global policy as a solution for both water shortage and low agricultural productivity (Owens et al., 2022). A paradox existing in many areas of irrigation worldwide is the so-called rebound effect, where infrastructure modernization destined to "economize water" aims to increase the efficiency of irrigation systems, but changing to channeled or pressurized drip systems often results in a loss of water and increased water consumption (Grafton et al., 2018). Irrigation with wastewater can make an excellent contribution to reducing the demand for water, recycling nutrients, improving the health of the soil and reducing the amount of pollution entering water courses (Hanjra et al., 2012).

3.6.4 Cluster 4: Precision Agriculture and Efficiency of Irrigation Systems

Water safety has an important role in socioeconomic development, in managing the ecosystem and environmental sustainability (Veettil et al., 2022). Agriculture is vulnerable to drought, indicating that the growing climate crisis requires sustainable agricultural production (Lee et al., 2022). Application of correct government policies for suitable management of water in agriculture and granting installations to mechanize irrigation can make water use more efficient in this economic sector and, to some extent, reduce the effects of the crisis (Yousefi et al., 2022). Water safety is the basis of food safety. The shortage of water resources will lead to variable production of food, which could originate a real food crisis. Water management is the key to ensuring more food can be produced for a growing population (Kang et al., 2017).

3.6.5 Cluster 5: Sustainability

The state of water safety is intimately linked to the sustainable development of society and ecosystems (Kang et al., 2017). The observed and projected consequences of climate change in the flow of water in the Pyrenees threaten the long-term sustainability of water resource systems downstream, especially those needing a great amount of irrigation (Haro-Monteagudo et al., 2022). Applying systems destined to increase irrigation efficiency and optimize crop production, observing irrigated cultivation with low water use, is the best approach to achieve environmental and economic sustainability (Gohari et al., 2017; Sinha, 2018). Implementing a combination of policies, such as changing growing patterns, increasing crop productivity, and controlling the retention of underground water, not only increases the safety of surface and underground water but also reduces the use of water for food and energy production, in this way ensuring the sustainability of water resources (Keyhanpour et al., 2021).

The above analysis of the clusters results in designing a research framework proposal (see Fig. 22 below), setting out from Cluster 1: Climate change, which is directly related to Cluster 2: Effects of climate change, which both precede Cluster 5: Sustainability. As for indirect relations, it is a question of integrating and exploring the implications of Cluster 1: Climate change, in relation to Cluster 3: Water crops irrigation, as well as in adjustment of Cluster 4: Precision farming and water efficiency, which interact ultimately with Cluster 5: Sustainability.

4 Discussion of the Results

As confirmed by this SLR, both the number of publications and the number of citations on the issue have increased in recent years, with the former reaching a peak in 2021 and the latter in 2022.

The most prominent authors and studies in this area of research are the following: Dalin et al., (2017); Kang et al., (2017); Costanza et al., (2007); Hanjra et al., (2012); Fang et al., (2010); De Silva et al., (2007); Salmon et al., (2015); Rockström

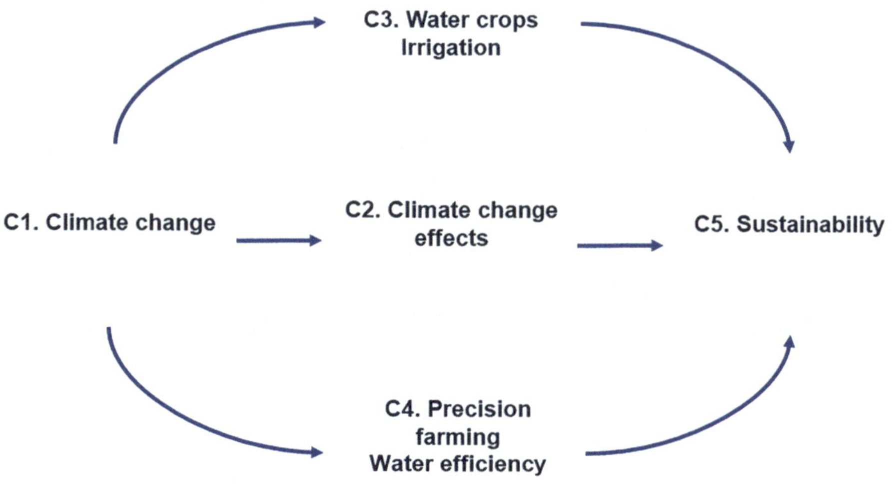

Fig. 22 Research framework proposal. (Source: Own elaboration)

et al., (2014); Brauman et al. (2013); and Archer et al., (2010a, 2010b), in decreasing number of citations.

The country publishing most articles with at least five citations is the USA, followed by the People's Republic of China, Australia, Spain, and Iran.

The main keywords are climate change, irrigation, sustainability, agriculture, water productivity, evapotranspiration, groundwater, water footprint, water management, and food security.

In turn, the main journals are "agricultural water management," "science of the total environment," and "water resources research," the countries with most being the USA, the People's Republic of China, and Australia.

Bibliographic coupling, elaborated by Kessler (1963), measures the relation between two articles based on the number of common references cited by both. A test was drawn up to confirm bibliographic coupling for documents with at least 25 citations, obtaining 51 articles and identifying 3 main ones (Costanza et al., 2007; Dalin et al., 2017; Kang et al., 2017). The most cited journal is agricultural water management, and the most cited countries are the USA, the People's Republic of China, and Australia.

Co-citation analysis measures the relation between two articles based on the number of publications in which they appear cited concomitantly. Despite having connections, some articles, according to the citnet program, do not belong to any of the clusters identified, namely, Ahmadaali et al. (2018), Archer et al. (2010a, 2010b), Santikayasa et al. (2014), Dalin et al. (2017), Laghari et al. (2012), and Nüsser et al. (2019).

There are also articles that despite being considered relevant by the program, due to their number of citations, do not present connections with any of the clusters identified. Highlighted here are Strzepek et al. (1999), Bouwer (2002), Costanza et al. (2007), De Silva et al. (2007), Brauman et al. (2013), Rockström et al. (2014), García-López et al. (2019), Sarindizaj and Zarghami (2019), and Baccar et al. (2021).

5 Conclusion

This SLR corroborates the view that water resource management is essential to ensure sustainable development of agricultural activities. However, water saving, to face up to future water shortages in systems with a great need of agricultural irrigation, is not enough to ensure the future sustainability of water resources. This warrants special attention by policymakers and academics, to encourage behavior and technological innovations that can save water to face prolonged periods of drought, given climate change and other extreme environmental events, to avoid situations of water resource stress.

In the scope of the studies selected for this SLR, there is a clear gap in existing knowledge about situations characterized by a rapid growth in the demand for food products and water shortage problems worldwide. This can be due to the recent increased media coverage of the effects of climate change, seen above all in prolonged periods of drought followed by short periods of intense precipitation. These abnormal occurrences also show that the soil is not prepared to absorb large amounts of water in a short period of time, which in turn leads to a greater frequency of flooding worldwide. This results in a growing need to reinforce storage areas and infrastructure, as mechanisms to conserve water in these periods, to anticipate the effects of climate change which will certainly cause more severe and frequent droughts.

Concerning the research questions formulated in this SLR, the following conclusions are drawn:

RQ1: How does climate change affect the sustainability of water management systems?

In the literature of reference, the growing coverage of the effects of climate change, especially in the form of longer periods of drought followed by short periods of intense precipitation, shows that soils are not prepared to absorb large amounts of water in a short period, which leads to more frequent flooding worldwide.

RQ2: What are the effects of climate change on the sustainability of irrigation systems?

To anticipate the impacts of climate change, which will cause more severe and frequent droughts, there is a growing need to reinforce storage areas, as well as infrastructure and water conservation techniques, including in the soil.

As for the limitations of this SLR, highlighted is the limited number of previous studies on the issue of water resource management oriented towards sustainable agricultural irrigation, in the face of climate change. Added to this is the predominance of case studies, rather than studies contrasting and comparing different situations, which makes it difficult to generalize findings in earlier studies.

To address the above limitations, a future line of exploration would be to prioritize in research and policy action the issue of water resource management oriented towards sustainable agricultural irrigation, focusing on micro-irrigation systems as smart instruments to prevent and combat the effects of climate change.

Also for the future, this SLR indicates the need for more wide-ranging studies on water resource management, in order to create mechanisms to combat, efficiently and sustainably, the adverse conditions arising from climate change and the lack of effective policies to address social challenges, regarding food production and consumption, as well as the associated environmental costs and benefits.

References

Abalos, D., Sanchez-Martin, L., Garcia-Torres, L., Willem, J., & van Groenigen, & Antonio Vallejo. (2014). Management of irrigation frequency and nitrogen fertilization to mitigate GHG and NO emissions from drip-fertigated crops. *Science of the Total Environment, 490*, 880–888.

Ahmadaali, J., Barani, G. A., Qaderi, K., & Hessari, B. (2018). Analysis of the Effects of Water Management Strategies and Climate Change on the Environmental and Agricultural Sustainability of Urmia Lake Basin, Iran. *Water, 10, 160.*

Al-Agele, H. A., Nackley, L., & Higgins, C. W. (2021). A pathway for sustainable agriculture. *Sustainability (Switzerland), 13*(8), 4328.
Allen, V. G., Baker, M. T., Segarra, E., & Brown, C. P. (2007). Integrated Irrigated Crop–Livestock Systems in Dry Climates. *Agronomy Journal, 99*(2), 346–360.
Apaydin, H., Feizi, H., Sattari, M. T., Colak, M. S., Shamshirband, S., & Chau, K. W. (2020). Comparative Analysis of Recurrent Neural Network Architectures for Reservoir Inflow Forecasting. *Water, 12, 1500.*
Araya, A., Prasad, P. V. V., Zambreski, Z., Gowda, P. H., Ciampitti, I. A., Assefa, Y., & Girma, A. (2020). Spatial analysis of the impact of climate change factors and adaptation strategies on productivity of wheat in Ethiopia. *Science of the Total Environment, 731*, 139094.
Archer, D. R., Forsythe, N., Fowler, H. J., & Shah, S. M. (2010a). Indus Basin water resources sustainability. Sustainability of water resources management in the Indus Basin under changing climatic and socio economic conditions. Indus basin water resources sustainability. *Hydrology and Earth System Sciences Discussions, 7*, 1883–1912.
Archer, D. R., Forsythe, N., Fowler, H. J., & Shah, S. M. (2010b). Sustainability of water resources management in the Indus Basin under changing climatic and socio economic conditions. *Hydrology and Earth System Sciences, 14*(8), 1669–1680.
Baccar, M., Bergez, J. E., Couture, S., Sekhar, M., Ruiz, L., & Leenhardt, D. (2021). Building climate change adaptation scenarios with stakeholders for water management: A hybrid approach adapted to the South Indian water crisis. *Sustainability, 13*(15), 8459.
Böhlke, J. K. (2002). Groundwater recharge and agricultural contamination. *Hydrogeology Journal, 10*(1), 153–179.
Bonetti, S., Sutanudjaja, E. H., Mabhaudhi, T., Slotow, R., & Dalin, C. (2022). Climate change impacts on water sustainability of south African crop production. *Environmental Research Letters, 17*(8), 084017.
Bouwer, H. (2002). Integrated water management for the 21st century: Problems and solutions. *Journal of Irrigation and Drainage Engineering, 128*(4), 193–202.
Brauman, K. A., Siebert, S., & Foley, J. A. (2013). Improvements in crop water productivity increase water sustainability and food security—A global analysis. *Environmental Research Letters, 8*(2), 024030.
Chen, D., Luo, Z., Webber, M., Chen, J., & Wang, W. (2014). Emergy evaluation of the contribution of irrigation water, and its utilization, in three agricultural systems in China. *Frontiers of Earth Science, 8*(3), 325–337.
Costanza, R., Graumlich, L., Steffen, W., Crumley, C., Dearing, J., Hibbard, K., Leemans, R., Redman, C., & Schimel, D. (2007). Sustainability or collapse: What can we learn from integrating the history of humans and the rest of nature? *Ambio, 36*(7), 522–527.
Cotterman, K. A., Kendall, A. D., Basso, B., & Hyndman, D. W. (2018). Groundwater depletion and climate change: Future prospects of crop production in the central High Plains aquifer. *Climatic Change, 146*(1–2), 187–200.
D'Odorico, P., Carr, J., Dalin, C., Dell'Angelo, J., Konar, M., Laio, F., Ridolfi, L., Rosa, L., Suweis, S., Tamea, S., & Tuninetti, M. (2019). Global virtual water trade and the hydrological cycle: Patterns, drivers, and socio-environmental impacts. *Environmental Research Letters, 14*(5), 053001.
Dalin, C., Wada, Y., Kastner, T., & Puma, M. J. (2017). Groundwater depletion embedded in international food trade. *Nature, 543*(7647), 700–704.
De Brito, S., & Leitão, J. (2021). Mapping and defining entrepreneurial ecosystems: A systematic literature review. *Knowledge Management Research & Practice, Taylor & Francis Journals, 19*(1), 21–42.
De Silva, C. S., Weatherhead, E. K., Knox, J. W., & Rodriguez-Diaz, J. A. (2007). Predicting the impacts of climate change—A case study of paddy irrigation water requirements in Sri Lanka. *Agricultural Water Management, 93*(1–2), 19–29.
Edenhofer, O., Sokona, Y., Minx, J. C., Farahani, E., Kadner, S., Seyboth, K., Adler, A., Baum, I., Brunner, S., Kriemann, B., Schlömer, J. S. W. M. S., von Stechow, C., & Scientist, T. Z. S. (2014). *Climate change 2014 mitigation of climate change working group III contribu-*

tion to the fifth assessment report of the intergovernmental panel on climate change edited by. Cambridge University Press.

Elliott, J., Deryng, D., Müller, C., Frieler, K., Konzmann, M., Gerten, D., Glotter, M., Flörke, M., Wada, Y., Best, N., Eisner, S., Fekete, B. M., Folberth, C., Foster, I., Gosling, S. N., Haddeland, I., Khabarov, N., Ludwig, F., Masaki, Y., Olin, S., Rosenzweig, C., Ruane, A. C., Satoh, Y., Schmid, E., Stacke, T., Tang, Q., & Wisser, D. (2014). Constraints and potentials of future irrigation water availability on agricultural production under climate change. *Proceedings of the National Academy of Sciences of the United States of America, 111*(9), 3239–3244.

Ezenne, G. I., Jupp, L., Mantel, S. K., & Tanner, J. L. (2019). Current and potential capabilities of UAS for crop water productivity in precision agriculture. *Agricultural Water Management, 218*, 158–164.

Falkenmark, M. (2013). Growing water scarcity in agriculture: Future challenge to global water security. *Philosophical Transactions of the Royal Society A: Mathematical, Physical and Engineering Sciences, 371*(2002), 20120410.

Fang, Q. X., Ma, L., Green, T. R., Yu, Q., Wang, T. D., & Ahuja, L. R. (2010). Water resources and water use efficiency in the North China plain: Current status and agronomic management options. *Agricultural Water Management, 97*(8), 1102–1116.

Farooq, N., & Gheewala, S. H. (2020). Assessing the impact of climate change on sugarcane and adaptation actions in Pakistan. *Acta Geophysica, 68*(5), 1489–1503.

Finley, J. W., & Seiber, J. N. (2014). The nexus of food, energy, and water. *Journal of Agricultural and Food Chemistry, 62*(27), 6255–6262.

Fujihara, Y., Tanaka, K., Watanabe, T., Nagano, T., & Kojiri, T. (2008). Assessing the impacts of climate change on the water resources of the Seyhan River basin in Turkey: Use of dynamically downscaled data for hydrologic simulations. *Journal of Hydrology, 353*(1–2), 33–48.

Gado, T. A., Hsu, K., & Sorooshian, S. (2017). Rainfall frequency analysis for ungauged sites using satellite precipitation products. *Journal of Hydrology, 554*, 646–655.

García-López, J., García-Ruiz, R., Domínguez, J., & Lorite, I. J. (2019). Improving the sustainability of farming systems under semi-arid conditions by enhancing crop management. *Agricultural Water Management, 223*, 105718.

Golfam, P., & Ashofteh, P. S. (2022). Performance indexes analysis of the reservoir-hydropower plant system affected by climate change. *Water Resources Management, 36*(13), 5127–5162.

Gornall, J., Betts, R., Burke, E., Clark, R., Camp, J., Willett, K., & Wiltshire, A. (2010). Implications of climate change for agricultural productivity in the early twenty-first century. *Philosophical Transactions of the Royal Society B: Biological Sciences, 365*(1554), 2973–2989.

Gohari, A., Mirchi, A., & Madani, K. (2017). System dynamics evaluation of climate change adaptation strategies for water resources management in Central Iran. *Water Resources Management, 31*, 1413–1434. https://doi.org/10.1007/s11269-017-1575-z

Grafton, R. Q., Williams, J., Perry, C. J., Molle, F., Ringler, C., Steduto, P., Udall, B., Wheeler, S. A., Wang, Y., Garrick, D., & Allen, R. G. (2018). The paradox of irrigation efficiency. *Science, 361*(6404), 748–750. https://doi.org/10.1126/science.aat9314

Guan, Y., Yang, X., Wang, Y., Zhang, N., & Chu, C. (2021). Assessing the health impacts attributable to PM2.5 and ozone pollution in 338 Chinese cities from 2015 to 2020. *Environmental Pollution, 286*, 117498. https://doi.org/10.1016/j.envpol.2021.117498

Hanjra, M. A., Blackwell, J., Carr, G., Zhang, F., & Jackson, T. M. (2012). Wastewater irrigation and environmental health: Implications for water governance and public policy. *International Journal of Hygiene and Environmental Health, 215*(3), 255–269.

Haro-Monteagudo, D., Palazón, L., & Beguería, S. (2022). Enhancing global hydrological models with local knowledge to support water management: A proof-of-concept in the Ebro River basin. *Science of the Total Environment, 830*, 154687. https://doi.org/10.1016/j.scitotenv.2022.154687

Higgins, J., & Green, S. (2008). Front matter. In *Cochrane handbook for systematic reviews of interventions* (pp. i–xxi).

Hordofa, A. T., Leta, O. T., Alamirew, T., & Chukalla, A. D. (2022). Response of Winter Wheat Production to Climate Change in Ziway Lake Basin. *Sustainability, 14*(20), 13666.

Hu, C., Delgado, J. A., Zhang, X., & Ma, L. (2005). Assessment of groundwater use by wheat (Triticum aestivum L.) in the Luancheng Xian region and potential implications for water conservation in the northwestern North China plain. *Journal of Soil and Water Conservation, 60*(2), 80–88.

Kang, S., Hao, X., Taisheng, D., Tong, L., Xiaoling, S., Hongna, L., Li, X., Huo, Z., Li, S., & Ding, R. (2017). Improving agricultural water productivity to ensure food security in China under changing environment: From research to practice. *Agricultural Water Management, 179*, 5–17.

Kessler, M. M. (1963). Bibliographic coupling between scientific papers. *American Documentation, 14*(1), 10–25.

Keyhanpour, Z., Sadeghi, S. H., & Jafari, H. R. (2021). Sustainable food production governance through system dynamics: A case study in Iran. *Journal of Cleaner Production, 320*, 128878. https://doi.org/10.1016/j.jclepro.2021.128878

Laghari, A. N., Vanham, D., & Rauch, W. (2011). The Indus basin in the framework of WRM. The Indus basin in the framework of current and future water resources management. The Indus basin in the framework of WRM. *Hydrology and Earth System Sciences Discussions, 8*, 2263–2288.

Laghari, A. N., Vanham, D., & Rauch, W. (2012). The Indus basin in the framework of current and future water resources management. *Hydrology and Earth System Sciences, 16*(4), 1063–1083.

Latif, S. D., Ahmed, A. N., Sathiamurthy, E., Huang, Y. F., & El-Shafie, A. (2021). Evaluation of deep learning algorithm for inflow forecasting: A case study of durian Tunggal reservoir, peninsular Malaysia. *Natural Hazards, 109*(1), 351–369.

Lee, S., Kim, H., Kim, J., & Lee, J. (2022). A bibliometrics review of hotspots in water footprint research based on co-words network analysis. *Frontiers in Environmental Science, 10*, 1027936. https://doi.org/10.3389/fenvs.2022.1027936

Leitão, J., Pereira, D., Gonçalves, Â., & Oliveira, T. (2023). Digitalizing the pillars of hybrid civic universities: A bibliometric analysis and new taxonomy proposal. *Journal of Open innovation: Technology, Market, and Complexity, 9*(1), 100026. https://doi.org/10.1016/j.joitmc.2023.100026

Liu, W., Ye, T., Jägermeyr, J., Müller, C., Chen, S., Liu, X., & Shi, P. (2021). Future climate change significantly alters interannual wheat yield variability over half of harvested areas. *Environmental Research Letters, 16*(9), 094045. https://doi.org/10.1088/1748-9326/ac1fbb

Loukas, A., Mylopoulos, N., & Vasiliades, L. (2007). A modeling system for the evaluation of water resources management strategies in Thessaly, Greece. *Water Resources Management, 21*, 1673–1702.

Lyon, S. W., Fischer, B. M. C., Morillas, L., Conejo, J. R., Sánchez-Murillo, R., Serrano, A. S., Frentress, J., Cheng, C. H., Garcia, M., & Johnson, M. S. (2022). On the Potential of Biochar Soil Amendments as a Sustainable Water Management Strategy. *Sustainability, 14, 7026*.

Mancosu, N., Snyder, R. L., Kyriakakis, G., & Spano, D. (2015). Water scarcity and future challenges for food production. *Water, 7*(3), 975–992.

Mateos, L. (2016). Irrigation scheduling in Spain: Experiences and approaches. *Water, 8*(6), 229. https://doi.org/10.3390/w8060229

Megdal, S. B., Eden, S., & Shamir, E. (2017). Water governance, stakeholder engagement, and sustainable water resources management. *Water, 9*(3), 190.

Mekonnen, M. M., & Hoekstra, A. Y. (2016). Sustainability: Four billion people facing severe water scarcity. *Science Advances, 2*(2), e1500323.

Morrison, J., Morikawa, M., Murphy, M., & Schulte, P. (2009). *Growing risks for businesses & investors water scarcity & climate change*. Authored by the Pacific Institute A Ceres Report.

Mostafa, S. M., Wahed, O., El-Nashar, W. Y., El-Marsafawy, S. M., & Abd-Elhamid, H. F. (2021). Impact of climate change on water resources and crop yield in the middle Egypt region. *Journal of Water Supply: Research and Technology-AQUA, 70*(7), 1066–1084.

Nam, W. H., Choi, J. Y., & Hong, E. M. (2015). Irrigation vulnerability assessment on agricultural water supply risk for adaptive management of climate change in South Korea. *Agricultural Water Management, 152*, 173–187.

Nüsser, Marcus, Juliane Dame, Sitara Parveen, Benjamin Kraus, Ravi Baghel, & Susanne Schmidt. 2019. «Cryosphere-fed irrigation networks in the northwestern Himalaya: Precarious livelihoods and adaptation strategies under the impact of climate change». 39(2):R1–11. https://doi.org/10.1659/MRD-JOURNAL-D-18-00072.1

Olesen, J. E., & Bindi, M. (2002). Consequences of climate change for European agricultural productivity, land use and policy. *European Journal of Agronomy, 16*(4), 239–262.

Owens, A., Kviten, K., Zarella, O., Caruso, B. A., Conrad, A., & Patrick, M. (2022). Water, sanitation, and women's empowerment: A systematic review and qualitative metasynthesis. *PLOS Water, 1*(6), e0000026. https://doi.org/10.1371/journal.pwat.0000026

Paul, M., Dangol, S., Kholodovsky, V., Sapkota, A. R., Negahban-Azar, M., & Lansing, S. (2020). Modeling the impacts of climate change on crop yield and irrigation in the Monocacy River watershed, USA. *Climate, 8, 139.*

Pereira, D., Leitao, J. C. C., Gaspar, P. D., Fael, C., Falorca, I., Khairy, W., Wahid, N., El Yousfi, H., Bouazzama, B., Siering, J., Hansmann, H., Zascerinska, J., Camilleri, S., Busuttil, F., Borg, M., Mizzi, J., Micallef, R., & Cutajar, J. (2023a). Exploring irrigation and water supply technologies for smallholder farmers in the mediterranean region. *Sustainability, 15*, 6875. https://doi.org/10.3390/su15086875

Pereira, D., Leitão, J., Oliveira, T., & Peirone, D. (2023b). Proposing a holistic research framework for university strategic alliances in sustainable entrepreneurship. *Heliyon, 9*(5), e16087.

Pereira, D., Leitao, J. C. C., Gaspar, P. D., Fael, C., Falorca, I., Khairy, W., Wahid, N., El Yousfi, H., Bouazzama, B., Siering, J., Hansmann, H., Zascerinska, J., Camilleri, S., Busuttil, F., Borg, M., Mizzi, J., Micallef, R., & Cutajar, J. (2023). Exploring irrigation and water supply technolo gies for smallholder farmers in the mediterranean region. *Sustainability, 15*, 6875. https://doi.org/10.3390/su15086875

Rahimi-Moghaddam, S., Kambouzia, J., & Deihimfard, R. (2019). Optimal genotype × environment × management as a strategy to increase grain maize productivity and water use efficiency in water-limited environments and rising temperature. *Ecological Indicators, 107*, 105570.

Risal, P., Muniyapillai, A., & Dhanasekaran, S. (2022). Water, sanitation, and hygiene (WASH) practices among households in rural and urban field practice areas of a tertiary care center in Perambalur, India. *Cureus, 14*(11), e31213. https://doi.org/10.7759/cureus.31213

Rockström, J., Falkenmark, M., Allan, T., Folke, C., Gordon, L., Jägerskog, A., Kummu, M., Lannerstad, M., Meybeck, M., Molden, D., Postel, S., Savenije, H. H. G., Svedin, U., Turton, A., & Varis, O. (2014). The unfolding water drama in the Anthropocene: Towards a resilience-based perspective on water for global sustainability. *Ecohydrology, 7*(5), 1249–1261.

Rodell, M., Velicogna, I., & Famiglietti, J. S. (2009). Satellite-based estimates of groundwater depletion in India. *Nature, 460*(7258), 999–1002.

Salmon, J. M., Friedl, M. A., Frolking, S., Wisser, D., & Douglas, E. M. (2015). Global rain-fed, irrigated, and paddy croplands: A new high resolution map derived from remote sensing, crop inventories and climate data. *International Journal of Applied Earth Observation and Geoinformation, 38*, 321–334.

Santikayasa, I. P., Babel, M. S., Shrestha, S., Jourdain, D., & Clemente, R. S. (2014). Evaluation of water use sustainability under future climate and irrigation management scenarios in Citarum River Basin, Indonesia. *International Journal of Sustainable Development & World Ecology, 21*(2), 181–194. https://doi.org/10.1080/13504509.2014.884023

Sarindizaj, E. E., & Zarghami, M. (2019). Sustainability assessment of restoration plans under climate change by using system dynamics: Application on Urmia Lake, Iran. *Journal of Water and Climate Change, 10*(4), 938–952.

Shrestha, A. B., & Nepal, S. (2016). *Springs, storage towers, and water conservation in the mid-hills of Nepal*. International Centre for Integrated Mountain Development (ICIMOD).

Sinha, J. (2018). Assessment of the impacts of climatic variability and anthropogenic stress on hydrologic resilience to warming shifts in peninsular India. *Journal of Hydrology, 561*, 943–954. https://doi.org/10.1016/j.jhydrol.2018.04.006

Sauer, T., Havlík, P., Schneider, U. A., Schmid, E., Kindermann, G., & Obersteiner, M. (2010). Agriculture and resource availability in a changing world: The role of irrigation. *Water Resources Research, 46*(6), 6503.

Scanlon, B. R., Faunt, C. C., Longuevergne, L., Reedy, R. C., Alley, W. M., McGuire, V. L., & McMahon, P. B. (2012). Groundwater depletion and sustainability of irrigation in the US High Plains and Central Valley. *Proceedings of the National Academy of Sciences of the United States of America, 109*(24), 9320–9325.

Sidhu, H. S., Jat, M. L., Singh, Y., Sidhu, R. K., Gupta, N., Singh, P., Pankaj Singh, H. S., & Jat, & Bruno Gerard. (2019). Sub-surface drip fertigation with conservation agriculture in a rice-wheat system: A breakthrough for addressing water and nitrogen use efficiency. *Agricultural Water Management, 216*, 273–283.

Strzepek, K. M., Major, D. C., Rosenzweig, C., Iglesias, A., Yates, D. N., Holt, A., & Hillel, D. (1999). New methods of modeling water availability for agriculture under climate change: The u.s. CORNBELT1. *JAWRA Journal of the American Water Resources Association, 35*(6), 1639–1655.

Teklay, A., Dile, Y. T., Asfaw, D. H., Bayabil, H. K., & Sisay, K. (2022). Evaluation of static and dynamic land use data for watershed hydrologic process simulation: A case study in Gummara watershed, Ethiopia. *Catena, 213*, 106213. https://doi.org/10.1016/j.catena.2022.106213

Vaghefi, S. A., Abbaspour, K. C., Faramarzi, M., Srinivasan, R., & Arnold, J. G. (2017). Modeling crop water productivity using a coupled SWAT–MODSIM model. *Water, 9, 157.*

Valverde, P., Serralheiro, R., de Carvalho, M., Maia, R., Oliveira, B., & Ramos, V. (2015). Climate change impacts on irrigated agriculture in the Guadiana river basin (Portugal). *Agricultural Water Management, 152*, 17–30.

Varela-Ortega, C., Blanco-Gutiérrez, I., Swartz, C. H., & Downing, T. E. (2011). Balancing groundwater conservation and rural livelihoods under water and climate uncertainties: An integrated hydro-economic modeling framework. *Global Environmental Change, 21*(2), 604–619.

Veettil, A. V., Mishra, A. K., & Green, T. R. (2022). Explaining water security indicators using hydrologic and agricultural systems models. *Journal of Hydrology, 607*, 127463. https://doi.org/10.1016/j.jhydrol.2022.127463

Vico, G., & Porporato, A. (2010). Traditional and microirrigation with stochastic soil moisture. *Water Resources Research, 46*(3), 1.

Vico, G., & Porporato, A. (2011). From rainfed agriculture to stress-avoidance irrigation: I. A generalized irrigation scheme with stochastic soil moisture. *Advances in Water Resources, 34*(2), 263–271.

Vörösmarty, C. J., Green, P., Salisbury, J., & Lammers, R. B. (2000). Global water resources: Vulnerability from climate change and population growth. *Science, 289*(5477), 284–288.

Wang, X. J., Zhang, J. Y., Ali, M., Shahid, S., He, R. M., Xia, X. H., & Jiang, Z. (2016). Impact of climate change on regional irrigation water demand in Baojixia irrigation district of China. *Mitigation and Adaptation Strategies for Global Change, 21*(2), 233–247.

Wang, Y., Zhou, Y., Franz, K. J., Zhang, X., Qi, J., Jia, G., & Yang, Y. (2022). Irrigation plays significantly different roles in influencing hydrological processes in two breadbasket regions. *Science of the Total Environment, 844*, 157253.

Ward, F. A., Crawford, T., O'Connell, M., & Gutzler, D. S. (2019). Integrated modeling to assess climate change impacts on water and agriculture in the Rio Grande Basin. *Frontiers in Environmental Science, 7*, 58. https://doi.org/10.3389/fenvs.2019.00058

Woznicki, S. A., Pouyan Nejadhashemi, A., & Parsinejad, M. (2015). Climate change and irrigation demand: Uncertainty and adaptation. *Journal of Hydrology: Regional Studies, 3*, 247–264.

Yousefi, K. P., Belleflamme, A., Goergen, K., & Kollet, S. (2022). Impact of deep learning-driven precipitation corrected data using near real-time satellite-based observations and model forecast in an integrated hydrological model. *Frontiers in Water, 4*, 1439906. https://doi.org/10.3389/frwa.2024.1439906

Zhao, M., & Boll, J. (2022). Adaptation of water resources management under climate change. *Frontiers in Water, 4*, 185.

Water Security in Central Asia: Towards Sustainable Development and Resilience

Iskandar Abdullaev and Shavkat Rakhmatullaev

Abstract Climate change will likely generate additional risks to the sustainable operation of water infrastructure and the provision of related services (irrigation, water supply, and hydropower) in arid Central Asia. This article discusses the current situation in water security realities in the context of climate change. Dam safety, groundwater, and return flows will be discussed as vital adaptation and mitigation options. Energy efficiency is an essential element for resiliency at the operational level of water management. In addition, this article sheds light on successful examples of management of shared irrigation infrastructure on transboundary watercourses, which is a prerequisite for regional stability and prosperity. This research contributes to the construction of an agenda on the process of climate change and adaptative water resources management in Central Asia.

Keywords Water resources · Climate change · Energy efficiency · Irrigation · Dam safety · Central Asia

1 Introduction

The region is particularly vulnerable to the impacts of climate change due to its arid and semiarid environment, high levels of poverty, and reliance on natural resources for livelihoods (ABD, 2023; Abdullaev & Rakhmatullaev, 2022; WWAP, 2020).

I. Abdullaev (✉)
Centre for International Development and Environmental Research (ZEU), Justus Liebig University, Giessen, Germany

S. Rakhmatullaev
Tashkent University of Irrigation and Melioration, Tashkent, Uzbekistan

J. C. Correia Leitão et al. (eds.), *Sustainable Water Management and Irrigation Systems*, Natural Resource Management and Policy 62,
https://doi.org/10.1007/978-3-031-93587-9_3

According to the World Bank, up to 2.4 million internal and external climate migrants may appear in Central Asia countries (CAC) by 2050, becoming the new norm (Clement et al., 2021). Hotspots of climate change are expected to emerge along the southern border of Kazakhstan, in areas around the Fergana Valley, some regions of eastern Turkmenistan, and southern Uzbekistan along the Amu Darya River, where water and land availability and crops are predicted to decline (UNCCD, 2022). Migration can be in-country and cross-border, pushing people from rural settlements to big cities.

A review of national migration policies indicates that climate change considerations are yet to be included. Achieving a resilient society where people can adapt and migrate with dignity toward areas of higher opportunity is integral to meeting national sustainable development goals. Inaction will be expensive, and intelligent climate investments are necessary for resilience in irrigation, drinking, and wastewater subsectors with improved early warning systems in the context of disaster risk (Adelphi and CAREC, 2017).

Water scarcity creates new challenges for many countries, including economic, environmental, and social losses in Central Asia (Table 1).

The region's agriculture sector is particularly vulnerable, with declining crop yields and increasing water scarcity. For example, manual labor jobs in agriculture are at high risk due to heat stress and productivity losses. This is leading to a decline in rural livelihoods and increased poverty. In urban areas, climate change significantly impacts public health, with increasing heatwaves and air pollution levels (Climate Risk Country Profiles: Kazakhstan, Kyrgyz Republic, Tajikistan, Turkmenistan and Uzbekistan 2021, WB and ADB). Low productivity of water use

Table 1 Estimated water scarcity losses (Authors' compilation)

#	Water scarcity impacts	Assessments
1	Agricultural losses due to crop failure, land abandonment	• 15–20% crop losses (Bekchanov & Lamers, 2016) • Around 10% of irrigated areas were abandoned (Rakhmatullaev & Abdullaev, 2014)
2	Sectoral losses: Energy, industries, irrigation	• Reduced hydropower production in water-scarce river basins (Didovets et al., 2021) • Water balance will be altered by Afghanistan factor (GET, 2023; EDB, 2023)
3	Social costs: Migration, education, health	• Around 30% of the population will be migrating internally (CAREC Institute, 2022) • Medical costs will be increased by 10–15% (Bekturganov et al., 2016)
4	Environmental losses: Land degradation, biodiversity	• Reduced biodiversity, drying out of natural lakes (Karthe et al., 2018). • Desertification (UNCCD, 2022) • Aral Sea and wetlands (Zan et al., 2022; Wang et al., 2023)
5	Political costs	• Potential water disputes and conflicts (Berndtsson & Tussupova, 2020)

in agriculture continues to absorb a disproportionate share of land, water, and labor resources (Rakhmatullaev et al., 2018). Diversifying crop production (still predominantly wheat and cotton) enables the spread of risk and increases trade opportunities.

One of the most significant impacts of climate change in Central Asia is the melting of glaciers, which provide a critical water source for the region. According to a study by Aizen et al. (2016), the Tien Shan Mountain range spans Kyrgyzstan, Kazakhstan, and Uzbekistan and has lost 27% of its ice volume since the 1960s. This has reduced water availability, particularly during the summer when the demand for irrigation is high. As a result, farmers in the region face declining crop yields and food insecurity.

The Afghanistan factor (construction of Qushtepa irrigation canal—285 km) will impact the water balance of the Amudarya River basin, especially during dry years for downstream reaches with its water development plans.

The decline in water availability is also having an impact on Central Asia's energy sector. The region relies heavily on hydropower, especially in the Kyrgyz Republic and Tajikistan, but reduced water availability is affecting the capacity of power plants, particularly in the summer months (CAREC Institute, 2022). This has led to power shortages and increased reliance on fossil fuels, further exacerbating the region's carbon footprint.

For example, extreme rainfall events will lead to flooding, affecting physical assets and turbidity levels and, thus, water quality. Increased surface temperatures can lead to algal bloom, again affecting water quality, increasing evapotranspiration, and reducing water supply.

Climate change is also having a significant impact on Central Asia's biodiversity. The region is home to diverse flora and fauna, including endangered species such as the snow leopard. However, climate change alters ecosystems, resulting in the loss of habitats and species. For example, the decline in water availability is affecting wetland ecosystems, which are critical habitats for migratory birds.

The Notre Dame Global Adaption Initiative (ND-GAIN) Index ranks 181 countries using a score calculating their vulnerability to climate change and other global challenges and their readiness to improve resilience (ND-GAIN, 2023). The more vulnerable a country is, the lower its score, while the more ready it is to improve its resilience, the higher it will be. Figure 1 depicts a snapshot of the CAC standings in the global rankings from 2005 to 2021.

As we can see, Kyrgyzstan and Uzbekistan have progressed their rankings for resilience, yet all countries still face adaptation challenges. Although the CAC is part of the international conventions on climate change, national adaptation policy frameworks still require more fine-tuning reflected in the Climate Risk Country Profiles: Kazakhstan, Kyrgyz Republic, Tajikistan, Turkmenistan, and Uzbekistan (WB and ADB, 2021a, 2021b, 2021c, 2021d, 2021e). Weak institutional capacities and only declarative political commitments cannot change the situation.

All nations face the same challenges related to sustainability in providing irrigation, drinking water, and energy services. At an operational level, resilient service delivery requires building and improving the capacity of local water and energy

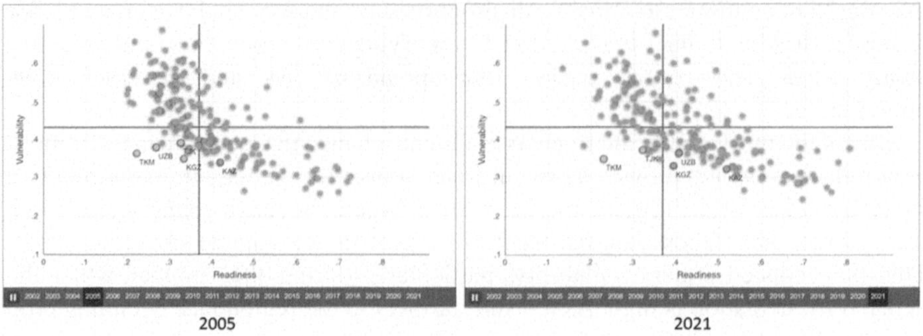

Fig. 1 Comparative resilience of countries for 2005 and 2021 (ND-GAIN, 2023)

service utilities to assess and identify climate risks, adapt their current practices to sustainable operation, and adequately respond to climate change challenges with proper plans and financing (Abdullaev, 2019).

The main goal of this paper is to discuss cross-cutting issues of water scarcity in the context of sectoral issues such as energy efficiency in the water sector (irrigation and drinking water), groundwater and return flow management, dam safety, and modernization. Transboundary cooperation is also examined through examples of bilateral shared irrigation infrastructure.

2 Importance of Energy Efficiency in the Water Sector

Reducing the energy footprint of the water sector is integral to efficient resource use and sustainable management of water systems. The delivery of reliable water services requires significant energy inputs. Therefore, improving efficiency is necessary to manage costs and mitigate potential scarcity risks. Uzbekistan plans to increase the area under drip irrigation from 44,000 ha to 600,000 ha by 2030 in Uzbekistan (MoWR, 2023). While these methods have greater efficiency in water application and saving water, higher energy inputs is required for pressurizing water through pumps.

The energy needs for irrigation and drainage (I&D) are higher, especially in semiarid and arid environments. On average, about 18% of the total irrigated areas in Central Asia are watered by electrically pump-lifted systems covering about 1.84 million ha (Rakhmatullaev & Abdullaev, 2014). During peak growing seasons, monthly irrigation energy demand for irrigation can be up to 20 percent of national energy requirements (Table 2). About half of the total irrigated lands are equipped with horizontal and vertical drainage facilities, yet the majority of them have fallen into disrepair and need rehabilitation.

In addition, approximately 500,000 ha of the area is irrigated from groundwater resources via irrigation wells, which requires approximately 30 percent more energy than surface water irrigation. Energy efficiency programs will be thus an integral

Table 2 Some characteristics of electrically pumped I&D systems in CAC (Dukhovny et al., 2007; Rakhmatullaev & Abdullaev, 2014; WB and FAO, 2022; EDB, 2023)

	Kazakhstan	Kyrgyz Republic	Tajikistan	Turkmenistan	Uzbekistan
Total irrigated land ('000 ha)	2.065[a]	1.021[a]	742	1.990	4.198
Total irrigated land by pump-lift ('000 ha)	41.3	51.1	296	318	1.133
% irrigated land equipped by drainage	29	14	47	51	66
% of pump-lifted irrigated lands	2	5	40	16	27
Total number of pumping stations (big, inter-farm and on-farm)	61	149	1054	3032	14,246
Number of vertical drainage systems/pumps	1988	64	1912	254	4431
Total number of irrigation boreholes/wells	136	383	1030	1365	5022
Total electricity used in irrigation (GWh/year)	2	0.2	1.5	2	8
% of electricity use by irrigation	3	1	11	10	20

[a] The whole of Kazakhstan and the Kyrgyz Republic, not only the Aral Sea Basin parts

part of mitigation measures. The lower water levels in waterways and irrigation canals may increase the demand for electricity to run and operate pump-lifted irrigation infrastructure sustainably.

Population growth and urbanization will place greater demand for drinking water and sanitation (WSS), with more stringent wastewater quality standards as the governments implement a universal access agenda. WSS services are typically the largest energy consumers in municipalities, often accounting for a third of the total energy consumed in the cities and towns (Fig. 2).

Outdated and oversized pumps, which lead to high operational costs and power outages, are often cited as the cause of intermittent service. For example, for Uzbekistan, in 2017, the water utilities consumed around 640 million kWh of electricity, which represents around 1.2% of the country's total electricity consumption. Energy costs for the utilities, at 2017 prices, were around US$19.5 million, which is 20–50% of the utilities' total operating costs. Furthermore, such costs will rise as suppressed energy tariffs rise over the coming years (WB, 2022).

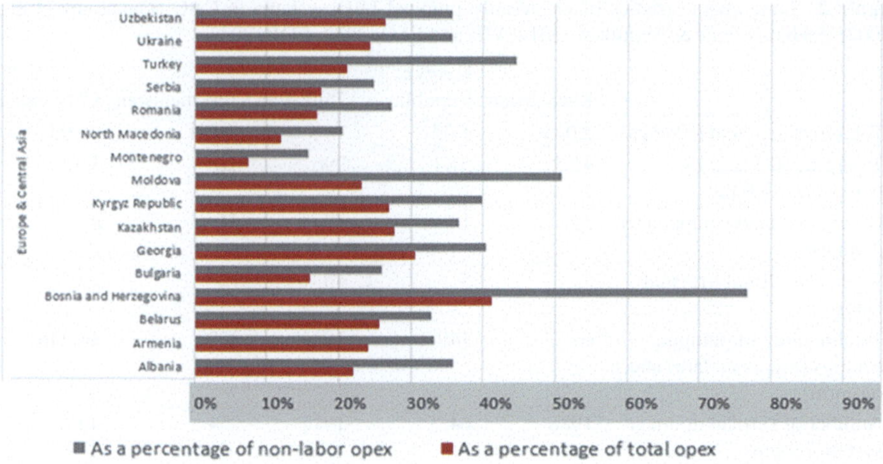

Fig. 2 Energy costs as a share of operating costs in selected countries, average for all available years (source: IBNET database)

3 Groundwater and Drainage Flow Resources

3.1 Groundwater Resources

Under climate change, groundwater will be both a vulnerable resource and a pivotal source for water users (UNESCO, 2022). There was a sufficient quantity and quality of surface water resources for different uses in the past (Rakhmatullaev et al., 2010). The water infrastructure financing came from the soviet state budget and, therefore, was well maintained. Then, national strategic development programs prioritized increasing irrigated areas via the construction or extension of irrigation systems or greater use of groundwater resources.

Very complicated hydrogeological conditions characterize the Central Asian region (Rakhmatullaev et al., 2012). Groundwater resources can be classified according to their recharge processes into two main classes: (1) groundwater formed under natural conditions in the mountain zone and catchment areas by infiltration of rainfall (autochthonous groundwater) and (2) groundwater formed from the infiltration losses from irrigated areas (allochthonous groundwater). In the region, groundwater basins occur along the alluvial valleys of the rivers. The mountains are the main groundwater recharge areas characterized by a complex hydrogeology. In the arid lowlands, local and shallow aquifers predominate often with saline groundwater.

In Central Asia, the total amount of potential groundwater reserves is estimated at 123 billion cubic meters (BCM/year), only 23% of which is currently utilized. Domestic and drinking purposes utilize around 38%, 42% for irrigation, and 8% for industry across CACs (Fig. 3).

The data suggests that over half of groundwater is used for irrigation across all countries. For example, the national inventory of wells only covers merely a quarter

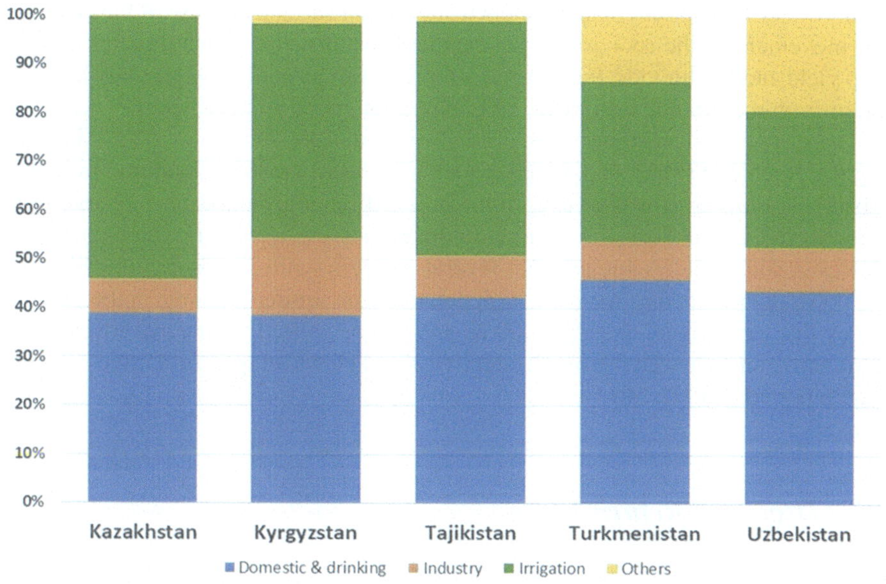

Fig. 3 Groundwater use (%) by sectors across countries of Central Asia (various sources)

of existing and unregistered wells in Uzbekistan. This situation can be easily interpolated into other countries. In the region, groundwater wells should be equipped with water-efficient technologies. Sticker control on groundwater use must be introduced in order to avoid overexertion of resources.

The main challenges in groundwater management in Central Asia are overexploitation, contamination, and inadequate monitoring and regulation. Overexploitation of groundwater resources is a significant problem in the region, with many aquifers being depleted at an alarming rate due to unsustainable use for irrigation and other purposes. This has led to declining water tables, land subsidence, and other negative impacts on the environment and local communities.

Groundwater contamination is another major challenge in the region, with industrial and agricultural activities being the primary sources of pollution. The region's outdated infrastructure and inadequate monitoring and regulation have also contributed to the problem. CACs must adopt a more integrated and sustainable approach to groundwater governance and management to address these challenges. This includes better monitoring and regulation of groundwater resources, promoting water-efficient technologies and practices, and investing in modern infrastructure to improve water quality and reduce losses.

In arid regions, the discharge of collector-drainage water into river systems, its informal reuse, and the chemigation of agriculture have led to regional pollution of unconfined groundwater resources by salts, nitrates, and pesticides. Insufficient drainage has caused secondary salinization of croplands, waterlogging problems, and groundwater pollution (OECD, 2020). The most critical situation is when the groundwater table is less than 2 meters from the soil surface because of the observed salt movement to the root zone by a capillary mechanism, which directly impacts crop yield and livelihoods. Furthermore, there is a weak control of groundwater pollution from the mining sector (i.e., uranium tailing pods and ore heaps).

There are technical solutions in groundwater management, such as the Managed Aquifer Recharge (MAR) approach, which temporarily stores ("banking") water in subsurface horizons for later use. In the past, artificial groundwater recharge technologies in Central Asia were limited to theoretical studies and research modeling. However, some studies suggest that several aquifers (with a capacity of about 1.65 BCM) were identified as having a high potential for aquifer recharge in the Fergana Valley that can irrigate over 500,000 ha and can be shifted from canal irrigation to conjunctive surface water-groundwater irrigation. However, practical uptake has not been reported (IWMI, 2013).

3.2 *Drainage/Return Flows*

The intensity of irrigation requires artificial drainage to control waterlogging and salinization. Drainage also prevents the accumulation of salts in the crop root zone, which could negatively impact crop growth and yields. CACs have about 5.35 million ha, combining surface drainage and vertical and horizontal subsurface drainage (Dukhovny et al., 2007). It is essential that CACs put drainage, wastewater management, and disposal within the context of resilience towards water scarcity realities.

According to expert estimates, about 30–35 billion m^3 of drainage/return flows is generated annually in the Aral Sea basin in high-water years with rapid development of I&D systems, which compromise about 30% of the total water withdrawal (Yakubov et al., 2011; OECD, 2020). About 50% of total return flows are discharged into rivers; 33% discharges into lowland depressions, creating artificial water bodies/ponds; and only 10–15% is reused for irrigation purposes (Fig. 4).

The main constraining factor for drainage flows is salt level (known as mineralization parameter) due to washing of salts from irrigated lands and chemization of agriculture. In upstream countries, Kyrgyzstan and Kazakhstan, about a third of return flows are reused for irrigation against low uptakes in downstream countries. Yet, the volume of return flows is mostly generated in Uzbekistan (24 BCM/year, i.e., 80% out of total in CAC) and in Turkmenistan (5 BCM/year, i.e., 16%) (Yakubov et al., 2011). Prior to the 1960s, mineralization of drainage flows was reported as less than 1 g/L in downstream reaches, yet in the 2010s, it increased up to 2 g/L, especially high values reported during March and April during salt leaching periods.

Drainage/flows are strategic assets under climate change, yet water quality and energy needed for water treatment are worrisome aspects. In general, drainage flows are used for irrigation and salt-leaching purposes. However, desalinization/treatment practices feasibility studies should be carried out for economic rationale, especially in Bukhara, Kashkadarya, Navoi provinces of Uzbekistan, and several provinces of Turkmenistan.

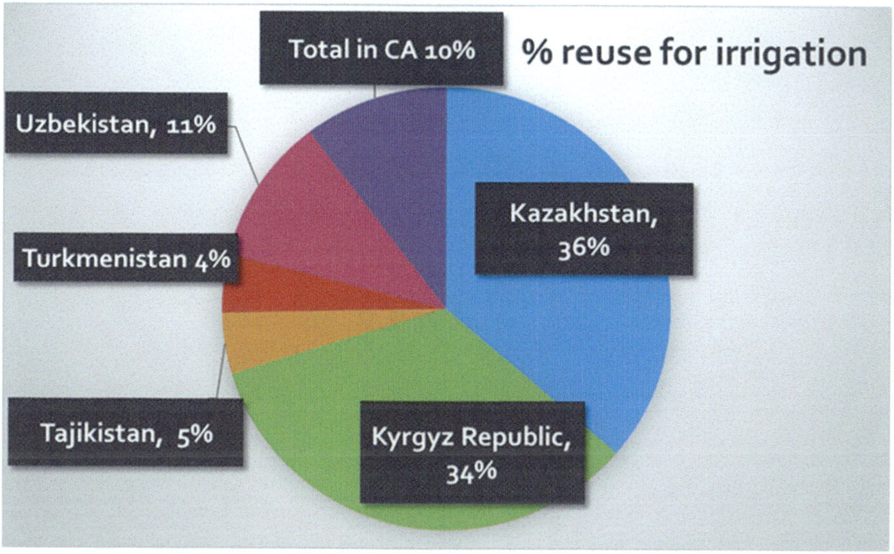

Fig. 4 Collector-drainage flows reuse in irrigation in CACs (Adopted from Yakubov et al., 2011)

4 Water Supply and Sanitation

CACs share common challenges in the WSS sector, such as aging and unsustainable infrastructure, insufficient capital investments, and a lack of institutional capacity (CAREC Institute and UNICEF, 2021). Moreover, the WSS sector is expected to become more vulnerable to the anticipated impacts of climate change (water availability) and population growth. National statistics report high piped drinking water connections, although there is no assurance that the quality of water delivered to the household meets national standards, especially in rural areas (Table 3).

The coverage of sewer connections is declining in countries, both in urban and rural areas. Sanitation usually takes the form of latrines and septic tanks. Generally, many wastewater treatment plants function only nominally (UNICEF and WHO, 2019). There is no accurate measurement, registration, or monitoring of water production and consumption. The water losses are high, and water metering is low in Central Asian countries.

High non-revenue water (NRW) rates can be attributed to technical losses from leakages and nonpayment (or underpayment). Extended and frequent service interruptions are also common in rural and urban areas. Moreover, staff in the WSS sector are often disincentivized due to low remuneration rates and insufficient training and professional development initiatives. These deficiencies hinder the performance of duties and promote corruption and negligence.

In Tajikistan, only 10% of existing rural sewerage systems function in an orderly way, 5% are partially functioning, and 85% do not function. As a result, raw sewage

Table 3 Technical characteristics of water supply and sewerage systems in CAC (Source: various)

Country	Households with piped water connections			Households with sewer connections			Non-revenue water (loss)	Fee collection rate
	Nationwide	Urban	Rural	Nationwide	Urban	Rural		
Uzbekistan	76%	89%	63%	16%	27%	5%	25–40%	85%
Kazakhstan	70%	88–93%	57–83%	58%	84–88%	11%	20–30%	64%
Kyrgyz Republic	43%	60%	25%	–	35%	10%	25–40%	25–50%
Turkmenistan	54%	–	–	–	62%	2%	30–55%	70%
Tajikistan	39%	–	–	–	45.2%	1.7%	20–40%	58%

is often discharged directly into waterways, which creates a persistent risk of waterborne diseases.

Due to intermittent water supply, pressure, and quality problems, connected households typically use multiple water sources to meet their needs, including purchasing more expensive water from tanker deliveries or obtaining untreated water from private shallow wells or irrigation canals. Lower-income households often use irrigation canals, rivers or ponds, and hand pumps in the yards, another sign of disparities in access to safe drinking water. In urban areas, service availability varies between 6 and 24 h per day, while in rural areas, service availability varies between 2 and 12 h per day.

Metered customers receive a volumetric tariff, while households without meters are billed flatly. Customers without meters receive bills that are much higher than their actual consumption. This leads to a lack of incentives for service providers to install meters and their resistance to the Government's efforts to increase the level of metering. On the other hand, consumers paying a flat tariff have no incentive to conserve, while utilities cannot charge a cost-reflective tariff, making them less sustainable.

5 Dam Safety and Modernization

Dams, with their strategic storage (associated water reservoirs), will be increasingly crucial to enhance resilience under the context of the food-water-energy nexus. About 70% of dams and dykes have passed a 50-year milestone in the region (UNECE, 2007). However, the existing dams do not perform up to their optimum levels due to backlog in maintenance, damaged structures, inadequate instrumentation and monitoring, deficient reservoir operation practices, and inadequate regulatory and operational safety measures (UNECE, 2021).

Moreover, sedimentation in reservoirs is increasingly causing a reduction in the storage capacity of dams and increasing the risk of failures. Notably, a recent study indicates that Central Asia dams will lose 26% of their volume by 2050 due to sedimentation only (Perera et al., 2023). In Uzbekistan, recent reports indicate that the average annual reservoir volume loss is estimated at 0.5% (Rakhmatullaev et al., 2013). The total volume capacities of major reservoirs have decreased by about 20%, and the dead storage capacities of the nation's dams have decreased by up to 75% in some reservoirs.

In general, the safety level of large water infrastructure facilities (especially hydropower) in countries is sound, but the safety monitoring system needs to be improved. However, the technical condition of small-scale and low-head water facilities, owned/managed by private, community-based water users associations, is generally rated as unsatisfactory. Moreover, many of these facilities are without clear ownership and, as such, have not been maintained and are without significant repairs for a long time. Dam operators need early warning systems, effective risk management strategies, and modern monitoring and surveillance instruments for sustainable operation and preparing themselves for emergencies.

The countries recognize that water infrastructure safety on transboundary watercourses is a priority area of cooperation. This is reflected in the regional cooperation agreements (Almaty 1992 agreement); regional cooperation program—Aral Sea Basin Program; multilateral and bilateral agreements; and national legislative frameworks. However, this work on regional dam safety is stalled because of disagreement on economic responsibility and liability under failure/incident cases. A draft umbrella agreement could help harmonize legislation, technical regulation, inter-sectorial cooperation, consideration of safety requirements, scientific cooperation, prevention, localization and elimination of emergencies, and technical audits.

In 2019, three countries, Kazakhstan, Kyrgyz Republic, and Uzbekistan, confirmed their consent to continuation. Only Uzbekistan has completed its internal procedures before signing the regional treaty. This framework agreement could become the basis of regional and basin agreements, containing the main directions, forms, and mechanisms of possible mutually beneficial cooperation in dam safety assurance. In addition, this will make it possible to attract international financing for loans and might be attractive for commercial insurance markets.

6 Shared Water Infrastructure

CACs have technical and historical experience in joint management, operation, maintenance (MOM), and rehabilitation of various hydraulic infrastructures as a part of the Soviet era (SIC ICWC, 2008). The Post-Soviet period also indicates successful cooperation examples: the joint operation of hydrotechnical structures on the Chu and Talas rivers between Kazakhstan and Kyrgyz Republic; the Orto-Tokoy (Kasansay) and Andijan (Kempirabad) reservoirs between the Kyrgyz Republic and Uzbekistan; the Farkhad dam and operation of the Kairakkum reservoir between Uzbekistan and Tajikistan; repair and rehabilitation of the Sultan Sanjar dam of the Tuyamuyun hydropower complex between Turkmenistan and Uzbekistan; and declarations by Uzbekistan and Tajikistan for potential joint construction and operation of hydropower plants in the Zarafshan river basin.

Countries of the region engage in transboundary collaboration over a shared hydraulic infrastructure between themselves and neighboring countries such as Afghanistan, Iran, China, and Russia (Table 4).

Kazakhstan is actively involved in transboundary cooperation with Central Asian countries, China, and Russia. Tajikistan has distinguished experience with Iran and Russia on concessional/joint ventures for MOM of hydropower. Turkmenistan and Uzbekistan practice a unique financing mechanism such as "land lease" in transboundary collaboration. In summary, the region has different regulatory and institutional arrangements over MOM of shared water infrastructure, indicating that countries are cooperating. However, scale-up is much needed for regional stability and sustainable water management.

Table 4 Transboundary collaboration on shared water infrastructure between CAC and neighboring countries (authors' compilation)

	Regulatory	Institutional
Kazakhstan	• Agreement on Chu-Talas River Basin with Kyrgyz Republic (2000) • RoadMap on construction of Kambarata HPP with Kyrgyzstan and Uzbekistan (2023) • Draft framework agreement on transboundary watercourses with Uzbekistan (under development). • Framework agreement on transboundary watercourses with China including 24 river basins (2001) • Framework agreement on transboundary watercourses with Russia including 20 river basins (2007) • Basin water organization "Syrdarya" (1992)	• Joint water Commission for Chu-Talas River Basin with Kyrgyz Republic • Working group with Uzbekistan • Joint water commission with China • Joint directorate on Dostluk Hydrocomplex with China. • Joint water commission with Russia
Kyrgyz Republic	• Agreement on Chu-Talas River Basin with Kazakhstan (2000) • RoadMap on construction of Kambarata HPP with Kazakhstan and Uzbekistan (2023) • Agreement on Andijan/Kempirabad dam with Uzbekistan (2022) • Agreement between the Ministry of Water Resources of Uzbekistan and water Service of Kyrgyz Republic on water cooperation (2022) • Agreement on Orto-Tokoy/Kasansay Dam with Uzbekistan (2017) • Draft framework agreement on transboundary watercourses with Tajikistan (under development) • Basin water organization "Syrdarya" (1992)	• Joint water Commission for Chu-Talas River Basin with Kazakhstan • Joint water Commission for Orto-Tokoy/ Kasansay dam with Uzbekistan • Joint water Commission for Andijan/ Kempirabad dam with Uzbekistan • Proposed joint water commission with Tajikistan
Tajikistan	• Agreement on Farhad hydropower complex with Uzbekistan (2018) • Agreement on Pyanj-Amu Darya River Basin with Afghanistan (2010) • Draft framework agreement on transboundary watercourses with Kyrgyz Republic (under development) • Agreement on Sangtuda-1 HPP with Russian (2009) • Agreement on Sangtuda-2 HPP with Iran (2005) • Agreement with Uzbekistan on construction of two hydropower plants in Zarafshan River Basin (2021) • Basin water organizations "Amudarya" and "Syrdarya" (1992)	• Working Group for Farhad hydropower complex with Uzbekistan • Working group with Afghanistan • Proposed joint commission with Kyrgyz Republic • Concessional/commercial arrangement with Russian and Iranian companies

(continued)

Table 4 (continued)

	Regulatory	Institutional
Turkmenistan	Agreements with Uzbekistan • On fee-based land use (1996a) • In water cooperation (1996b) • On crossing the Uzbek-Turkmen border by persons servicing the infrastructure located on the territories of border regions (2004) • On cooperation for operation, repair and rehabilitation of infrastructure facilities of Uzbekistan and Turkmenistan located in the border areas of the Parties (2008) • On the Joint Uzbek-Turkmen Intergovernmental Water Commission (2021) • Amendment to 1996 agreement on fee-based land use (2021) • On the management, protection and sustainable use of water resources of the Amudarya River (2022). • Agreement on Dostluk/Friendship Dam with Iran (2007). • BWO Amudarya and BWO Syrdarya (1992).	• Joint water commission with Uzbekistan • Joint directorate on Dostluk/friendship dam with Iran
Uzbekistan	• Draft framework agreement on transboundary water cooperation with Kazakhstan (under development) • RoadMap on construction of Kambarata HPP with Kazakhstan and Kyrgyz Republic (2023) • Agreement between the Ministry of Water Resources of Uzbekistan and water Service of Kyrgyz Republic on water cooperation (2022) • Agreement on Andijan/Kempirabad dam with Kyrgyz Republic (2022) • Agreement on Orto-Tokoy/Kasansay Dam with Kyrgyz Republic (2017) • Agreement on Farhad hydropower complex with Tajikistan (2018) • Agreement with Tajikistan on construction of two hydropower plants in Zarafshan River Basin (2021) • Agreements with Turkmenistan (1996a, 1996b, 2004, 2008, 2021a, 2021b, 2022). • Basin water organizations "Amudarya" and "Syrdarya" (1992).	• Joint commission on Orto-Tokoy/Kasansay dam with Kyrgyz Republic • Joint commission on Andijan/Kempirabad dam with Kyrgyz Republic • Working group on Farhad HPP with Tajikistan • Joint water commission with Turkmenistan

6.1 Bilateral Water Commissions

Countries established joint intergovernmental water commissions with a mandate to streamline transboundary cooperation in general and for specific water infrastructures, which means there is continuous networking and dialogue. However, those institutions need capacity building and strengthening.

Recent news on the Andijan (/Kempir-Abad) dam with its associated water reservoir (located in Kyrgyz Republic and Uzbekistan) has been in the headlines of most of the newspapers in Kyrgyzstan and Uzbekistan. Two governments signed an intergovernmental agreement on border demarcation. Accordingly, the Kyrgyz Republic will receive 19,000 ha of land as compensation against 4500 ha of land under the Andijan reservoir will go to Uzbekistan. However, local communities in the Kyrgyz Republic opposed this potential deal even though it was declared that a joint venture (50/50 share) would be established to manage this critical water infrastructure jointly.

The main reason for contestation is the absence of community-based organizations as a part of any formal joint water commissions or working groups between countries (Table 5). New stakeholders and public actors will become critical in meeting energy and water-related challenges. In contrast, the traditional actors (water technocratic bureaucrats) are reluctant to engage with "outsiders" and are afraid to lose power and influence.

6.2 BWO (Basin Water Organizations)

Mandates of regional Basin Water Authorities (BWAs) of "Amudarya" and "Syrdarya," as executive bodies for intestate water allocation, are limited given that national authorities manage some water-important hydropower systems with

Table 5 Stakeholder groups represented in joint water commissions across the region

Stakeholder groups	Kazakhstan	Kyrgyz Republic	Tajikistan	Turkmenistan	Uzbekistan
Water management authorities	+	+	+	+	+
Agriculture	–	–	+	+	+
Energy	–	–	+	+	+
Foreign affairs	–	–	+	+	+
Environment	–	–	–	+	+
Local governments	–	–	–	+	+
Other government (epidemiology, border, emergency, hydrometeorology, finance)	–	–	+	–	+
Community-based organizations	–	–	–	–	–

reservoirs. There is no official document where countries of the region would recognize the interstate status of hydraulic structures located on transboundary watercourses.

6.3 Basin Councils

A comprehensive examination of national legal and institutional frameworks shows that since the mid-2000s, the national Water Codes of Kazakhstan, Kyrgyz Republic, and Tajikistan have been amended to introduce consultative advisory platforms (Abdullaev & Rakhmatullaev, 2015). Basin councils are being advocated as formalized platforms to foster inter-sectorial dialogue and support local participatory processes inside a river basin/watershed. Basin councils are established inside operating national water management organizations (WMOs) to enhance their facilitating role and change the behavioral response of other stakeholders to national WMOs, overarching towards balancing the decision-making power among all stakeholders.

Under USAID's funded projects, 19 small basin councils were created from 2011 to 2020 and are functioning in the region, 10 being bilateral (Yakubov, 2021). At the level of small transboundary basins, the challenges emerging in adjacent territories can be addressed at the joint meetings of the small basin councils. This is one of the most effective tools to resolve conflicts and problems related to shared management. Decisions to initiate joint actions in the basin are taken at the BC meetings. Historically, cleaning and minor repairs of irrigation networks in Central Asia have always been carried out using the joint voluntary efforts of local communities (called *ashhar, kashar, khashar*, etc.).

The experience and results of piloting such institutional mechanisms show the feasibility of introducing community participation in water management at national and transboundary levels.

Kazakhstan initiated the creation of basin councils for all major river basins in 2005, and currently, all eight major hydrographic basins of the country have been established. In 2006, the Kyrgyz Republic adopted the Water Code that envisaged the establishment of basin councils, which is still in progress (Abdullaev & Rakhmatullaev, 2016a). Afghanistan, Turkmenistan, and Tajikistan made relevant changes to their water legislation in 2009, 2016, and 2020, respectively, while these institutional changes are still under discussion in Uzbekistan.

Development partners supported countries to successfully showcase transboundary basin councils in selected small transboundary river basins. However, countries somewhat limit practical application due to the absence of sustainable financing, although such institutions are legally recognized in national water legislation.

7 Financing

The water sector (irrigation and WSS) is underfinanced globally, increasing water-related risks (Briscoe, 1998; Camdessus, 2003). According to ADB, Central Asia's annual infrastructure investment needs at the baseline scenario are equivalent to $33 billion or 6.8% of regional GDP (Gross Domestic Product), considering all types of infrastructure, including energy, transport, telecommunications, water, and sewerage (ADB, 2020). The inclusion of measures making this infrastructure more resilient to the impacts of climate change adds additional costs, increasing annual investments to $38 billion or 7.8% of regional GDP.

According to recent estimations, almost 60–75% of the water sector financing in CACs is spent on capital investment/construction and O&M of water infrastructure (CAREC Institute, 2023). Therefore, an important issue is the prioritization of infrastructure financing. The current setting of the water infrastructure is meant for large-scale, mechanized, and collective farming systems. Although the main canals and major water infrastructure are still intact and perform well, numerous diversion and delivery water infrastructures are outsized, delivering excessive water to much smaller lands segmented by small and individualized farms. Therefore, financing the old and outdated water infrastructure may not be adequate for the current agricultural setting.

All countries face the same challenges related to sustainability in the provision of water services. Moreover, the engagement of the private sector has proved to be a cumbersome endeavor not only in irrigation but paradoxically in WSS, as most water infrastructure is still under state/municipality jurisdiction. However, costs for the O&M of these systems are assigned to different organizations and supported through various financial mechanisms. Water user fees have been introduced in all CACs. However, service fees are symbolic and do not cover the operational costs of water systems. The primary sector problems are low public financing, low collection rates of user fees, and reluctance of users to pay owing to unreliable water services and delivery.

Agriculture policies and land ownership issues also hinder private sector engagement in irrigation. The agricultural water management in CACs is a social function of the state. Therefore, the state is not planning to introduce a complete agricultural market system or the full cost recovery arrangements for water delivery. None of the region's countries has a competitive land market for private sector participation in land resources management (CAREC Institute, 2020).

In irrigation, irrigation service fee collection is still far too low (up to 10% for full cost recovery of the O&M) alone. The countries are trying to increase service collection rates by introducing volumetric water metering (Uzbekistan), increasing the role of water user organizations (Kyrgyzstan) and water management organizations (Tajikistan), giving the right to retain part of the fees and introducing a more commercial model of water supply (Kazakhstan). Currently, payments received from water users associations and other irrigation water users for the service delivery cover only about 30% of MOM costs in all five countries (FAO and WB, 2022).

Since water is essentially a public commodity, water tariffs for irrigation are politically and socially sensitive. The "willingness to charge" in countries is low because of social concerns regarding the affordability of irrigation services by the population. Yet, material evidence suggests that farmers are willing to pay for higher irrigation tariffs if the service satisfied their needs.

Although the public-private partnership (PPP) framework has been justified in many ways to stimulate private sector involvement in large infrastructure projects, the rationale for why and how this can be practically implemented is unexplored in the region. In Uzbekistan, the irrigation sector has the highest number of PPP projects, with 157 (CAREC Institute, 2023). Typically, PPPs have been signed between farm clusters and local enterprises with irrigation authorities for the joint MOM of the pumps and pump structures, irrigation wells, and irrigation canals. In addition, the government of Uzbekistan plans to enter into joint development and management agreements for small reservoirs and dams with the private sector.

7.1 Water Use Tax in Uzbekistan

Since 1997, there has been a so-called water-use tax on water withdrawal from surface and groundwater sources. These revenues are allocated to local governments, but they have not been earmarked for financing the irrigation sector up to date. However, the government plans to allocate about 25% of these revenues to irrigation organizations starting in 2024. Payers are legal entities, individual entrepreneurs, and private farmers. Tax rates per m^3 of water depend on the type of water source and economic activity. In 2015 and 2019, new water-tax categories were introduced for enterprises producing nonalcoholic beverages and industrial enterprises for car-washing stations, respectively.

Communal utilities can withdraw water to produce drinking water for the population free of charge; they only have to pay for water resources used for their own operational needs. This implies that technical water losses do not enter into their operating costs. Hydropower plants are also tax-exempted for water withdrawals with salt-leaching practices. Revenues from the water resources tax amounted to UZS 140.4 billion (US$18 million) in 2018, UZS 391.4 billion (US$36.9 million) in 2021, and UZS 688 billion (US$62 million) (WB, 2022). A tax rate of UZS 80 (exchange rate USD1 = UZS12,200) per one cubic meter is established for water used for irrigating agricultural land and fish breeding, including farms.

8 Addressing Challenges and Sustaining Water Security

CACs should introduce stronger water rights and water distribution principles within regional and national legal documents. Strong focus must be given to the prevention of major water risks under climate change—emergence of higher demand and reduced water availability which will lead to the regular water shortages. Important to introduce regular, long-term planning of water resources for basins, subbasins under different climate scenarios would be the most important tool to reduce the impact of climate change.

The water planning in all rivers and basins of the Central Asia is key for the development of sustainable and long-term, climate adaptive water management systems (Abdullaev, 2012). Currently, due to the absence of the long-term planning in all levels, interstate, basin, subbasin, and territories are major problem of water sector. There is no linkage between state of the water resources with development plans (national, territorial).

It is equally important to focus on improving the quality and density of climate and water monitoring systems in the crucial transboundary systems (Abdullaev & Rakhmatullaev, 2016b). The successful adaptation also depends on information openness and digital means as tools for major analysis and communication on climate-related risk in the water sector.

The water dependency in Central Asia is very high for almost all economic development sectors. Moreover, the population's lifestyles are highly water-dependent (Abdullaev & Atabaeva, 2012). The unsystematic and unjust water distribution results in a high water deficit in the tail end of the systems, such as rivers, canals, and settlements. On top of human-made water deficit, climate change will further deteriorate the situation. The poorer areas are most affected due to unequal water distribution. Water management is a social process. Therefore, active engagement, participation, and the voice of water users are vital in preventing water problems/issues.

Moreover, introducing market mechanisms in the water sector and formulating stronger water rights systems are part of the resilience to climate change. Activities must be implemented at water systems' transboundary, national, and operational levels (Fig. 5 and Table 6).

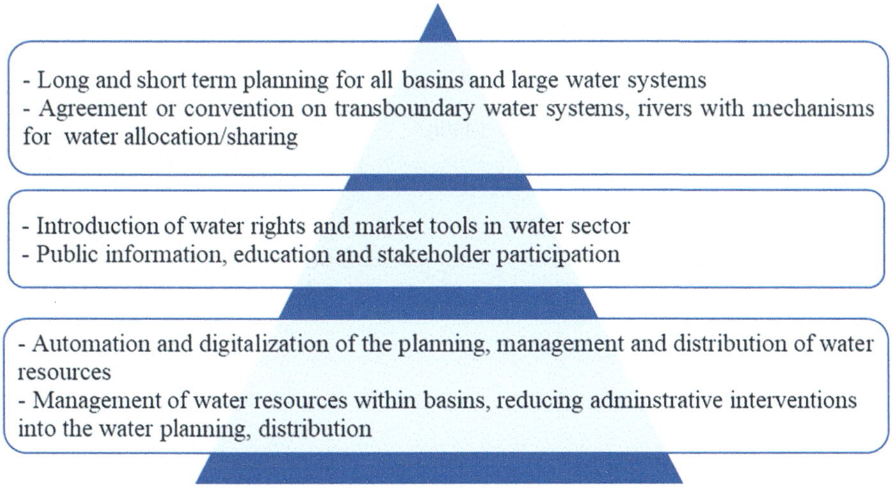

Fig. 5 Activities at different levels of water systems to cope with climate risks

Table 6 Policy options for adaptation measures in water sector (compiled by authors)

Domains	Obstacles/limits	Options
Legislation, regulation and judiciary	Undeveloped or fragmented legal and normative framework for climate change adaptation	• Sector specific adaptation plans • River basin plans • Asset management plans
Institutional and Educational	• Poor institutional knowledge/ know-how • Capacities in design, planning • Poor inter-sectoral coordination. • Absence of curriculum in higher educational system	• Increasing awareness among irrigation/water stakeholders • Incentives for research and development • Basin councils-advisory consultative platforms • Introduction of climate change adaptation into theCurriculum (specialization) of local universities • Professionalize joint water commissions on dispute resolution, arbitration, and negotiation skills
Technical and technological	• High costs of technologies • Integration of renewable/energy efficiency programs into existing water and sewerage infrastructure • Absence of instrumentation/ diagnostics and survey technologies and equipment • Publicly limited data	• Partnership with international producers for domestic production of some technologies • Use of modern GIS, GPS, remote sensing (satellite imagery) technologies • Modern methodologies for diagnostics and surveys • Reforestation and landscape • Automation, digitalization of databases and maps • Where feasible, convert pup-fed schemes to gravity-fed
Fiscal and market	• Low electricity tariffs/subsidies for irrigation • Undeveloped financial mechanisms • Nascent insurance liability programs	• Public-private partnership • Irrigation service fees • Ecosystem services • Increase tariffs for cost recover at O&M level • Independent energy audit system • Attractive bank loans • Government import tax subsidies to local manufacturers/suppliers • Develop and adopt insurance liability programs and mechanisms

The annual economic losses due to land salinity and its abandonment are estimated to be more than hundreds of million dollars. Balanced agricultural cropping systems utilizing climate-smart varieties are prerequisites for the resiliency of countries in terms of food security and foreign revenues. Supporting reforestation can generate multiple benefits not only by providing forage, food, and construction materials but also by improving watershed management and preventing soil erosion.

It is of strategic importance that safety assessments and technical audits be carried out for all water- and energy-related facilities. Harnessing the innovation of solar irrigation pumping would require a nuanced approach to address specific objectives in each local context (Cakmak et al.2004).

Water pricing is one, albeit often unpopular, mechanism for the promotion of more efficient water use, but now is the time for redesigning water pricing for I&D and WSS at least for cost recovery level for O&M in the region. Governments must therefore introduce incentives within the water sector to make it attractive to private and international investors for potential PPPs.

The increased participation of stakeholders and actors requires open-source, easy-to-access information and user-friendly data management systems to be in place to support decision-making and early warning objectives. Information and communication technologies are essential assets for modeling, early warning systems, streamlining billing, and monitoring purposes (Abdullaev & Rakhmatullaev, 2014).

Countries need enhanced knowledge on designing disaster risk-financing strategies and catastrophe insurance pools/bonds for dam safety programs. Dam operators and owners report that there are no third-party civil liability insurance schemes for damages resulting from accidents in the region. Stakeholder analysis shows that public and community-based groups not properly represented in joint water commissions might create potential mistrust of local communities, regardless of good intensions of riparian governments.

References

Abdullaev, I. (2012). *Socio-technical aspects of water resources management in Central Asia*. Lambert Academic.

Abdullaev, I., & Atabaeva, S. (2012). Water sector in Central Asia: Slow transformation and potential for cooperation. *International Journal of Sustainable Society, 4*(1/2), 103–112.

Abdullaev, I., & Rakhmatullaev, S. (2014). Data management for integrated water resources management in Central Asia. *Journal of Hydroinformatics, 16*(6), 1425–1440.

Abdullaev, I., & Rakhmatullaev, S. (2015). Transformation of water management in Central Asia: From state-centric, hydraulic mission to socio-political control. *Environment and Earth Science, 73*, 849–861.

Abdullaev, I., & Rakhmatullaev, S. (2016a). River basin management in Central Asia: Evidence from Isfara Basin, Fergana Valley. *Environmental Earth Science, 75*, 677. https://doi.org/10.1007/s12665-016-5270-9

Abdullaev, I., & Rakhmatullaev, S. (2016b). Setting up the agenda for water reforms in Central Asia: Does the nexus approach help? *Environmental Earth Science, 75*, 870. https://doi.org/10.1007/s12665-016-5409-8

Abdullaev, I., & Rakhmatullaev, S. (2022). Water security and climate change challenges in the transition economies of Central Asia. In O. F. Tanrisever & B. H. Sakal (Eds.), *Water, energy and environment in Eurasia* (pp. 191–222). Cappadocia University Press. https://doi.org/10.35250/kun/9786054448227

Abdullayev, R. (2019). *Development of water sector in Tajikistan, National report*. USAID-CAREC Smart Waters Project.

ADB. (2020). *Asian water development outlook 2020. Advancing Water Security Across Asia and the Pacific*. Asian Development Bank. https://www.adb.org/sites/default/files/publication/663931/awdo-2020.pdf

ADB. (2023). *CAREC 2030: Supporting regional actions to address climate change. A scoping study*. Asian Development Bank. https://doi.org/10.22617/SPR230126-2

Adelphi and CAREC. (2017). Rethinking water in Central Asia. In *The costs of inaction and benefits of water cooperation*. SDC. https://carececo.org/Rethinking%20Water%20in%20Central%20Asia.pdf

Aizen, E. M., Aizen, V. B., Takeuchi, N., Mayewski, P. A., Grigholm, B., Joswiak, D. R., & Schwikowski, M. (2016). Abrupt and moderate climate changes in the mid-latitudes of Asia during the Holocene. *Journal of Glaciology, 62*(233), 411–439.

Bekchanov, & Lamers, J. P. A. (2016). Economic costs of reduced irrigation water availability in Uzbekistan (Central Asia). *Regional Environmental Change, 16*, 2369–2387.

Bekturganov, Z., Tussupova, K., Berndtsson, R., Sharapatova, N., Aryngazin, K., & Zhanasova, M. (2016). Water related health problems in Central Asia—A review. *Water, 8*(6), 219. https://doi.org/10.3390/w8060219

Berndtsson, R., & Tussupova, K. (2020). The future of water Management in Central Asia. *Water, 12*, 2241. https://doi.org/10.3390/w12082241

Briscoe, J. (1998). The Changing Face of Water Infrastructure Financing in Developing Countries. *International Journal of Water Resources Development, 15*, 301.

Cakmak, B., Beyribey, M., & Kodal, S. (2004). Irrigation water pricing in water user associations, Turkey. *International Journal of Water Resources Development, 20*, 113–124.

Camdessus, M. (2003). Financing water for all. In *Report of the World Panel on Financing Water Infrastructure*.

CAREC Institute. (2020). *Climate Vulnerability, Infrastructure, Finance and Governance in the CAREC Region*. CAREC Institute. https://www.carecinstitute.org/publications/climate-vulnerabilityinfrastructure-finance-and-governance-in-carec/

CAREC Institute. (2022). *Water–agriculture–energy nexus in Central Asia through the lens of climate change*. CAREC Institute. https://www.carecinstitute.org/publications/report-on-water-agriculture-energy-nexus-in-central-asia-through-the-lens-of-climate-change/

CAREC Institute. (2023). *Water infrastructure in Central Asia: Promoting sustainable financing and private capital participation*. Central Asia Regional Economic Cooperation Institute, Urumqi, Xinjiang, the PRC. https://www.carecinstitute.org/wp-content/uploads/2023/02/CI-Water-Financing-Report.pdf

CAREC Institute and UNICEF. (2021). *The development of sustainable water and sanitation Systems in Rural Areas of the CAREC region with a focus on China, Mongolia, Tajikistan and Uzbekistan. Research Report*. CAREC Institute and UNICEF. https://www.carecinstitute.org/wp-content/uploads/2021/07/CI-UNICEF-WASH-report-July-2021.pdf

Clement, V., Kanta, R., de Sherbinin, A., Bryan, J., Susana, A., Jacob, S., Nian, S., & Elham, S. (2021). *Groundswell part 2: Acting on internal climate migration*. The World Bank.

Didovets, I., Lobanova, A., Krysanova, V., Menz, C., Babagalieva, Z., Nurbatsina, A., Gavrilenko, N., Khamidov, V., Umirbekov, A., Qodirov, S., Muhyyew, D., & Hattermann, F. F. (2021). Central Asian rivers under climate change: Impacts assessment in eight representative catchments, journal of hydrology. *Regional Studies, 34*, 100779.

Dukhovny, V., Umarov, P., Yakubov, H., & Madramootoo, C. A. (2007). Drainage in the Aral Sea basin. *Irrigation and Drainage, 56*, S91–S100. https://doi.org/10.1002/ird.367

EDB. (2023). Efficient irrigation and water conservation in Central Asia. In E. Vinokurov, A. Ahunbaev, S. Chuyev, A. Adakhayev, & T. Sarsembekov (Eds.), *Reports and working papers 23/4*. Almaty, Kazakhstan, Eurasian Development Bank.

FAO, & WB. (2022). Modernizing Irrigation in Central Asia: Concept and Approaches. In R. Dankova, M. Burton, M. Salman, A. K. Clark, & E. Pek (Eds.), *Directions in Investment, No. 6*. United National Food and Agriculture Organization and The World Bank. https://doi.org/10.4060/cb8230en

German Economic Team (2023). *Qush-tepa canal to significantly impact Uzbekistan's agriculture*. Issue No 25.

IWMI. (2013). *Managed aquifer recharge: The solution for water shortages in the Fergana Valley* (p. 51. (IWMI research report 151)). International Water Management Institute. https://doi.org/10.5337/2013.205

Karthe, D., Chifflard, P., & Bueche, T. (2018). Hydrogeography—linking water resources and their management to physical and anthropogenic catchment processes. *Die Erde; Zeitschrift der Gesellschaft für Erdkunde zu Berlin., 149*. https://doi.org/10.12854/erde-2018-412

ND-GAIN (2023) *Notre Dame Global Adaptation Initiative website*. Retrieved September 26, 2023, from https://gain.nd.edu/our-work/country-index/methodology/

OECD (2020). Overview of the use and management of water resources in Central Asia: A discussion document. Organisation for Economic Co-operation and Development.

Perera, D., Williams, S., & Smakhtin, V. (2023). Present and future losses of storage in large reservoirs due to sedimentation: A country-wise global assessment. *Sustainability, 15*(219). https://doi.org/10.3390/su15010219

Rakhmatullaev, S., & Abdullaev, I. (2014). Central Asian irrigation sector in a climate change context: Some reflections. *Journal of Water and Climate Change, 05*(3), 341–356.

Rakhmatullaev, S., Huneau, F., Kazbekov, J., Le Coustumer, P., Jumanov, J., El Oifi, B., Motelica-Heino, M., & Hrkal, Z. (2010). Groundwater resources use and management in the Amu Darya river basin (Central Asia). *Environmental Earth Sciences, 59*, 1183–1193.

Rakhmatullaev, S., Huneau, F., Kazbekov, J., Celle-Jeanton, H., Motelica-Heino, M., Coustumer, P., & Jumanov, J. (2012). Groundwater resources of Uzbekistan: An environmental and operational overview. *Open Geosciences, 4*, 67–80.

Rakhmatullaev, S., Huneau, F., Celle-Jeanton, H., Le Cousumer, P., Motelica-Heino, M., & Bakiev, M. (2013). Water reservoirs, irrigation and sedimentation in Central Asia: A first-cut assessment for Uzbekistan. *Environment and Earth Science, 68*(4), 985–998.

Rakhmatullaev, S., Abdullaev, I., & Kazbekov, J. (2018). Water-energy-food-environmental nexus in Central Asia: From transition to transformation. In S. S. Zhiltsov et al. (Eds.), *Water resources in Central Asia: International context, The handbook of environmental chemistry* (pp. 103–120). Springer. https://doi.org/10.1007/978-3-030-11205-9

SIC ICWC. (2008). *Intergovernmental agreements on transboundary cooperation in Eastern Europe, Caucasus and Central Asia* (Legal volume no 19. Scientific information center, Interstate Commission for Water Coordination). Tashkent.

The World Bank Group and the Asian Development Bank (2021a) *Climate Risk Country Profile: Turkmenistan*.

The World Bank Group and the Asian Development Bank (2021b). *Climate Risk Country Profile: Kyrgyz Republic*.

The World Bank Group and the Asian Development Bank (2021c) *Climate Risk Country Profile: Kazakhstan*

The World Bank Group and the Asian Development Bank (2021d) *Climate Risk Country Profile: Tajikistan*

The World Bank Group and the Asian Development Bank (2021e). *Climate Risk Country Profile: Uzbekistan.*

UNCCD. (2022). *The nexus between land degradation, climate change and migration in Central Asia*. United Nations Convention to Combat Desertification.

UNECE. (2007). *Dam safety in Central Asia: Capacity building and regional cooperation*. United Nations Economic Commission for Europe. http://cawater-info.net/review/pdf/unece_water_series_publication5_r.pdf

UNECE. (2021). *Water infrastructure safety in Central Asia: Capacity building and regional cooperation. Nations economic Commission for Europe*. Switzerland. https://unece.org/sites/default/files/2021-07/Dam%20Safety%20Review_RU.pdf

UNICEF, & WHO. (2019). *Progress on household drinking water, sanitation and hygiene 2000-2017. Special focus on inequalities*. United Nations Children's Fund (UNICEF) and World Health Organization (WHO).

United Nations. (2022). *The United Nations world water development report 2022: Groundwater: Making the invisible visible*. UNESCO.

Wang, X., Zhang, J., Wang, S., Ge, Y., Duan, Z., Sun, L., et al. (2023). Reviving the Aral Sea: A hydro-eco-social perspective. *Earth's Futures, 11*, e2023EF003657. https://doi.org/10.1029/2023EF003657

World Bank. (2022). Uzbekistan public expenditure review. In *Better value for money in human capital and water infrastructure*. World Bank.

WWAP (UNESCO World Water Assessment Programme). (2020). *The United Nations world water development report 2020: Water and climate change*. UNESCO.

Yakubov, M. (2021). Small Basin governance scheme: Linking water sector reforms and governance in the Aral Sea basin. *International Journal of Water Resources Development, 38*, 783. https://doi.org/10.1080/07900627.2021.1926936

Yakubov, K. I., Yakubov, M. A., & Yakubov, S. K. (2011). *Collector-drainage flow of Central Asia and assessment of its use for irrigation*. Monograph.

Zan, C., Liu, T., Huang, Y., Bao, A., Yan, Y., Ling, Y., Wang, Z., & Duan, Y. (2022). Spatial and temporal variation and driving factors of wetland in the Amu Darya River Delta, Central Asia. *Ecological Indicators, 139*, 108898. https://doi.org/10.1016/j.ecolind.2022.108898

Challenges and Trends in Water System Management

Hermilio Vilarinho and Ana S. Camanho

Abstract Access to safe drinking water, sanitation, and hygiene services is paramount for human health; yet progress towards universal access remains inadequate, leaving billions at risk. Water infrastructure systems, intricate and essential, face management complexities exacerbated by territorial spread and operational costs. Failures in these systems pose profound social and legal repercussions, necessitating effective asset management and maintenance strategies. Against this backdrop, this chapter advocates for accelerated efforts to enhance water utility management, emphasising the criticality of safe water supply and wastewater treatment for public health and environmental sustainability. Through an overview of the multifaceted challenges of sustainable water management and the imperative for improved managerial practices, this research underscores the pivotal role of water management in fostering resilience and meeting the needs of present and future generations.

Keywords Water sector · Water utilities · Sustainability · Water systems

This chapter has appeared previously in the author's dissertation: [https://repositorio-aberto.up.pt/handle/10216/153852]. Used with permission.

H. Vilarinho (✉)
INEGI, Faculdade de Engenharia, Universidade do Porto, Porto, Portugal
e-mail: hvilarinho@inegi.up.pt

A. S. Camanho
Faculdade de Engenharia, Universidade do Porto & INESC TEC, Porto, Portugal
e-mail: acamanho@fe.up.pt

J. C. Correia Leitão et al. (eds.), *Sustainable Water Management and Irrigation Systems*, Natural Resource Management and Policy 62,
https://doi.org/10.1007/978-3-031-93587-9_4

1 Introduction

Access to safe drinking water, sanitation, and hygiene services is critical for human health. However, current progress towards achieving universal access to these services is slower than required, and billions of people are at risk of being left without them. The challenges to sustainable water management are vast and complex, with increasing demand due to population growth, urbanisation, and pressure from agriculture, industry, and energy sectors. Decades of misuse, poor management, and contamination have led to water stress and degraded water-related ecosystems, with far-reaching impacts on human health, economies, and food and energy supplies. Urgent action is needed to reverse this trend and ensure sustainable access to water resources for all.

According to the United Nations World Water Assessment Programme Report (WWAP, 2016), more than 40% of the global active workforce, which is over one billion jobs, heavily rely on water. These water-intensive jobs are present in various sectors, such as agriculture, forestry, inland fisheries, mining, power generation, water supply and sanitation, as well as manufacturing industries like food, pharmaceuticals, and textiles. Moreover, over one-third of the world's total active workforce, representing another billion jobs, are moderately water-dependent. These jobs are found in sectors such as construction, recreation, transportation, and manufacturing and transformation industries such as wood, paper, rubber/plastics, and metals.

The United Nations has set forth the 2030 Agenda for Sustainable Development, which requires countries to implement a structured process of tracking and assessing their progress towards achieving the goals and targets. This involves using standardised global indicators that provide a framework for measuring and evaluating performance. Monitoring and review processes are vital for promoting transparency, accountability, and continuous improvement towards sustainable development. Goal 6 of the 17 Sustainable Development Goals (SDGs) established by the United Nations (UN) focuses on water and sanitation, aiming to provide safe and accessible drinking water and sanitation for everyone. A closer look at the United Nations' SDGs reveals that all of them, in a certain way, prioritise the sustainable use and consumption of water, making Goal 6 the centrepiece for achieving all SDGs. As such, all SDGs can be considered interconnected with Goal 6. This relationship is illustrated using a water-centric figure (Fig. 1) presented by Makarigakis and Jimenez-Cisneros (2019). The figure effectively displays how water is central to achieving sustainable development across all sectors.

The UN Sustainable Development Goals Report (United Nations, 2022) shows that there has been an increase in the proportion of the global population with access to safely managed drinking water and sanitation services. However, there are still significant challenges to overcome. As of 2020, 2 billion people still lack access to safe drinking water, with 1.2 billion people lacking even basic service, and the majority of those without access to drinking water live in rural areas and least developed countries. Similarly, access to safely managed sanitation services remains a

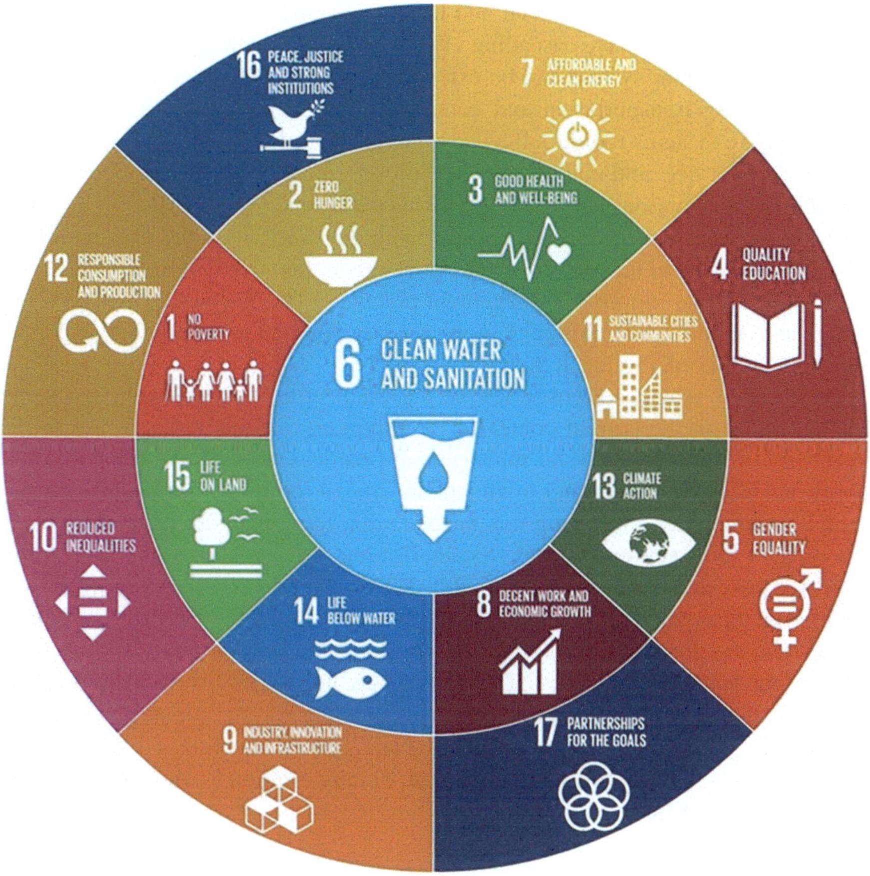

Fig. 1 The water centric 17 Sustainable Development Goals (Makarigakis & Jimenez-Cisneros, 2019)

challenge, with 2.8 billion people lacking access as of 2020. While the world is on track to eliminate open defecation by 2030, the increase in access to hand washing facilities with soap and water has been minimal. Despite improvements, over 829,000 people still die each year from diseases caused by unsafe water, inadequate sanitation, and poor hygiene practices. To achieve universal coverage by 2030, the rate of progress would need to increase fourfold. While a vast collection of "sustainable approaches" have been conceptualised and implemented in the water industry since the publication of UN SDGs, there is still room for improvement to maximise the tangible benefits they provide (Silva, 2022).

Water and wastewater systems are complex and essential infrastructure systems for societies, consisting of pipes, valves, pumps, and reservoirs. However, managing the reliability of these complex networks spread over large territorial areas, while keeping operational costs under control, is a challenging task. Interruptions to these systems can have significant social and legal implications, as failures can happen unexpectedly and may have cascading effects on other systems. The consequences of failures in a water system can be extensive, causing damage to adjacent structures, disrupting transportation and commerce, and impacting urban life. These asset-intensive systems require effective management of asset portfolios to sustain their level of service, and maintenance and preservation initiatives must be in place to minimise interruptions and breakage events (Mazumder et al., 2018).

Given the essential role of water systems in our daily lives and the challenges faced by these critical infrastructure networks, it is imperative that we prioritise and accelerate efforts to improve management practices. The various aspects of water supply and sanitation utilities range from cost-efficiency to service quality and asset management, in order to reach their critical goals.

This chapter provides an overview of the water utility management issues and their relevance in the current context. The importance of this research lies in the urgent need for improved water management practices, as safe and reliable water supply and effective wastewater treatment are crucial for public health and environmental sustainability. Following this introduction, the remainder of the chapter is structured in four sections. A brief characterisation of water supply and wastewater utilities is provided in Sect. 2. Section 3 discusses the pressing issues in water system management, such as climate change, population growth, infrastructure deterioration, and regulatory compliance, which require a multidisciplinary approach for effective management strategies. These issues create challenges, still discussed in Section 3, that need to be addressed by management in various perspectives such as cost-efficiency, service quality, and asset management. Section 4 examines the emerging trends in water utility management. Finally, Section 5 summarises the key conclusions drawn from the chapter.

2 Water Supply and Wastewater Utilities

This section aims to characterise water utilities for water supply and sanitation and describe their main elements.

Urbanisation has a profound impact on the natural water cycle, leading to a more intricate hydrological cycle in urban areas due to various human activities and interventions. This complex hydrological cycle in urban areas is commonly known as the urban water cycle (UWC), and it comprises several components and pathways, as depicted in Fig. 2. According to Marsalek et al. (2008), the UWC is a valuable framework for studying the water balance in urban areas. The high level of interaction among the elements of the UWC, as seen in Fig. 2, underscores the need to understand these interactions for effective urban planning and integrated management of urban water systems. Pinto et al. (2023) emphasise that integrated management of the UWC is crucial for ensuring the sustainability and resilience of urban water systems. To achieve this, the system must be viewed as a whole, considering

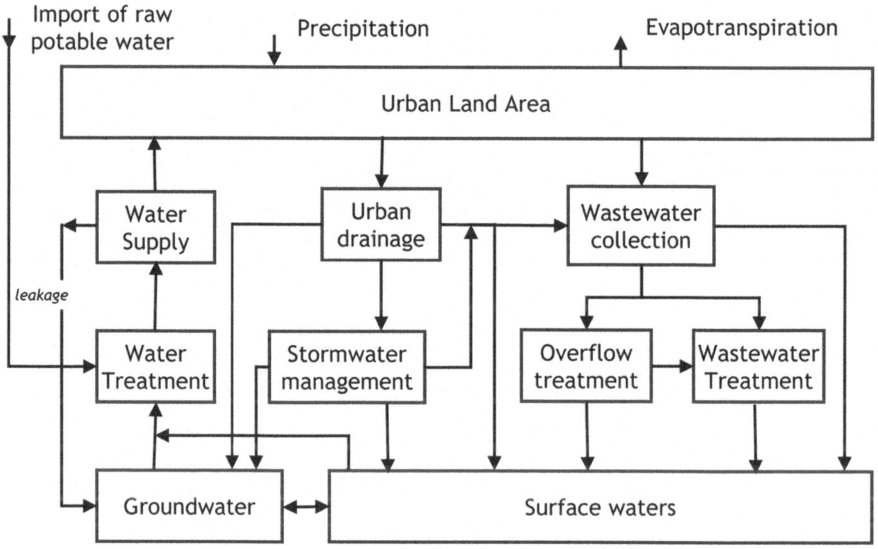

Fig. 2 Urban water cycle (Marsalek et al., 2008)

the interdependencies of the various elements. Although these interactions can create complexity, they also present opportunities for synergies and gains in integrated management. For instance, treated wastewater can be used for non-potable purposes, reducing the demand for freshwater resources. Franco-Torres et al. (2021) outline that with the adoption of a new approach that emphasises the coordinated management of water systems, it is possible to repurpose stormwater as potable water, assess leaky sewers as source of pollution for water supply, and take wastewater as a valuable resource for replenishing groundwater aquifers. As discussed by McConville (2023), there is a growing trend in the sanitation industry to view wastewater as a resource rather than waste. This has resulted in proposals and implementations of separate collection systems for various wastewater fractions. Separating wastewater by levels of contamination can simplify the treatment processes since the volumes requiring advanced treatment will be only a fraction of the total wastewater volumes.

Although the need for integrating the activities of the UWC has been widely discussed, the structure of the water industry varies significantly across the globe in terms of activities undertaken by individual businesses, size, customer base, private sector involvement, competition, regulation, and oversight. Water businesses typically perform a range of activities including water collection, transfer, treatment, distribution, sewerage collection and treatment, irrigation, and drainage.

Various factors, such as the source of water, geography, geology, topography, customer type, demand, and density, influence the activities undertaken by water businesses. According to Abbott and Cohen (2009), in many small- to medium-sized markets, water businesses operate as vertically integrated monopolies, while larger metropolitan areas may present several vertically integrated entities coexisting with separate local distribution networks.

Abbott and Cohen (2009) highlight the significant capital costs associated with water supply networks as a major barrier to competition. Compared to energy distribution systems where distribution infrastructure costs represent only 40% of the supplied electricity, the cost of water and distribution systems accounts for two-thirds of the water supply, making it expensive to duplicate and contributing to the creation of natural monopoly conditions. These factors make it difficult for new players to enter the market and offer competitive prices, which can ultimately lead to higher prices and reduced innovation.

Water supply and wastewater services can be managed jointly or as separate 'management entities'. Examples of typical operational stages found in infrastructures of water supply and wastewater utilities can be seen in Fig. 3a, b, respectively.

As seen in Fig. 3a, water supply utilities collect raw water from various sources such as rivers, lakes, ground water, underground aquifers, and reservoirs. Most of these water sources cannot be directly consumed, so they must be treated. The choice of an appropriate treatment method depends on various factors, with the raw water quality being the primary driver. Filtration, coagulation, and disinfection are some of the possible treatment methods that can be employed to achieve suitable water quality (Adedeji & Hamam, 2020).

Pumping is utilised to facilitate the movement of water, including to locations at higher elevations, by adding energy to the flow and raising pressure levels. They are applicable in various sections of a water supply system, such as conveying untreated water from the source to a treatment facility, pressurising water to be distributed to the system, and enhancing pressure in segments of the system that experience low pressure (i.e. booster pumps). Adduction systems are usually employed to convey raw water to treatment and treated water to storage tanks.

Storage tanks are used for several purposes, such as meeting variable demands, balancing operating pressure, and serving as a reserve for emergency and firefighting needs.

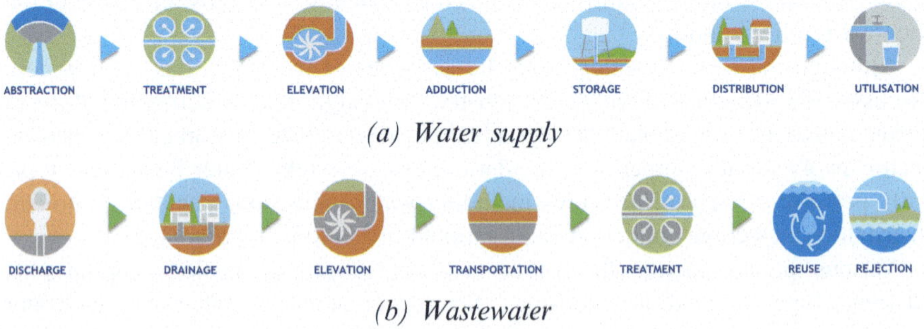

Fig. 3 Typical operational stages of water and wastewater utilities (ERSAR, 2021). (**a**) Water Supply. (**b**) Wastewater

Distribution networks consist of various pipe types, such as water transmission mains that transfer potable water from storage facilities to different parts of the system, including supply pressure zones or district metered areas. Distribution pipes, which have a smaller diameter than transmission mains, distribute potable water within local areas. Supply/customer connection pipes connect individual end-user properties to the water distribution system (Farmani & Sweetapple, 2023).

In wastewater systems, as displayed in Fig. 3b, the drainage or collection of wastewater is primarily done through piped sewer systems. Depending on the terrain and size of the area being served, pumping stations can be required in the collection and transportation systems. Conventional wastewater treatment uses physical, biological, and chemical processes to purify the incoming wastewater in response to increasingly stringent treatment standards. Initially, treatment focused on removing suspended solids and pathogens, known as primary treatment. The development of secondary treatment processes in the 1960s was in response to the need for reduction of dissolved organic matter. In the 1970s, tertiary treatment processes emerged for phosphorus and nitrogen removal, with widespread adoption in the 1990s. Today, the majority of high-income countries' wastewater treatment plants have at least a secondary treatment, if not a tertiary treatment. However, it is important to note that treatment levels vary between countries.

The final stage of the sanitation service chain involves either rejection or reuse of treated products. Conventional urban sanitation systems in high- and middle-income countries generally produce effluent water and sewage sludge, which are often released into local waterways. However, reuse of treated wastewater is becoming more common, especially in water-stressed areas. In Europe, only slightly over 2% of the treated urban wastewater effluent is currently being reused. Nevertheless, the European Commission is taking steps to encourage and facilitate water reuse by developing rules and incentives. In contrast to Europe, Singapore, a country facing water scarcity, has made significant progress in water reuse. Treated wastewater is used for direct non-potable use and indirect potable use, meeting over 40% of its total water demand. Meanwhile, reuse of organic matter and nutrients from sanitation systems is more prevalent worldwide compared to treated wastewater reuse (McConville, 2023).

3 Critical Challenges in Management of Water Utilities

To provide a comprehensive understanding of the management of water utilities, it is crucial to highlight the major challenges that must be tackled. This section aims to provide an overview of the significant challenges that water utilities encounter in achieving sustainable and effective management. Subsection 3.1 outlines the most pressing issues in the current scenario, while subsections 3.2, 3.3, and 3.4 delve into specific challenges related to cost-efficiency, service quality, and asset management, respectively. Finally, subsection 3.5 describes the recent approaches developed to address these challenges.

3.1 *Pressing Issues for Sustainable and Effective Management of Water Utilities*

According to Haasnoot et al. (2011), the future of water management is demanding and sometimes unpredictable due to factors such as climate change, population growth, and regulatory pressing issues. These factors create uncertainties that may require transdisciplinary water management strategies, making it essential to identify the most sustainable strategy to adopt (Larsen et al., 2016). Water-related challenges are commonly expressed by professionals using the phrase "too much, too little, or too polluted." This succinct statement captures the three primary issues related to water management that pose significant challenges in various contexts (Makarigakis & Jimenez-Cisneros, 2019).

Some urgent concerns currently faced by water utilities are as follows:

3.1.1 Climate Change

Changes in weather patterns and increasing frequency of extreme weather events can impact water availability and quality. Climate change impacts water system management particularly in terms of planning, building, and maintenance. The rise in global temperatures is leading to extreme weather cycles, such as floods and droughts, while sea levels are also increasing, resulting in droughts in some regions that negatively affect water sources. Additionally, the combination of decreasing precipitation rates and increasing extraction rates is putting stress on groundwater supplies. With two-thirds of the world's mega-cities located in climate-vulnerable regions, water utilities must improve water resource management and infrastructure to provide safe water to residents. Diversifying water sources through sustainable practices like groundwater extraction, water trading, conservation, and the use of recycled or desalinated water is crucial to managing these challenges (Danilenko et al., 2010; Vairavamoorthy et al., 2008; Wilby, 2007).

3.1.2 Population Growth

In just the last century, the global population has grown by approximately four times, and during that same time, human water usage has increased significantly, with agricultural, industrial, and municipal usage growing approximately five, 18, and 10 times, respectively (Nazemi & Madani, 2018). In the years to come, water demand is projected to rise notably due to the global population growth, which could result in water scarcity (Pahl-Wostl et al., 2016). Population growth is concentrated in regions with less abundant water, including Sub-Saharan Africa, the Middle East, and Central Asia. A forecast suggests that by 2050, over 50% of the world population will live in regions that will face water scarcity at least once a year (Ortega-Ballesteros et al., 2021). Sanitation systems are under similar pressure to adapt to the increasing demand. As the demand in the water sector increases, putting pressure on existing water systems, water utilities must find ways to meet this growing demand while ensuring efficient use of resources.

3.1.3 Regulatory Compliance

According to Marques and Pinto (2018), the establishment of regulatory agencies is deemed necessary to protect the welfare of the general public, given the complexity of regulated sectors which make it difficult to ensure their protection. The creation of these agencies is often a part of the reform process with the aim to enhance the accountability of service providers, to establish an independent pricing mechanism, and to utilise regulatory expertise. In order to achieve these goals, autonomous regulatory bodies are endowed with legislative, executive, and judicial powers, allowing them to supervise operators and enforce the necessary rules. Water utilities must adhere to those various regulations and standards related to water quality, infrastructure, and service provision. Meeting these requirements may entail significant efforts and investments in technology, personnel, and infrastructure.

3.1.4 Infrastructure Deterioration

Many cities and locations with ageing water infrastructure require immediate rehabilitation to meet the demand. Some drinking water transmission systems in Europe and the United States are over a century old, with some pipes dating back to the nineteenth century. Those laid after World War II typically have a life span of 75 to 100 years, indicating that many are reaching the end of their design life. Due to those conditions, there has been a 27% increase in water main breaks between 2012 and 2018 in the United States, translating to an estimated 250,000 to 300,000 breaks annually or a break every 2 minutes (Ortega-Ballesteros et al., 2021; ASCE, 2021a). Regarding the wastewater systems, typically their pipes have a lifespan of 50 to 100 years. As collection systems age, they deteriorate, causing groundwater and stormwater to infiltrate the networks through cracks and joints. When collection systems become overwhelmed, sanitary sewer overflows may occur. In North America, approximately 28% of mains are older than 50 years (ASCE, 2021b). Given this scenario, infrastructure renewal and replacement ranking have been considered as the most pressing issue in the sector according to the American Water Works Association Water Industry report for ten consecutive years (ASCE, 2022). The results of decades of little or no pipe replacement or refurbishment are low pressure and high leakage. Low pressure is a symptom of systems not being able to cope with expansion. Leakage alone is estimated to have an equivalent cost of USD 1 billion per year in selected African cities (Rouse, 2013). Global water losses of 126 billion cubic meters per year, worth USD 39 billion, reflect inadequate infrastructure preservation. Saving half of these losses could provide water to 90 million people (Molinos-Senante et al., 2022).

Effective management is of paramount importance to ensure the efficient and sustainable provision of water services by utilities. Management is faced with various challenges in the current scenario, which arise from the constraints listed above that require immediate attention. Although water utilities face a multitude of challenges (Hasit et al., 2012; EPA, 2017), some of the most significant include achieving cost-efficiency, maintaining high service quality, and effective asset management.

3.2 The Cost-Efficiency Challenge

Achieving sustainable water use requires a focus on efficiency due to the high cost of operation and maintenance in water utilities. However, some pricing modes, such as egalitarian pricing, hinder efficiency and negatively impact sustainability. Bithas (2008), in a study that examined several pricing modes, advocates that full-cost prices are necessary to promote social equity in the long run and ensure sustainable water use. Water managers face the challenge of dealing with operational costs that are sometimes not covered by prices. Among these costs, energy, which is a significant operational cost and increasing over time, plays a crucial role in water supply as all stages of water supply utilities require energy. According to Coelho and Andrade-Campos (2014), the global energy consumption for water distribution is around 7% of the total energy generated. Pumping station operations are the primary consumers of energy in water supply systems, accounting for 50–80% of the total energy consumed. In Europe only, the estimated cost of energy for pumping operations is between 2.9 and 3.9 billion e/year (Reis et al., 2023).

At a broader level concerning cost-efficiency, there are ongoing discussions regarding the most effective structural arrangements for the water supply and wastewater sectors. This discourse includes integrating different scopes, such as water and wastewater or bulk and retail operations, identifying opportunities for economies of scale, and determining management systems and ownership (Abbott & Cohen, 2009; González-Gómez & García-Rubio, 2008).

According to a literature review conducted by Berg and Marques (2011), the question of the optimal size of water utilities is a commonly discussed topic in the water sector. However, the results of the research on this issue are inconclusive and vary depending on the sample and output mix analysed. The existing studies suggest that there are economies of scale up to a certain point, but the implications for optimal utility size are unclear. Consolidation of geographically separate utility operations may reduce managerial overhead costs, but coordination costs may rise, making decisions slower with larger bureaucracies. Thus, determining the optimal size of utility operators cannot be generalised, and location-specific factors are more important when evaluating consolidation or decentralisation.

More research is needed to assess efficiency gains associated with economies of scope, such as joint provision of drinking water supply with sanitation, or other activities like piped gas, electricity, or urban waste. Vertical integration within water and sanitation utilities, i.e. providing wholesale and retail services simultaneously, can result in some advantages for drinking water supply. Economies of density, both in production and customer density, have been identified in the literature.

Berg and Marques (2011) also discuss the issue of ownership in water utilities and identified lack of consensus in the literature. While some economic theories suggest that the private sector performs better, a review of 47 studies found that 18 concluded that private utilities are more efficient, 12 found public utilities to be more efficient, and 17 were inconclusive. There is some agreement that private utilities tend to have higher labour productivity but also higher capital expenses. Historically, the public sector has tended to under invest in infrastructure and has lower labour productivity.

3.3 The Service Quality Challenge

Ensuring safe drinking water and reliable supply and sanitation services is a critical aspect of water utility management. Water utilities must invest in water and wastewater treatment technologies and establish effective water quality monitoring programs to safeguard public health. It involves maintaining water quality, preventing contamination, managing water resources, and ensuring adequate storage and distribution systems. Making sure that the drinking water supplied by public systems is of good quality is a crucial aspect of public health policies (Roeger & Tavares, 2018).

Moreover, the quality of utilities' services is not limited to the physical quality of water and wastewater. It also involves ensuring that these services are accessible and reliable. A frequently used measure of utility performance is the uninterrupted provision of services, with a goal to provide available service 24 hours a day, 7 days a week (known as continuous supply), being regarded as the ideal standard (Rawas et al., 2020). As stated by Kumar et al. (2013), assuring reliable and safe water services is a highly intricate task, particularly in developing countries where piped water supply is irregular, with low pressure, high leak rates, and poor maintenance practices along the supply chain, leading to a significant gap between supply and demand. This inadequate water supply is also unequally distributed among different consumer categories. Despite being the most crucial factor in public health, a clean water supply is jeopardised by source contamination, deteriorating infrastructure of water distribution systems, leakages, cross-contamination, and unsanitary practices. Diarrhea is the most frequent cause of illness and death globally, with 88% of these deaths attributable to a lack of access to safe drinking water and adequate hygiene facilities.

Addressing these complex issues requires the implementation of comprehensive strategies that encompass infrastructure investment, improved maintenance practices, and equitable distribution of water resources. By prioritising the quality of services provided, water utilities play a crucial role in advancing human society. Their efforts contribute to ensuring universal access to clean and reliable water, a fundamental necessity for the well-being and prosperity of all individuals.

3.4 The Asset Management Challenge

Effective management is essential for maintaining the infrastructure and assets required for the provision of water services. Water utilities are asset-intensive systems, and asset management practices are vital for ensuring their long-term sustainability. This involves developing and implementing asset management plans to optimise the use of existing infrastructure, identify maintenance and repair needs, and plan for future investments. Regular maintenance, repair, and replacement of ageing assets are needed to ensure an uninterrupted water supply. Many water systems were built decades ago and have reached the end of their design life. As a result, water utilities must invest in upgrading and replacing ageing infrastructure to ensure reliable and efficient service provision. The neglect of infrastructure preservation, especially water mains and sewers, is a prevalent problem in many developed regions. These systems have long lifetimes and are difficult to inspect, resulting in significant service disruptions before any visible deterioration occurs. Water charges have been too low to save enough money for necessary replacements. While system managers acknowledge the need for refurbishment, the political difficulty of increased government subsidies or charges has hindered the implementation of renovation programmes. Only in the United States, the investment in water infrastructure is estimated in USD 1.7 trillion up to 2050. There is a similar backlog of investment in existing systems in developing countries. Many of the systems were constructed in colonial times, and although the cities have grown since that time, water supply systems have not been able to keep pace with urban development (Molinos-Senante et al., 2022).

The importance of enhancing the resilience of infrastructure, particularly in the context of extreme weather events, is steadily increasing. According to Hallegatte et al. (2019), investing $1 in enhancing the resilience of critical infrastructure can yield $4 in return. Hence, there is a need for appropriate actions to be taken towards water-related assets, considering their entire lifecycle and future generations. This involves developing the capability to (1) withstand the initial impact of hazardous events, (2) minimise the adverse effects of such events, (3) quickly adapt to the resulting changes, and (4) enhance water-related assets to enhance preparedness against forthcoming threats and variations (Rezvani et al., 2022).

A common issue with water infrastructure is the transportation and protection of treated water through the distribution system. Many current systems have poorly maintained and inadequately sized infrastructure, resulting in low or no pressure in the network, intermittent water supply, and contamination. Some systems have oversized pipes and pumps, leading to higher costs and poorer water quality due to excess capacity. This is often due to outdated sizing or declining water demands. Furthermore, innovative methods for detecting leaks such as analysing water usage data with advanced metering, pressure loggers, data mining and inspection with traditional acoustic methods, thermal imaging, robotic inspection, and pressure management can assist water suppliers in identifying distribution system issues and managing water losses. Other than that, repairing and replacing underground water infrastructure is costly and complicated, as it may affect other properties, traffic, and buildings. Various techniques are required to repair the different types of pipes in a distribution system, making the process even more challenging. Therefore, there is an urgent need for innovations to efficiently reduce and control water loss and extend the life of pipes (Milman et al., 2021).

3.5 Contributions to Address the Challenges

In recent years, the scientific community has dedicated considerable efforts towards finding solutions and alternatives to help utilities overcome the challenges they face in managing and operating their systems. According to a review by Ortega-Ballesteros et al. (2021), there has been a significant increase in awareness among researchers in this field, leading to a growing number of scientific publications on this topic since the start of the current century.

Moreover, in the past 20 years, there has been a shift in thinking about urban water management, which reflects a larger cultural shift towards valuing natural processes and systems, as opposed to relying solely on technological and mechanistic solutions. This new paradigm represents a departure from the traditional approach to managing water in cities. New management frameworks have been developed to tackle current challenges. According to Franco-Torres et al. (2021), different scholars have launched novel ideas on frameworks for water management, including new styles of governance, circular use of resources, or modular ecosystem-based infrastructures.

The International Organisation for Standardisation (ISO) has been also promoting actions aimed at improving the management of water systems. In 2001, the ISO technical committee ISO/TC 224 was launched to develop guidelines for the activities related to water supply and wastewater systems. Six years later, ISO/TC 224 released three new ISO standards addressing water services: ISO 24510 (ISO, 2007a) focusing on the service provided to users of both water and wastewater systems, ISO 24511 (ISO, 2007b) dealing with management practices of wastewater utilities, and ISO 24512 (ISO, 2007c) addressing management practices of drinking water utilities.

The ISO/TC 224 standards provide a framework to improve the governance of water services and promote quality and efficiency. They facilitate dialogue among stakeholders such as consumers, water authorities, utilities, research departments, and laboratories. The standards enable a mutual understanding of responsibilities and tasks and provide methods and tools to define objectives and specifications and assess performance. They can be applied voluntarily in both industrialised and developing countries and, in any setting, whether the utility operator is public or private. The ISO/TC 224 standards address several key areas, including defining common terminology for stakeholders, clarifying consumer expectations and service elements, developing optimised management actions, proposing measurable quality criteria and performance indicators, and enabling local-scale comparisons of observed results with targets set by water authorities. These standards promote benchmarking among water utilities and can help protect the environment while meeting the needs of water users (ISO, 2004).

ISO has taken another important step in promoting effective asset management by publishing a family of standards in 2014. Comprising of ISO 55000, ISO 55001, and ISO 55002, this framework provides a comprehensive approach for managing any type of assets, whether it be railway sleepers, brand reputation, or telecommunications networks (ISO, 2014a, 2014b). ISO 55001, specifically, lays out the requirements for effective asset management and creates a system designed to optimise performance, mitigate risks, and minimise costs throughout the entire asset lifecycle. Organisations can seek certification for compliance with this standard, similar to other ISO standards like ISO 9001 for quality management systems or ISO 14001 for environmental management systems, upon successful implementation of the requirements. It is worth noting that these management systems have a common foundation in the Deming cycle (PDCA: Plan-Do-CheckAct), originating in the field of quality management. According to Sousa and Meireles (2022), the PDCA cycle serves as the inspiration and basis for the sequence of Leadership, Planning, Support, Operation, Performance, and Improvement, which is shared among all ISO management standards.

The United Nations has recognised the crucial role of effective water management in achieving sustainable development. To this end, UNESCO, the scientific arm of the United Nations, launched the International Hydrological Programme (IHP) to raise awareness among communities and decision-makers about the significance of water-related issues. Since 1975, IHP has developed a comprehensive set of 17 worldwide initiatives aimed at creating scientific and technological tools for evidence-based decision-making, promoting international cooperation through networking, enhancing the science-policy interface, and focusing on education and training for building human capital at the local, regional, and global levels (Makarigakis & Jimenez-Cisneros, 2019).

The World Bank Group is actively supporting the latest advancements in water management practices. In 2014, the Bank established the Water Global Practice, which serves as an integrated platform for financing, expertise, and implementation efforts. By combining the Bank's global knowledge with local investments, this initiative generates greater capacity for implementing transformative solutions that promote sustainable growth in countries. As part of this effort, the Bank launched the Utility of the Future (UoF) programme, which offers a novel approach to planning and implementing reforms aimed at making water utilities more sustainable (Lombana Cordoba et al., 2022).

The European Union (EU) recognises that water has value beyond its usefulness as a resource for humans and has taken steps to establish policies for the preservation of aquatic ecosystems as natural capital. The Water Framework Directive (2000/60/EC) was a significant milestone in this direction, establishing a new approach to water management that considers its role in supporting ecosystems. The EU's more recent action plan for sustainability, the Biodiversity 2030 strategy, recognises the importance of restoring and preserving nature, including water resources.

As part of this plan, measures such as reducing water pollution, promoting efficient and circular water use, raising awareness of the natural capital provided by biodiversity and ecosystem services, and adapting to climate change will help to protect and restore ecosystems. It is important for the global community to make a collective effort to protect water resources in order to achieve sustainable growth (European Commission, 2020).

4 Emerging Trends in Water Utility Management

This section presents some latest trends in water utility management. With the water sector rapidly evolving, it is crucial to remain up-to-date with emerging trends that are being adopted to address the previously discussed challenges in water utility management. Some of those trends include:

4.1 *Digitalisation*

The integration of advanced Industry 4.0 technologies such as artificial intelligence, the Internet of Things (IoT), and Big Data can significantly improve the performance and efficiency of the water sector. The industry is also witnessing the emergence of innovative technologies like smart water systems, which utilise real-time data analysis and sensors to monitor and manage water distribution networks, resulting in increased efficiency and reduced waste. Furthermore, the implementation of artificial intelligence and machine learning algorithms can optimise water treatment processes and predict potential system failures (Silva, 2022).

To achieve a "Smart City" approach to water resource management, infrastructure monitoring, data collection, and analysis and decision support systems integrating IoT, Big Data, and blockchain technologies are necessary. Data analysis and the use of computational intelligence techniques are critical in this approach for monitoring water consumption, predicting pipe failures, and forecasting water demand. Decision support systems must incorporate multiple tools to address various issues such as ensuring water quality, availability, and infrastructure maintenance. Anomaly detection tools generate notifications and alarms to signal abnormalities such as water contamination, distribution network malfunctions, or cyberattacks on water treatment facilities. Water demand models and simulation tools are employed to evaluate the effect of new rules on reducing water consumption or forecasting water demand for residential areas under development. Infrastructure maintenance planning tools should include pipe failure predictors to reduce the costs associated with unexpected pipe breaks.

Future developments should focus on integrating different types of data, creating decision support systems, and standardising methodologies to create large-scale systems for water infrastructure management (Hangan et al., 2022).

4.2 Increased Focus on sustainability and Circular Economy

The increasing concern about water scarcity in various areas has prompted utilities to explore methods of minimising water waste and encouraging water conservation. This encompasses the promotion of water reuse, the collection of rainwater, and other unconventional water sources. Moreover, utilities are adopting more sustainable practices in their day-to-day operations, which entails lowering energy consumption and greenhouse gas emissions (Silva, 2022).

In the study performed by Larsen et al. (2016), various strategies to improve water productivity in the Urban Water Cycle are discussed, including reducing wastewater, reusing lower-quality water, and regenerating high-quality water from used water. Globally, only 1.7% of water supply is currently reused, primarily for irrigation purposes. The authors also suggest source separation of waste as a promising trend, which can be implemented at both the household and device level, such as using a recycling shower that recirculates shower water in real time. The practice of separating greywater not only conserves water but also can recover energy and nutrients. Examples of successful source separation initiatives include China's 40 million domestic biogas reactors and the almost 100,000 urine-diverting dry toilets in peri-urban areas of eThekwini, South Africa.

4.3 Infrastructure Risk Management

Regulatory authorities have shifted towards promoting an integrated risk management approach to minimise the risk of water system infrastructure failures. This involves implementing several protective barriers such as source water protection, redundant treatment designs, and continuous monitoring of distribution systems. The adoption of the World Health Organisation's Water Safety Plans (WSPs) by over 90 countries is an example of this approach. The WSPs require the proactive management of all water system assets from source water to tap, resulting in a decrease in diarrheal diseases, improved communication between water system stakeholders, and better overall management of water system assets. However, implementing risk management approaches necessitates institutional and cultural change. Water suppliers need to undertake an honest evaluation of their current systems and practices proactively instead of reacting to risks to be effective (Milman et al., 2021).

4.4 Capacity-Building Efforts

Historically, ownership has been considered the key factor in improving the management of water utilities. However, the privatisation movement in the 1980s resulted in inconsistent outcomes, with some privatised systems performing no better or even worse than public ones. In the 1990s and 2000s, interest shifted towards public-private partnerships or corporatisation initiatives, where state-owned systems were kept under public control but managed as independent entities. Despite a variety of different structures appearing, success rates were inconclusive.

It has now been recognised that good management of water utilities is not solely dependent on ownership, and there is a need to look beyond this factor to understand the reasons behind the success or failure of these systems. Recently, attention has shifted towards the institutional and regulatory environments in which these utilities operate, with "governance" being a key point of discussion.

The focus is on creating an environment that enables water systems to operate effectively, which includes ensuring accountability and autonomy, providing oversight before new systems are developed, and designing appropriate incentives for water system management. Additionally, there is a focus on improving the internal capacity and culture within water systems. Studies have shown that successful water suppliers have a strong knowledge base, human and organisational capacities, a willingness to learn, a focus on customers and business, effective measurement practices, and a culture of continuous improvement.

The emphasis on improving water management through the enabling environment is expected to continue, with a specific focus on enhancing capacity-building efforts. This is already happening with support from national and subnational governments and international donors who are investing resources in water system management. Key activities include training, creating and promoting self-assessment tools, providing guidance and best practices, and providing financial support to water suppliers to improve their capacity. Capacity-building efforts will also likely continue to facilitate the development of water system partnerships, which can help build capacity through resource leveraging, taking advantage of economies of scale, and filling gaps in specific operational or management responsibilities such as billing, water quality testing, or shared operators (Milman et al., 2021).

Despite a century of efforts to develop successful paradigms, management of piped water systems remains a significant challenge. Improving the institutional and regulatory environment and promoting capacity-building efforts and partnerships are key to addressing the challenges associated with managing those systems.

5 Conclusion

This chapter has highlighted the crucial importance of management practices within the water utilities sector, emphasising its central role as both a subject of research and an area for practical application. Drawing upon a robust body of literature in this field, the chapter has explored the challenges facing water utilities, including the need to balance cost-efficiency with service quality and to manage assets effectively.

Furthermore, the chapter has emphasised the ever-changing nature of the water utilities landscape, where shifting demographics, evolving regulatory frameworks, and emerging technological advancements continually shape the operational environment. In navigating this dynamic landscape, water utilities must adopt a forward-thinking approach characterised by adaptability, resilience, and a commitment to ongoing improvement.

The study has underscored the intrinsic connection between effective water utility management and broader societal goals, such as public health, environmental sustainability, and economic development. As caretakers of a crucial public resource, water utilities hold a significant responsibility to ensure the provision of safe, dependable, and fair water services to all sections of society. This necessitates not only a strong focus on operational excellence but also a clear commitment to involving stakeholders, promoting openness, and maintaining accountability.

In conclusion, this chapter serves as a call to action for collective efforts to address the pressing challenges facing the water utilities industry. By leveraging insights from both academic research and practical experience, stakeholders can forge innovative pathways towards a more resilient, sustainable, and inclusive water future. Through collaborative endeavours, informed decision-making, and a shared commitment to excellence, we can overcome the obstacles ahead and realise the transformative potential of effective water utility management.

References

Abbott, M., & Cohen, B. (2009). Productivity and efficiency in the water industry. *Utilities Policy, 17*(3–4), 233–244.

Adedeji, K. B., & Hamam, Y. (2020). Cyber-physical systems for water supply network management: Basics, challenges, and roadmap. *Sustainability, 12*(22), 9555.

ASCE. (2021a). *2021 Report card for America's infrastructure: Drinking water*. American Society of Civil Engineers. https://infrastructurereportcard.org/cat-item/drinking-waterinfrastructure/

ASCE. (2021b). *2021 report card for America's infrastructure: Wastewater*. American Society of Civil Engineers. https://infrastructurereportcard.org/cat-item/wastewaterinfrastructure/

ASCE. (2022). *2022 State of the water industry*. American Society of Civil Engineers. https://www.awwa.org/Professional-Development/Utility-Managers/

Berg, S., & Marques, R. (2011). Quantitative studies of water and sanitation utilities: A literature survey. *Water Policy, 13*, 591–606.

Bithas, K. (2008). The sustainable residential water use: Sustainability, efficiency and social equity. The european experience. *Ecological Economics, 68*(1–2), 221–229.

Coelho, B., & Andrade-Campos, A. (2014). Efficiency achievement in water supply systems—a review. *Renewable and Sustainable Energy Reviews, 30*, 59–84.

Danilenko, A., Dickson, E., & Jacobsen, M. (2010). *Climate change and urban water utilities: Challenges and opportunities*. World Bank.

EPA. (2017). *Effective utility management: A primer for water and wastewater utilities*. United States Environmental Protection Agency. https://www.epa.gov/sustainable-waterinfrastructure/effective-utility-management-primer-water-and-wastewater-utilities

ERSAR. (2021). Relatório anual dos serviços de águas e Resíduos em Portugal (2021). In *Caracterização do setor de águas e resíduos* (Vol. 1). ERSAR. https://www.ersar.pt/pt/sitepublicacoes/Paginas/edicoes-anuais-do-RASARP.aspx

European Commission (2020). Biodiversity strategy for 2030. https://ec.europa.eu/environment/strategy/biodiversity-strategy-2030_en

Farmani, R., & Sweetapple, C. (2023). Traditional systems of drinking water delivery: Technical aspects and sources. In *Routledge handbook of urban water governance* (pp. 34–43). Routledge.

Franco-Torres, M., Rogers, B. C., & Harder, R. (2021). Articulating the new urban water paradigm. *Critical Reviews in Environmental Science and Technology, 51*(23), 2777–2823.

González-Gómez, F., & García-Rubio, M. A. (2008). Efficiency in the management of urban water services. What have we learned after four decades of research. *Hacienda Pública Española/ Revista de Economía Pública, 185*(2), 39–67.

Haasnoot, M., Middelkoop, H., Van Beek, E., & Van Deursen, W. (2011). A method to develop sustainable water management strategies for an uncertain future. *Sustainable Development, 19*(6), 369–381.

Hallegatte, S., Rentschler, J., & Rozenberg, J. (2019). *Lifelines: The resilient infrastructure opportunity*. World Bank Publications.

Hangan, A., Chiru, C.-G., Arsene, D., Czako, Z., Lisman, D. F., Mocanu, M., Pahontu, B., Predescu, A., & Sebestyen, G. (2022). Advanced techniques for monitoring and management of urban water infrastructures—An overview. *Water, 14*(14), 2174.

Hasit, Y., Yeager, F., & Matichich, M. (2012). Industry survey builds on 10 attributes of effective utility management. *Opflow, 38*(9), 8–9.

ISO. (2004). *ISO/TC 224 information brochure: Guidelines to improve governance and quality of water services*. International Organization for Standardization. https://www.iso.org/committee/299764.html?view=documents

ISO. (2007a). *ISO 24510:2007a. Activities relating to drinking water and wastewater services—Guidelines for the assessment and for the improvement of the service to users. Technical report*. International Organization for Standardization.

ISO. (2007b). *ISO 24511:2007b. Activities relating to drinking water and wastewater services—Guidelines for the management of wastewater utilities and for the assessment of wastewater services. Technical report*. International Organization for Standardization.

ISO. (2007c). *ISO 24512:2007c. Activities relating to drinking water and wastewater services—Guidelines for the management of drinking water utilities and for the assessment of drinking water services. Technical report*. International Organization for Standardization.

ISO. (2014a). *ISO 55000:2014a. Asset management—Overview, principles and terminology. Technical report*. International Organization for Standardization.

ISO. (2014b). *ISO 55001:2014b. Asset management—management systems—requirements. Technical report*. International Organization for Standardization.

Kumar, M. M., Manohar, U., Pallavi, M., & Anjana, G. (2013). Urban water supply and management. *Journal of the Indian Institute of Science, 93*(2), 295–318.

Larsen, T. A., Hoffmann, S., Lüthi, C., Truffer, B., & Maurer, M. (2016). Emerging solutions to the water challenges of an urbanizing world. *Science, 352*(6288), 928–933.

Lombana Cordoba, C., Saltiel, G., & Penalosa, F. P. (2022). Utility of the future, taking water and sanitation utilities beyond the next level. In *A methodology to ignite transformation in water and sanitation utilities*.

Makarigakis, A. K., & Jimenez-Cisneros, B. E. (2019). UNESCO'S contribution to face global water challenges. *Water, 11*(2), 388.

Marques, R. C., & Pinto, F. S. (2018). How to watch the watchmen? The role and measurement of regulatory governance. *Utilities Policy, 51*, 73–81.

Marsalek, J., Cisneros, B. J., Karamouz, M., Malmquist, P.-A., Goldenfum, J. A., & Chocat, B. (2008). *Urban water cycle processes and interactions: Urban water series-UNESCO-IHP* (Vol. 2). CRC Press.

Mazumder, R. K., Salman, A. M., Li, Y., & Yu, X. (2018). Performance evaluation of water distribution systems and asset management. *Journal of Infrastructure Systems, 24*(3), 03118001/1- 03118001/24.

McConville, J. R. (2023). Conventional systems for urban sanitation and wastewater management in middle-and high-income countries. In *Routledge handbook of urban water governance* (pp. 119–133). Routledge.

Milman, A., Kumpel, E., & Lane, K. (2021). The future of piped water. *Water International, 46*(7–8), 1000–1016.

Molinos-Senante, M., Maziotis, A., Sala-Garrido, R., & Mocholi-Arce, M. (2022). Estimating performance and savings of water leakages and unplanned water supply interruptions in drinking water providers. *Resources, Conservation and Recycling, 186*, 106538.

Nazemi, A., & Madani, K. (2018). Urban water security: Emerging discussion and remaining challenges. *Sustainable Cities and Society, 41*, 925–928.

Ortega-Ballesteros, A., Manzano-Agugliaro, F., & Perea-Moreno, A.-J. (2021). Water utilities challenges: A bibliometric analysis. *Sustainability, 13*(14), 7726.

Pahl-Wostl, C., Gupta, J., & Bhaduri, A. (2016). Water security: A popular but contested concept. In *Handbook on water security* (pp. 1–16). Edward Elgar Publishing.

Pinto, F. S., Bolognesi, T., & Gasson, C. (2023). Urban water cycle and services: An integrative perspective. In *Routledge handbook of urban water governance* (pp. 17–33). Routledge.

Rawas, F., Bain, R., & Kumpel, E. (2020). Comparing utility-reported hours of piped water supply to households' experiences. *npj Clean Water, 3*(1), 6.

Reis, A. L., Lopes, M. A., Andrade-Campos, A., & Antunes, C. H. (2023). A review of operational control strategies in water supply systems for energy and cost efficiency. *Renewable and Sustainable Energy Reviews, 175*, 113140.

Rezvani, S. M., de Almeida, N. M., Falcao, M. J., & Duarte, M. (2022). Enhancing urban resilience evaluation systems through automated rational and consistent decision-making simulations. *Sustainable Cities and Society, 78*, 103612.

Roeger, A., & Tavares, A. F. (2018). Water safety plans by utilities: A review of research on implementation. *Utilities Policy, 53*, 15–24.

Rouse, M. (2013). Policy brief: The urban water challenge. *International Journal of Water Resources Development, 29*(3), 300–309.

Silva, J. A. (2022). Implementation and integration of sustainability in the water industry: A systematic literature review. *Sustainability, 14*(23), 15919.

Sousa, V., & Meireles, I. (2022). Quality and asset management: Conceptual compatibility towards sustainable infrastructures management. *Total Quality Management & Business Excellence, 34*(5–6), 743–767.

United Nations. (2022). *The sustainable development goals report: 2022. Technical report*. United Nations. https://unstats.un.org/sdgs/report/2022/

Vairavamoorthy, K., Gorantiwar, S. D., & Pathirana, A. (2008). Managing urban water supplies in developing countries–climate change and water scarcity scenarios. *Physics and Chemistry of the Earth, Parts A/B/C, 33*(5), 330–339.

Wilby, R. L. (2007). A review of climate change impacts on the built environment. *Built Environment, 33*(1), 31–45.

WWAP. (2016). The United Nations world water development report 2016: Water and jobs. In *Technical report, UNESCO, Paris, France, UNESCO, Paris, France*. United Nations World Water Assessment Programme.

Integrated Aquaculture-Agriculture System: A Tool for Sustainable and Optimum Water Management

Magdy A. Salah El Deen and Wael Khairy

Abstract The sustainability of the aquaculture sector worldwide is at stake due to the predicted effects of climate changes on water resources availability and unsustainable aquaculture practices. The integration of aquaculture and agriculture (IAA) has long been implemented by rural communities in many Asian and African countries as part of integrated resources management for enhancing the efficiency of natural resources and the livelihood of the farming community. These systems are particularly relevant for developing countries where the capacity of the farming community for intensive aquaculture is limited. Some case studies are presented and discussed in this chapter. Aquaponics combines the components of a recirculating aquaculture system (RAS) and hydroponics into a single-loop recirculating system. This integration allows the plant to use nutrients from the water provided by the fish which improves water quality characteristics for fish use and emerges as a sustainable means of organic food production. However, income from aquaponic systems mainly comes from the sale of plants rather than fish. Egypt is the first in Africa and among the top ten countries in the world in the field of aquaculture and has substantial experience in raising freshwater fish, mainly tilapia. Therefore, it is feasible to contribute and provide straightforward and inexpensive technological

M. A. Salah El Deen
National Water Research Center (NWRC), Ministry of Water Resources and Irrigation, Kalubia, Egypt
e-mail: Magdy_Salah-Eldeen@nwrc.gov.eg

W. Khairy (✉)
Faculty of Engineering, Heliopolis University, Cairo, Egypt
e-mail: wael.khairy@hu.edu.eg

J. C. Correia Leitão et al. (eds.), *Sustainable Water Management and Irrigation Systems*, Natural Resource Management and Policy 62,
https://doi.org/10.1007/978-3-031-93587-9_5

solutions to various aquaculture practices. The present chapter provides an overview of the impact of climate change on water resources and the different integrated aquaculture agriculture (IAA) systems that have great potential to increase food productivity and reduce risks associated with water scarcity.

Keywords Aquaculture · Integrated farming system · Aquaponics · Climate change · Water productivity · More crop per drop

1 Introduction

The world population is expected to reach 8.1 billion by 2025 and 9.6 billion by 2050 (United Nations, 2017). Rapid population growth will increase global demand for food production, which is expected to increase by 60% by 2050. On the other hand, the per capita share of global agricultural land has gradually decreased from 0.44 to 0.25 ha over the last 50 years (FAO, 2013).

Population growth has long been linked to food security, climate change, and water resources. By 2050, the world's population is expected to grow to nearly 9 billion; thus, our need for food will increase by more than 100%. Currently, 1 in 8 people or 842 million people struggle with hunger every day. Furthermore, nearly one billion people in the world are food insecure, meaning they lack access to an adequate amount of affordable nutritious food (FAO, 2013).

Annually, 7130 km^3 of freshwater is required for global food production, and it is expected to increase to 13,500 km^3 by 2050 (Molden et al., 2007). The effects of climate change may threaten global food production by increasing water demand and reducing water availability and crop productivity (Hanjra & Qureshi, 2010). Therefore, increasing food production is one of the key challenges in feeding a growing global population. In some world regions where scarce water reserves are becoming scarcer, innovative approaches to growing food are crucial. Known for its heat and deserts, it is not surprising that the Near East and North Africa are located in these regions.

Agriculture is essential for food security to meet the growing global population. Globally, agriculture consumes more than 69–70% of annual water withdrawals and 90% in some arid countries (United Nations, 2017). While agriculture has evolved to meet these exponentially increasing food demands, farmers will need to increase food production by 70–100% to meet global nutrition needs. Therefore, there is a necessity to produce higher yields using the same (or less) available lands than they use today while relying on fewer natural water resources (United Nations, 2019).

On the other hand, with a global focus on environmentally and economically sustainable development of natural resources, it is logical to integrate appropriate farming practices, where possible, to enhance farm productivity and water-use efficiency (the concept of more crop per drop). A possible method to increase crops per drop would be through integrating aquaculture into the existing agriculture systems (IAAS). IAAS has the privilege of augmenting food productivity and minimizing the menace correlated with water scarcity (Ahmed et al., 2012). It is a form of intensification and sustainable development of agriculture, that is, producing more food from essentially the same area of land and water without (or less) environmental impacts (Godfray et al., 2010).

This chapter aims to (a) briefly explore the impact of climate change on water resources, aquaculture, and more crops per drop; (b) present and highlight the different types, challenges, and risks associated with the integrated aquaculture-agriculture systems (IAAS) as efficient solutions to maximize water use efficiency; and (c) assess the expected benefits from IAAS regarding water productivity and food security.

2 Impacts of Climate Change on Water Resources and Aquaculture

Regardless of all contradicting views, there is a general agreement that climate change is a fact and will have an impact on water resources, aquaculture, food production processes, and ecosystem balance as well as human well-being. These impacts include sea level rise, increase in global temperature, and frequency of extreme weather events and patterns (De Silva, 2013; United Nations, 2020; Elsheikh, 2021).

Therefore, it is vital to promote sustainable water management practices, support resilient aquaculture systems, and implement innovative agricultural techniques that improve water-use efficiency (more crop per drop) and contribute to global food security. Implementing efficient irrigation techniques (e.g., drip irrigation, soil moisture sensors, improved crop varieties, mulching and conservation tillage, etc.) and sustainable water management practices help minimize the strain of climate change on water resources and promote environmentally sustainable agriculture.

2.1 Impact of Climate Change on Water Resources

Climate change affects water availability which varies from one region to another and can cause damage to people's health and ecosystems. Climate change is a water issue and may disturb water resources' stability at local, regional, and national levels (Sukanya & Sabu, 2023). The following are some of the climate change impacts on water resources:

- Changes in precipitation patterns: Climate change can cause alterations in rainfall in terms of distribution and quantity, leading to more recurrent and drastic droughts in some regions and increased rainfall in others. These changes can disrupt water availability and cause water scarcity, affecting both freshwater sources and groundwater reserves both of which can disrupt water resources for various uses, including agriculture and aquaculture.

- Rising temperatures: Higher temperatures accelerate evaporation rates, reducing the amount of water available for different uses. This phenomenon contributes to the depletion of surface water bodies and reduces soil moisture, impacting agricultural productivity.
- Melting glaciers and snowpack: In many regions, water supplies are dependent on the gradual melting of glaciers and snowpack. With rising global temperatures, glaciers are melting at an accelerated rate, affecting the timing and quantity of water flows in rivers and streams.
- Saltwater intrusion: Rising sea levels can cause saltwater intrusion into coastal aquifers, contaminating freshwater sources used for drinking, irrigation, and aquaculture.

2.2 *Impacts of Climate Change on Aquaculture*

Many authors investigated and evaluated the impacts of climate change on aquaculture practices (De Silva, 2013; Maulu et al., 2021; Elsheikh, 2021). These impacts are expected to remain persistent and most probably irreversible with critical consequences on the aquaculture and concerned communities (Barange et al., 2018; Dabbadie et al., 2018). The following are some of the changes that the climate will have an impact. Moreover, the potential positive and negative effects of some climate effects on aquaculture are summarized in Table 1.

- Changes in water temperature: Aquatic species have specific temperature ranges in which they thrive. Climate change alters water temperature, leading to shifts in the distribution and abundance of various species of aquatic organisms. Fish and shellfish are sensitive to temperature changes, and variations beyond their optimal ranges can reduce growth rates, alter reproduction patterns, increase susceptibility to diseases, and potentially reduce aquaculture yields and survival (Troell et al., 2017; Guyondet et al., 2018).
- Altered water quality: Climate change affects water quality variables such as dissolved oxygen concentrations, pH levels, water salinity, nutrient concentrations, and the occurrence of harmful algal blooms. These changes can negatively impact the survival, health, and productivity of aquaculture species (Ngoan, 2018).
- Disease outbreaks: Warming water temperatures can facilitate the spread of pathogens and harmful bacteria, thereby increasing the risk of disease outbreaks among farmed fish and other aquatic organisms. This can result in significant economic losses for aquaculture facilities (Sae-Lim et al., 2017).
- Loss of habitat and biodiversity: Climate change disrupts ecosystems, leading to the loss of critical habitats for aquatic species. Destruction of mangroves, coral reefs, and seagrass beds due to rising sea temperatures and ocean acidification leads to shifts in species distribution and can affect the overall biodiversity and productivity of aquaculture systems (Kibria et al., 2017; Rutkayova et al., 2017).

Table 1 Potential effects of climate change on aquaculture systems

No.	Element	Negative effects	Positive effects
1.	Rising temperature	– Slow growth rate and viability of cold-water species – Water quality deterioration – Fragile immunity in cold water species – Slow down carbon absorption capacity in the oceans – Enhance thermal stratification – Increased severity of disease-causing agents in warm waters	– Prolong growing seasons of warm-water species – Promote the possibility of developing genetic traits
2.	Diseases and harmful algal blooms	– Slow growth and survival rates – Water quality deterioration – Increase disease outbreaks which lead to increased production costs – Increased incidence of unfamiliar diseases	– Possibility of eradicating cold-water pathogens – May develop disease-tolerant species – Possibility of identifying and developing new species
3.	Changes in rainfall/ precipitation patterns	– Droughts may cause water use competition and high production costs – Floods may lead to water quality deterioration, polluting the environment, and high loss of species – Destruction of production facilities	– In some regions, floods may develop suitable areas for various aquaculture practices – Drought can promote improved management of wastewater

- Ocean acidification: Increased carbon dioxide (CO_2) emissions in the atmosphere are not only warming the planet but also causing the oceans to absorb more CO_2. This leads to ocean acidification, which can hinder shell formation in mollusks, making them more vulnerable to predation, disrupting the food chain and potentially impacting the growth and survival of various species as well as other aquaculture operations (FAO, 2009).
- Rising sea levels: The warming climate is causing ice caps and polar glaciers to melt leading to sea levels rising. This puts coastal aquaculture facilities at risk of flooding, salinization of freshwater sources, and increased vulnerability to storms and erosion (Rutkayova et al., 2017).
- Extreme weather events: More frequent and intense storms, hurricanes, and typhoons can damage aquaculture facilities, disrupt production, and result in significant economic losses for the industry (Seggel et al., 2016).
- Economic impacts: Disruptions in aquaculture production due to climate change can result in financial losses for farmers, processors, and related industries, thus affecting livelihoods and regional economies.

2.3 *Adaptation Strategies*

To mitigate climate change impacts on water resources and the aquaculture industry, stakeholders are exploring various adaptation strategies. These include developing efficient water management systems, such as recirculating aquaculture systems (RAS), reducing water consumption and minimizing the impact of changing water availability, implementing water-saving technologies, and diversifying aquaculture species to increase resilience to changing environmental conditions such as temperature-tolerant or disease-resistant varieties (Elsheikh, 2021; Malau, 2021).

Improved monitoring and early warning systems are crucial to detect and respond to any changes in water quality, disease outbreaks, and acute weather events. Adopting integrated multi-trophic aquaculture systems can help optimize resource utilization and reduce environmental impacts. By combining different species in a symbiotic relationship, these systems can enhance water quality and reduce the need for external inputs (Elsheikh, 2021). Several researchers have indicated that the impacts of climate change on aquaculture may vary depending on many factors like geographical areas, economics, climatic zones, production systems, and farmed species (Barange et al., 2018; Zolnikov et al., 2019; Maulu et al., 2021). Table 1 summarizes the expected elements as well as the negative and positive impacts of climate change that affect the production and sustainability of the aquaculture sector.

Ultimately, climate change poses significant challenges to aquaculture, particularly regarding water resources. The industry must adapt and develop strategies to mitigate these impacts. Collaboration between policymakers, scientists, and aquaculture practitioners is vital to ensure long-term sustainability and implement climate-resilient strategies for the sustainable use of water resources and the advancement of integrated agriculture and aquaculture industries.

3 The Integrated Aquaculture-Agriculture Systems (IAAS)

Aquaculture is one of the fastest progressing sectors of food production, with an annual growth rate of 5.3% in the period 2001–2018, ensuring an average of 46% of fish production worldwide up from 25.7% in the year 2000. Therefore, it is expected to fill the gap in fish demands (Halwart, 2020; FAO, 2020). Integrated aquaculture-agriculture systems (IAAS) are innovative and sustainable farming systems that link the cultivation of aquatic organisms with other forms of agricultural crops in a symbiotic manner. Wastes generated from one system are recycled in IAAS as inputs to another system, thus decreasing pollution (Ahmed et al., 2019). Therefore, these systems enhance resource utilization and waste recycling and increase overall productivity.

From a broader context, the IAAS has been mentioned as part of managing integrated resources to improve natural resource efficiency and increase productivity, profitability, and sustainability (Dey et al., 2010; Ahmed et al., 2019). The following paragraphs briefly discuss water productivity in agriculture and aquaculture and the different types and advantages of IAAS.

3.1 Water Productivity

Water productivity, also called water use efficiency (WUE), is the amount of agricultural output (such as crop yield or industrial production) that is achieved per unit of water consumed or withdrawn. In other words, it is a measure of how efficiently water resources are utilized to generate a certain level of productivity or output (Molden et al., 2007; Kumar, 2021; Ibrahim et al., 2023).

Water productivity is a convenient concept to guide efficient water management implementations, improve irrigation techniques, boost sustainable agricultural practices, and confront water scarcity challenges, particularly in regions where water resources are finite. High water productivity implies that more output is generated for every drop of water used, which is essential for sustainable development, food security, and environmental conservation (Molden et al., 2010; Brauman et al., 2013). Water productivity relies on different aspects which comprise varieties of growing plants and crops, strategies of water application, farming techniques, soil quality, and other agricultural matters (Ali & Talukder, 2008; Kilemo, 2022).

3.1.1 Water Productivity in Agriculture

The concept of more crop per drop and sometimes even more crop per drop and clean energy unit is addressed as part of the perspective of the water-food-energy nexus (El-Bably & Abd El-Hafez, 2019). A possible method to increase crop per drop would be through integrating aquaculture into the existing agriculture systems (IAAS). IAAS has the privilege of augmenting food productivity and minimizing the menace correlated with water scarcity (Ahmed et al., 2012). It is a form of intensification and sustainable development of agriculture, that is, producing more food from essentially the same area of land and water without (or less) environmental impacts (Godfray et al., 2010).

Both water productivity and water use efficiency measure the effectiveness of water use. However, water productivity refers to a physical mass, yield produced per unit of water input, and/or an economic value of produce per unit volume of water (Fig. 1). It measures the efficiency of water use in achieving a particular production such as crop yield, fish biomass, domestic animal breeding, or useful algal production. Water productivity (WP) can be expressed in various units, such as kilograms of crop per cubic meter of water (kg m^{-3}), liters of milk per cubic meter of water (L/m^3), or dollars of revenue per cubic meter of water (\$/m^3). On the other hand, water use efficiency (WUE) also measures the efficiency of water use but focuses specifically on how effectively water is utilized by a system or process to achieve its

Fig. 1 Relation between water productivity (WP), crop yield (kg), and water consumption (m^3). (Source: El-Bably & Abd El-Hafez, 2019)

intended purpose. In agriculture, WUE typically refers to the ratio of crop yield (or biomass production) to the amount of water used for irrigation. It quantifies how efficiently water is converted into plant biomass. WP has a broader application than WUE (El-Bably et al., 2015; Molden & Sakthivadivel, 1999), which relates the production of mass to process depletion (evapotranspiration or transpiration for irrigated agriculture). WP is defined in terms of net biomass, depleted water, and process depletion (Molden, 1999). In agriculture, WP is often used to assess the efficiency of irrigation systems and crop water management practices. It helps farmers and policymakers evaluate the effectiveness of different water management strategies in maximizing crop yield/animal production while minimizing water consumption.

Among agricultural crops, the highest average water footprint (i.e., the volume of freshwater used throughout the entire production chain of a consumer item or service) has been recorded for rice production. Moreover, due to its popularity as a main food crop, rice is one of the highest consumers of blue water (i.e., surface water and groundwater) and green water (i.e., natural rainfall) in the world (Chapagain & Hoekstra, 2011; Ahmed et al., 2014).

As indicated in Table 2, water productivity varies among major agricultural crops. The water productivity range of rice averaged 0.74–1.10 kg/m^3, for maize is 0.56–1.59 kg/m^3, and for wheat is 0.94–1.10 kg/m^3. A wider range of water productivity for rice at 0.15–1.60 kg/m^3 and for wheat at 0.20–1.20 kg/m^3 was reported by Molden et al. (2010). Increasing water productivity in different agricultural practices is the main strategy to face the challenges of water scarcity and may abate the necessity for additional land and water resources.

Water productivity in agriculture can be enhanced by adopting modern irrigation strategies, managing water efficiently, increasing soil fertility, selecting suitable crops, improving farming systems, and increasing yield per volume of water used (Ali & Talukder, 2008; Molden et al., 2010; Brauman et al., 2013).

3.1.2 Water Productivity in Aquaculture

Aquaculture practice is considered non-exhaustive and nonconsumptive regarding water use and therefore is a water-efficient food production system and does not compete with irrigation (Ahmed & Granett, 2011; Abdul-Rahman et al., 2011). Whenever water is available, aquaculture can be integrated with other crop production thus potentially increasing water productivity (Ahmed et al., 2014). However, water productivity in aquaculture systems varies widely depending on location,

Table 2 Comparative water productivity among crops in different farming systems

Practice	Crop	Type of study	Average water productivity (kg/m^3)	References
Agriculture	Maize	Field exp.	0.56	Igbadun et al. (2006)
		Field exp.	1.01	Moayeri et al. (2011)
		Monitoring through remote sensing	1.59	Fapeng et al. (2013)
	Rice	Crop production and land use data	0.74	Cai et al. (2010)
		Data review	1.09	Zwart and Bastiaanssen (2004)
		Experimental study	1.10	Steduto et al. (2012)
	Wheat	Crop production and land use data	0.94	Cai et al. (2010)
		Data review	1.09	Zwart and Bastiaanssen (2004)
		Experimental study	1.10	Steduto et al. (2012)
Aquaculture	Fish	Assessment	0.21	Verdegem et al. (2006)
		Data review	0.36	Brummett (1997)
		Assessment	0.37	Verdegem et al. (2006)
Integrated Agriculture aquaculture (IAA)	Fish maize	Filed exp.[a]	2.31 (fish 0.22 + crop 1.91)	Abdul-Rahman et al. (2011)
	Fish vegetables	Filed exp.[a]	8.46 (fish 0.22 + crop 8.24)	Abdul-Rahman et al. (2011)
	Rice fish	Assessment	1.21	Ahmed and Granett (2011), Abdul-Rahman et al. (2011)[b]

[a] Fish stocking density 10/m^3 seems to be an extensive culture

[b] The studies of Ahmed and Granett (2011) and Abdul-Rahman et al. (2011) revealed that IAA increases water value index by 11% and water productivity by over 10% without any additional consumptive use of water. Based on these results, it appears that fish and rice farming will increase water productivity compared to rice farming by at least 10% from 1.1 to 1.21 kg/m^3

culture method, and fish species. Improving water productivity in aquaculture varies widely depending on the culture method and the implemented strategies to maximize the yield of aquatic organisms while minimizing water usage and waste. Improving aquaculture production will increase the economic water productivity of water especially when selecting high-value fish (Ali & Talukder, 2008).

Some key considerations for enhancing water productivity in aquaculture include water management, feed management, stocking density, recirculating aquaculture systems (RAS), integrated multi-trophic aquaculture (IMTA), polyculture and integrated systems, water quality and waste management, species, and site selection as well as monitoring and evaluation (Verdegem et al., 2006; Verdegem & Bosma, 2009; Molden et al., 2010). Efforts to improve water productivity in aquaculture can contribute to sustainable food production, reduction of environmental impacts, conservation of freshwater resources, and the sustainability of the aquaculture industry.

3.1.3 Water Productivity in Integrated Aquaculture-Agriculture System (IAAS)

Water productivity in integrated aquaculture-agriculture systems (IAAS) refers to the efficient use of water resources to produce both aquatic products (such as fish or shrimp) and agricultural products (such as crops) within the same integrated system. These systems aim to maximize the benefits derived from water use by optimizing the interaction between aquaculture and agriculture components. Water productivity in such systems involves measuring the combined output of both aquatic and agricultural products relative to the amount of water used (Verdegem et al., 2006).

Water productivity in well-managed pond systems is reported to be 0.21–0.37 kg/m^3 and in pond-based IAAS is 2.13 kg/m^3 for fish and maize production and up to 8.46 kg/m^3 for fish and vegetable production (Table 2). Although water productivity in rice-fish farming is not well documented, IAA is reported to increase water productivity by at least 10% (Abdul-Rahman et al., 2011). Moreover, studies showed that water productivity could be improved in non-integrated farming systems through the integration and diversification of crops (Ali & Talukder, 2008; Ahmed & Granett, 2011).

3.2 Categories of IAAS

The integrated aquaculture-agriculture systems (IAAS) can be tailored to suit different geographical and environmental conditions, making them versatile and adaptable to various regions. However, successful implementation needs proper management and awareness of the essential requirements of the integrated species involved. There are several systems of IAA practice, each combining different elements of aquaculture and agriculture. Some of these are as follows.

3.2.1 Integrated Crop-Aquaculture Systems

It includes either rice, horticulture, sericulture, or mushrooms integrated with the fish system (Ahmed et al., 2007; Fapeng et al., 2013; Tran et al., 2013). Integrated rice-fish farming is the most widely spread activity where fish (e.g., common carp, catfish, and tilapia) feed on weeds and insects particularly mosquitoes that carry malaria reducing the need for pesticides and herbicides (Liu et al., 2020). In return, the fish waste acts as a natural fertilizer for the rice plants. The flooded rice paddies also provide an ideal habitat for fish growth, increasing overall productivity and local farmers' economic income (Ahmed & Granett, 2011).

Fish farming in rice fields increases the rice yield by 8–26%, as a result of the presence of fish (Dey et al., 2010). Therefore, such an integrated system is better than rice monoculture in terms of high production and providing a more divergent and nourishing food supply. However, it requires about 26% more water than rice monoculture. In this context, the second aim of integrated fish and crop production applications is to eliminate aquaculture residues and provide organic fertilizers for agricultural crops (Phong et al., 2010; Cai et al., 2010; Tran et al., 2013).

Fish farming in rice fields is an extensive culture system that primarily relies on natural foods (e.g., plankton, periphyton, and benthos); accordingly, fish productivity is low (Table 3). To increase the production of the system, other livestock such as cattle are raised and fed rice straw; then dunk is used as fertilizer for homestead gardening. Duck rearing on rice fields is also practiced in many countries (Ahmed et al., 2012; Long et al., 2013).

Table 3 Productivity of rice (crop) and fish in IAA among different countries

Farming system	Country	Fish species	Productivity (kg/ha)		References
			Fish	Crop/rice	
Pond-based IAA	Bangladesh	Prawn, carp and mola	1983	1470	Ahmed et al. (2007)
	China		13,575	Not documented	Kangmin (2010)
	Egypt	Tilapia	350	5400	van der Heijden (2012)
	India		3000	800	Kumar and Ayyappan (1998)
	Malawi	Tilapia	2017	1976	Tran et al. (2013)
	Tanzania	Tilapia-catfish	2460	2460	Mulokozi (2021)
	Thailand		2307	Not documented	Pant et al. (2004)
	Vietnam	Common carp, silver carb, kissing gourami, tilapia, and catfish	474	1618	Phong et al. (2008)
Rice-fish[a] farming	Bangladesh	*C. carpio*, *B. gonionotus*, *O. niloticus*	259	5261	Ahmed and Garnett (2011)
	China	*Cyprinus carpio*	372	6290	Hu et al. (2013)
	Ghana	Nile tilapia (*Oreochromis niloticus*)	201	4410	Ofori et al. (2005)
	India	*C. catla*, *C. carpio*, *C. mrigala*, *L. rohita*	906	3629	Mohanty et al. (2004)
	India	Rohu, Catla, Silver carp, Common carp, and Mrigal	1230	5800	Dey et al. (2010)
	Indonesia		0.8–625	7800	Purba (1998)
	Nigeria		2376	1656	Akegbejo-Samsons (2010)
	Philippines		233	2750	Corales et al. (2004)
	Thailand		146–363	1630–2651	Deomampo (1998), Halwart and Gupta (2004)
	Vietnam	Carpio, *B. gonionotus*, *O. niloticus*	326 C	4209	Berg (2002)

[a] Fish production from rice fields is low because they only depend on natural foods (e.g., plankton, periphyton, and benthos)

3.2.2 Integrated Livestock-Fish Aquaculture Systems

In this system, livestock, such as poultry, pigs, or cattle, are raised in combination with aquaculture. Livestock manure is utilized to fertilize fish ponds, providing nutrients like nitrogen (N) and phosphorus (P), stimulating plankton growth, and providing natural food for the fish (Boxes 1 and 2). The fish pond water can also be used for livestock drinking water and irrigation for crops (Little & Edwards, 2004).

Box 1 Case Studies from Ethiopia

In a study conducted by Waktola et al. (2016) where components of the integration were fish (*O. niloticus*), livestock (poultry and cow), and vegetables (tomato and onion), the authors reported the following:

- Generally, there is no significant change in the physical and chemical characteristics of pond water, and it remained within the required level for the growth of Nile tilapia.
- Higher levels of nitrate and total phosphorus in the IAA ponds were recorded.
- Higher levels of soil organic carbon and organic matter were recorded in the IAA ponds compared to the non-IAA ponds.
- There were higher number and bigger size of tomato fruit and onion from the IAA plots than the control plots.
- The results of the expenditure and income analysis indicated that the integration of vegetable cultivation using fish pond water alone was more profitable than the conventional method of vegetable cultivation with the application of fertilizers.

In another study by Tugie et al. (2017) where components of the integration were fish (*O. niloticus, C. carpio, and C. gariepinus*), livestock (Lohman brown chicken), and vegetables (maize and onion), the authors reported the following:

- Total profit from the integrated system was 12,030 ETBirr (1 Birr = 0.44 USD), while the profit farmers get from the sole maize production from the same plot of land was 1300 ETB.
- The poultry manured ponds gave significantly better growth rates in both *O. niloticus* and *C. carpio* species.
- The overall farm production was 8 tons of fish/ha/year, 233 eggs/hen/year, 10,800 kg onion/ha/year.

The authors indicated in their studies that IAA farms were significantly more efficient compared to non-IAA farms. IAA farms also had higher total factor productivity, higher farm income per hectare, and higher returns to family labor.

Box 2 Case Studies from South Africa and Kenya

Prinsloo and Schoonbee (1987) conducted a study in South Africa where components of the integration were four species of carps, pecking ducks, and three types of vegetables (tomato, spinach, and lettuce); the authors reported the following:

- Ducks grew to an average weight of 2.6 kg in 55 days.
- Fish production exceeded 19.5 tons/ha/year with a substantial yield of vegetable production.
- There were ten batches of Pekin ducks at a density of 2500 ducks/ha of water over 6 months with an average yield of 32.184 tons/ha/year.
- The mean feed conversion factor (FCR) was 3.05.

In another study by Muendo et al. (2011), in Kenya, the authors showed that frms with IAA systems were found to have a drop by 23–35% in nutrient consumption rates, increased agricultural production by 2–26%, and an increase of overall farm food production by 22–70%.

Generally, the authors' assessment indicated that IAAS practices have increased farm productivity, increased household incomes, and improved food and nutritional security through the production and consumption of fresh fish, ducks, and food crops grown around fish ponds.

3.2.3 Integrated Multi-trophic Aquaculture (IMTA)

Integrated multitrophic aquaculture is a resource recycling system in which discharged nutrients from higher trophic levels become input to the lower trophic levels (Reiriz et al., 2013; Melaku & Natarajan, 2019). IMTA involves the cultivation of different species in the same aquaculture system to create a balanced ecosystem. For example, finfish are farmed alongside shellfish and seaweeds. The fish provide nutrients for the shellfish and seaweeds, while these species help in removing excess nutrients and waste from the water. This system is mainly practiced in developed countries where there is an intensive cage culture of finfish like salmon (Handa et al., 2012).

3.2.4 Aquaponics

Aquaponics is a popular highly efficient and sustainable IAAS technique where a recirculating aquaculture system (RAS) and hydroponics are merged to grow plants and fish in a closed system, thus increasing economic efficiency and water productivity (Atique et al., 2022).

The basic principle of aquaponics involves cultivating fish in a tank and using the nutrient-rich water from the fish tanks to nourish plants grown in a soil-less medium.

The fish waste contains ammonia, which is converted by beneficial bacteria into nitrates and nitrites. These nutrients serve as natural fertilizers for the plants, providing essential elements for their growth. The plants, in turn, act as a biofilter by taking up these nutrients, purifying the water, and returning it to the fish tanks. This symbiotic relationship creates a self-sustaining ecosystem where both fish and plants thrive (Fig. 2) with a large reduction in water use and thus will mitigate and alter the negative environmental impacts of the system (Krastanova et al., 2022).

Aquaponic systems support food security and are equally effective in places with drought or poor soil quality resulting in local food production, boosting the local economy, and reducing food transportation. Aquaponics have the advantage of being sustainable and environmentally safe compared to other aquaculture practices because no agrochemicals are required and there is a continuous natural supply of organic fertilizers (Table 4).

Current studies on aquaponic sustainability gave a lot of information on its applicability. These studies revealed that, although the system is simple, it can only be sustainably managed through the availability of enough knowledge and experience of the fish, bacteria, and plant components at the individual and system levels (Sharma et al., 2018; Goddek et al., 2019; Atique et al., 2022).

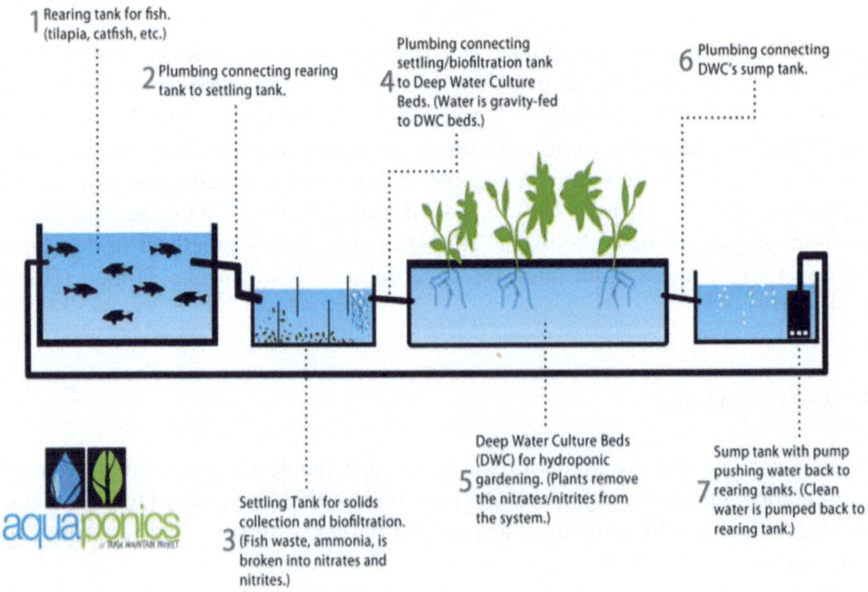

Fig. 2 General setup of an aquaponics system (Panigrahi et al., 2016)

Table 4 Comparison between hydroponics, aquaculture, and aquaponics

System	Advantages	Disadvantages
Hydroponics	• Produces a high crop yield in a small space • Highly water use efficient for crop production	• Using expensive fertilizers
Aquaculture	• Produce a large number of fish in a small space	• May fail if high fish stocking density • Fish produce toxic wastes that should be continuously removed
Aquaponics	• No pesticide, thereby environmentally safe and reducing carbon footprint • Plants get their automatic nutritional supply from the fish's water as they filter water for fish	• Management requires expertise in both growing fish and plants • High initial and operational costs compared to soil culture • The system requires high energy and maintenance • Numerous ways in which the entire system can break down

Table 5 Most common farmed fish in aquaponics

Scientific name	Common name	Most commonly farmed fish (ranked)	References
Oreochromis niloticus	Nile tilapia	First place	Love et al. (2014, 2015), Villarroel et al. (2016)
	Ornamental fish	Second place	Love et al. (2014, 2015)
Silurus glanis	European catfish	Third place	Love et al. (2014, 2015)

3.2.4.1 Species in Aquaponics

3.2.4.1.1 Aquaculture

Currently, some fish species are frequently used in aquaponic systems. In 2014, a study conducted in 44 countries revealed that the most prevalent fish species in aquaponic were tilapia (55%) followed by ornamental fish (koi, goldfish, and tropical fish) (48%) (Love et al., 2014). Other studies reported that catfish (25%), bluegill (15%), trout (10%), and some other aquatic animals (18%) are also used (Love et al., 2015; Villarroel et al., 2016). The most common farmed fish in aquaponics is presented in Table 5.

Tilapia is popular in the aquaponics industry due to its omnivorous nature, low cost, robustness, and rapid growth rate as it has a lower feed conversion ratio (FCR) of approximately 1.0 compared to large ruminants, whose FCR is 8.8. (Rakocy et al., 2004). Moreover, tilapia can thrive in various water conditions and a wide range of water temperatures (15–30 °C) and free ammonia (NH_3) concentrations (0.2–3.0 mg/L) for short period, making them adaptable to different climates and locations (Popma & Masser, 1999).

3.2.4.1.2 Vegetation (Plants)

Many types of plants can grow in an aquaponics system. However, the choice depends on the stocking density of the fish and the concentration of nutrients available in the aqueous solution. Nitrate requirements for plants vary during their growth phases depending on the surface area of the roots and the type of plant. For example, leafy vegetables require higher amounts of nitrate than fruit vegetables (Li et al., 2019).

The most favored and prominent plants in the aquaponic systems are herbs and spices for cooking purposes (e.g., basil, coriander, chives, parsley, purslane, and mint), leafy vegetables (e.g., lettuce, spinach, chard, Chinese cabbage, and watercress), and medicinal plants (e.g., aloe vera, chamomile, feverfew, lavender, ginger, and sage). These plants grow well in water with high nitrogen concentration, have a short-term growing period, and do not require high nutrient contents, and there is substantially a high consumer demand (Bailey & Ferrarezi, 2017). Moreover, in well-stocked and established aquaponic systems, vegetables with higher nutritional requirements such as tomatoes, peppers, and cucumber can perform well (Licamele, 2009).

Love et al. (2014, 2015), Villarroel et al. (2016), and Danner et al. (2019) have outlined the most commonly grown plant species in aquaponics (Table 6). Experimental data generally showed a higher yield of crops in aquaponics than in hydroponics or conventional agriculture (Buzby et al., 2016; Johnson et al., 2017).

3.2.4.2 Water Quality in Aquaponics

In aquaponic systems, a large production of plants and fish could be attained if there is a prominent balance between dissolved nutrients and the quality of water. Thus, proper monitoring and management of certain water quality variables are essential to have a functional system and obtain high performance in raising fish and plants. It is also necessary to have enough background on the effects of these parameters and how they interact.

Table 6 Most commonly grown crops in aquaponics

Scientific name	Common name	Crop	Most commonly grown crops (ranked)	References
Ocimum basilicum	Basil	Herb	First place	Love et al. (2014, 2015)
Lactuca sativa	Lettuce	Leaf vegetable	Second place	Villarroel et al. (2016)
Solanum lycopersicum	Tomato	Fruit vegetable	Second place Third place	Love et al. (2014), Villarroel et al. (2016)
	Salad greens	Leaf vegetable	Second place Third place	Love et al. (2014, 2015)
	Herbs	Leaf vegetable	Third place First place	Love et al. (2015) Villarroel et al. (2016)

Table 7 Water quality ranges for aquaponics systems

Parameters	Range	Parameters	Range
pH	6.5–7.0	Total ammonia-nitrogen (TAN) = NH_3 + NH_4-N	<2 mg/L
Water temperature (T)	17–30 °C	Nitrites (NO_2)	<1 mg/L
Dissolved oxygen (DO)	>4 mg/L	Nitrates (NO_3)	50–100 mg/L
Electrical conductivity (EC)	1100–2000 μS/cm	Flow	1–2 L/min
Total dissolved solids (TDS)	<1000 mg/L	Air temp.	18–30 °C
Salinity	0–2 ppt	Relative humidity	60–80%
Water hardness	50–150 mg/L as $CaCO_3$	Carbon dioxide (CO_2)	340–1300 ppm
Alkalinity	50–150 mg/L as $CaCO_3$	Light intensity	600–900 PPFD (PPFD is the photosynthetic photon flux density)

The ranges of the most important water quality parameters that affect the health and productivity of aquaponics systems are shown in Table 7. These parameters should be carefully monitored particularly key water quality parameters such as pH, temperature, and total ammonia nitrogen (TAN) to avoid sudden changes in their values and losses in productivity (Odema et al., 2017; Ibrahim et al., 2023).

4 Aquaculture in Egypt

Aquaculture in Egypt has been a significant and growing industry for several decades. The country's favorable climate, access to water resources, and historical reliance on the Nile River have all contributed to the development of aquaculture as an important economic sector. Aquaculture is considered as the only practical resource for decreasing the current gap between production and consumption of fish in Egypt. Egypt has continued to be a significant player in the global aquaculture industry. It is now number one in Africa, number six worldwide in aquaculture production, and number three in tilapia production globally (Shaalan et al., 2017; Adeleke et al., 2020; Wally & Akingbe, 2022). In Egypt, aquaculture is practiced in almost all different production systems that have been described in previous paragraphs. (Soliman & Yacout, 2016; Shaalan et al., 2017).

Egypt's total fish production increased by 17.64% from 1.7 million metric tons (MMT) in 2016 to 2.0 MMT in 2020 (Fig. 3). Egypt's total fish production increased by 17.6% from 1.7 million metric tons (MMT) in 2016 to 2.0 MMT in 2020 (Khairy et al., 2020). Aquaculture's share grew from 1.37 MMT in 2016 to 1.62 MMT in

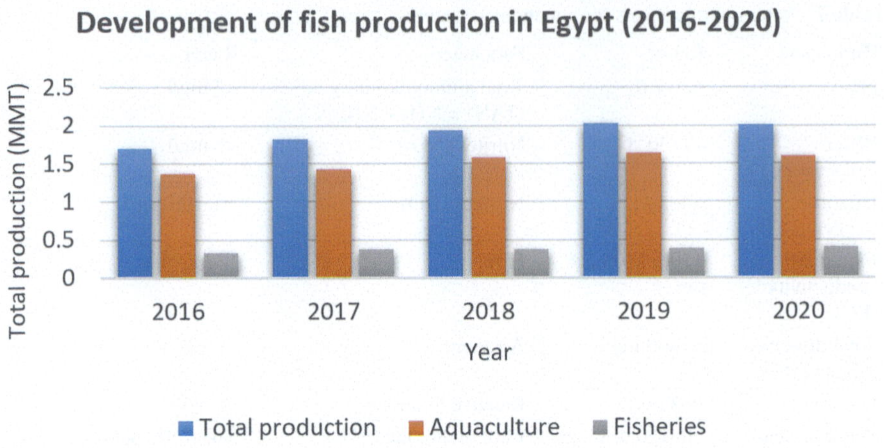

Fig. 3 Development of fish production in Egypt (2016–2020). (Source: GAFRD, 2020)

2020 representing about 81% of total fish production. Moreover, there is an increase of 18.2% growth from 2016, followed by production from lakes with 10%, seawater with 4.4%, freshwater with 3.8%, and finally rice fields with 0.8% of the total production (GAFRD, 2022; Wally & Akingbe, 2022).

4.1 IAA Systems in Egypt

According to Ibrahim et al. (2023), several commercial farms are implementing the IAAS and producing finfish particularly. The most common fish are Nile tilapia (*Oreochromis niloticus*), African catfish (*Clarias gariepinus*), carps, flathead gray mullet (*Mugil cephalilus*), European seabass (*Dicentrarchus labrax*), and gilthead seabream (*Sparus aurata*). The water source comes from groundwater reserves and/or agricultural drainage. The majority of these farms have applied integrated techniques that allow them to maximize water productivity by producing three different crops (fish, plant, and sheep) at a specific period. Some other farms have applied the recirculation aquaculture system (RAS) and produce tilapia in densities of 20–30 kg/m^3 to a market size of 250–400 g in 6–8 months. This production was attainable due to the suitable climate and available warm groundwater present throughout the year (Corner et al., 2020).

Moreover, brackish groundwater with a salt concentration of 25 g/L is also used in IAAS as Salicornia crops (halophytic vegetables with an array of potential health benefits) combined with intensive European seabass and gilthead seabream aquaculture. The yearly production of both fish species is about 100 tons/year (Corner et al., 2020).

Furthermore, aquaponics is a relatively new technique in Egypt for growing aquatic species and soilless plants in a single integrated system. Commercial aquaponics is still developing and practiced on a pilot scale. The predominant challenge is to balance the conditions required for the growth of multiple species which eventually may lead to a complex dynamic system (Table 4). In Egypt, some trials (Box 3) have been initiated by national institutions and the private sector to highlight that aquaponics could be an ally or alternative to traditional agriculture and land reclamation (Goda et al., 2015; El-Essawy et al., 2019; Corner et al., 2020). Although aquaponic presents a great opportunity, especially for developing countries like Egypt, it still requires further studies, research, and training to be prepared for unexpected challenges.

Box 3 Case Studies on Aquaponics in Egypt

The study conducted by Goda et al. (2015) aimed to highlight some of the technical, biological, social, and economic features of aquaponic systems in Egypt as a method of integrated aquaculture agriculture system (IAAS). Two systems using renewable energy were studied: (1) Integrated Multi-Trophic Aquaculture, IMTA–Nutrient Film Technique (NFT) System and (2) IMTA–Floating Raft System (FRS) in comparison with traditional soil culture system. The study included different fish species (e.g., tilapia, catfish, mullets, prawns, and clams) and plants (broccoli, cucumber, tomato, eggplant, and pepper).

The results revealed that IMTA–FRS and IMTA–NFT systems were able to bear costs and attain excess economic capacity of 53% and 47%, respectively. Moreover, the results confirmed the potentiality of these two systems to tolerate the burden of increased costs of production and bear the risks of lower fish and vegetable prices.

Another study was conducted by El-Essawy et al. (2019). The study discussed the possibility of applying the aquaponic system as an alternative to traditional agriculture in Egypt. A comparative pilot study examined two aquaponic systems (deep water culture system and integrated aquaculture agriculture system, IAAS). The crop quality (e.g., vitamins, heavy metals, and pesticide residues) of the two systems was compared to each other, as well as to that of the commercially organic crops available in the market. Results indicated that both systems produced safe high-quality organic food.

The economic feasibility analysis indicated that IAAS produced more crops with a wider variety of almost 20% less capital expenditure and operational expenditure costs. Fish production in both systems was similar, yet slightly higher in deep water culture. Although, in the short term, aquaponics entails relatively high capital and operational expenditure costs compared to traditional agriculture, however, it is more profitable in the long term, saving up to 85% of the water wasted due to traditional farming techniques.

5 Conclusion

This review highlighted the potential effects of climate change on aquaculture production and implications on the sector's sustainability. It is also discussed that the IAAS is a combined activity between aquaculture and agriculture to increase farm productivity, waste recycling, maximize resource use efficiency, and an effective means of using the same land resource to produce both agricultural products and animal protein.

Globally, aquaculture is considered one of the most efficient solutions to meet the increasing demands for aquatic products. However, this sector is under threat from climate change's negative and positive effects, although the negatives outweigh the positive ones.

Even though climate change poses a universal threat to food production, the correlated risks on aquaculture are predicted to vary according to geographical or climatic zones, national economy, rearing environment, production system scale, cultured species, and feed ingredient availability.

Because of the continued growth of the aquaculture sector while climate impacts become more evident, there is a necessity to adopt a comprehensive approach to anticipating the impacts of climate change on aquaculture and in addressing these impacts. Therefore, developing mitigation and adaptation strategies and action plans would be more effective.

Integrated aquaculture agriculture system (IAAS) is perceived as an efficient use of water to increase productivity and reduce the risks associated with water shortages. Therefore, IAAS is considered to produce "more crop per drop" and play a crucial role in food security.

Aquaponics depicts a promising practice for producing both fish and vegetables in ways that use less land and water. It also reduces the chemical and fertilizer inputs that are used in conventional food production. Aquaponics can be a cost-effective business, while the substantial challenges are the high initial and operational costs, available skill and knowledge from two separate fields, daily maintenance, limited plant selection, monitoring water quality for fish and plants, as well as marketing. There are multiple ways the entire system can fail.

The topic of water resources availability and management must be carefully addressed before the adoption of the IAAS in-country strategy. These encompass the efficient use and better management practices of blue and green water.

References

Abdul-Rahman, S., Saoud, I. P., Owaied, M. K., Holail, H., Farajalla, N., Haidar, M., & Ghanawi, J. (2011). Improving water use efficiency in semi-arid regions through integrated aquaculture/agriculture. *Journal of Applied Aquaculture, 23*(3), 212–230.

Adeleke, B., Andersson, D. R., Moodley, G., & Simon Taylor, S. (2020). Aquaculture in Africa: A comparative review of Egypt, Nigeria, and Uganda Vis-À Vis South Africa. *Reviews in Fisheries Science & Aquaculture, 29*(3). https://doi.org/10.1080/23308249.2020.1795615

Ahmed, N., & Granett, S. T. (2011). Integrated rice-fish farming in Bangladesh: Meeting the challenges of food security. *Food Security, 3*, 81–92. https://doi.org/10.1007/s12571-011-0113-8

Ahmed, N., Wahab, M. A., & Thilsted, S. H. (2007). Integrated aquaculture-agriculture systems in Bangladesh: Potential for sustainable livelihoods and nutritional security of the rural poor. *Aquaculture Asia, 12*, 15–23.

Ahmed, N., Muir, J. F., & Garnett, S. T. (2012). Bangladesh needs a "blue-green revolution" to achieve a green economy. *Ambio, 41*, 211–215.

Ahmed, N., Ward, J. D., & Saint, C. P. (2014). Can integrated aquaculture-agriculture (IAA) produce "more crop per drop"? *Food Security, 6*, 767–779. The final publication is available at link.springer.com

Ahmed, N., Thompson, S., & Glaser, M. (2019). Global aquaculture productivity, environmental sustainability, and climate change adaptability. *Environmental Management, 63*, 159–172.

Akegbejo-Samsons, Y. (2010). *Rice-fish production strategies in the coastal floodplains of Ondo State, Nigeria*. Second Africa Rice Congress. https://api.semanticscholar.org/CorpusID:55610328

Ali, M. H., & Talukder, M. S. U. (2008). Increasing water productivity in crop production—A synthesis. *Agricultural Water Management, 95*, 1201–1213.

Atique, F., Lindholm, L. P., & Pirhonen, J. (2022). Is aquaponics beneficial in terms of fish and plant growth and water quality in comparison to separate recirculating aquaculture and hydroponic systems? *Water, 14*, 1447. https://doi.org/10.3390/w14091447

Bailey, D. S., & Ferrarezi, R. S. (2017). Valuation of vegetable crops produced in the UVI commercial aquaponic system. *Aquaculture Reports, 7*(C), 77–82. https://doi.org/10.1016/j.aqrep.2017.06.002

Barange, M., Bahri, T., Beveridge, M. C. M., Cochrane, A. L., Funge-Smith, S., & Paulain, F. (2018). *Impacts of climate change on fisheries and aquaculture, synthesis of current knowledge, adaptation and mitigation options* (FAO Technical paper 627), Rome, Italy.

Berg, H. (2002). Rice monoculture and integrated rice-fish farming in the Mekong Delta, Vietnam—Economic and ecological considerations. *Ecological Economics, 41*, 95–107.

Brauman, K. A., Siebert, S., & Foley, J. A. (2013). Improvements in crop water productivity increase water sustainability and food security—A global analysis. *Environmental Research Letters, 8*(2), 1–7. https://doi.org/10.1088/1748-9326/8/2/024030

Brummett, R. E. (1997). Farming fish to save water. *Bioscience, 47*, 402.

Buzby, K., Waterland, N., Semmens, K., & Lin, L. S. (2016). Evaluating aquaponic crops in a freshwater flow-through fish culture system. *Aquaculture, 460*, 15–24. https://doi.org/10.1016/j.aquaculture.2016.03.046

Cai, X., Sharma, B. R., Matin, M. A., Sharma, D., & Gunasinghe, S. (2010). *An assessment of crop water productivity in the Indus and Ganges River Basins: Current status and scope for improvement*. International Water Management Institute, Rep Project CGIAR Challenge Program Water Food.

Chapagain, A. K., & Hoekstra, A. Y. (2011). The blue, green and grey water footprint of rice from production and consumption perspectives. *Ecological Economics, 70*(4), 749–758.

Corales, R. G., Juliano, L. M., Capistrano, A. O. V., Tobias, H. S., Dasalla, N. V., Canete, S. D., Casimero, M. C., & Sebastian, L. S. (2004). Palayamanan: A rice-based farming systems model for small-scale farmers. *Philippine Journal of Crop Science, 29*, 21–27.

Corner, R., Fersoy, H., Crespi, V. (Eds.). (2020). *Integrated agri-aquaculture in desert and arid lands: Learning from case studies from Algeria, Egypt and Oman* (Fisheries and Aquaculture Circular No. 1195). Cairo, FAO.

Dabbadie, L., Aguilar-Manjarrez, J. J., Beveridge, M. C. M., Bueno, P. B., Ross, L. G., & Soto, D. (2018). *Chapter 20: Effects of climate change on aquaculture: Drivers, impacts and policies*. FAO.

Danner, R., Mankasingh, U., Anamthawat-Jonsson, K., & Thorarinsdottir, R. (2019). Designing aquaponic production systems towards integration into greenhouse farming. *Water, 11*(10), 2123. https://doi.org/10.3390/w11102123

De Silva, S. S. (2013). Climate change impacts: Challenges for aquaculture. In *Proceedings of the Global Conference on Aquaculture 2010. Farming the Waters for People and Food* (p. 75110). FAO/NACA.

Deomampo, N. R. (1998). Farming systems, marketing and trade for sustainable aquaculture (pp. 201–217). Annex IV-5 of ADB/NACA, 1998. Aquaculture sustainability and the environment. *Report on a regional study and workshop on aquaculture sustainability and the environment. Bangkok, Thailand*, 492 p. Asian Development Bank and Network of Aquaculture Centers in Asia-Pacific.

Dey, M. M., Paraguas, F. J., Kambewa, P., Diemuth, E., & Pemsl, D. E. (2010). The impact of integrated aquaculture-agriculture on small-scale farms in Southern Malawi. *Agricultural Economics, 41*(1), 67–79. https://doi.org/10.1111/j.1574-0862.2009.00426.x

El-Bably, A. Z., & Abd El-Hafez, S. A. (2019). A strategic framework for sustainable agriculture development goals based on: Water, energy, food security nexus. *International Journal of Water Resources and Arid Environments, 8*(2), 118–131. ISSN 2079-7079, © PSIPW, 2019, Soils, Water and Environment Research Institute (SWERI), Agricultural Research Center, Egypt.

El-Bably, A. Z., Abd El-Hafez, S. A., Mahmoud, M. A., & Oud, S. A. H. (2015). A conceptual framework for water accounting: Addressing water balance, rotation, and economics crop. *International Journal of Water Resources and Arid Environments, 4*(2), 120–127.

El-Essawy, H., Nasr, P., & Sewilam, H. (2019). Aquaponics: A sustainable alternative to conventional agriculture in Egypt—A pilot-scale investigation. *Environmental Science and Pollution Research, 26*, 15872–15883. https://doi.org/10.1007/s11356-019-04970-0

Elsheikh, W. (2021). Effects of climate change on aquaculture production. *Eurasian Journal of Food Science and Technology, 5*(2), 167–173.

FAO. (2009). Food and Agriculture Organization. Climate change implications for fisheries and aquaculture, overview of current scientific knowledge, Rome, Italy.

FAO. (2013). *Food and Agriculture Organization. FAO statistical yearbook 2013*. World Food and Agriculture.

FAO. (2020). Food and Agriculture Organization. The state of world fisheries and aquaculture 2020. In *Brief. Sustainability in action*. Rome, Italy.

Fapeng, L., Chesheng, Z., Zongxue, X., Shanshan, J., & Jun, X. (2013). Remote sensing monitoring on regional crop water productivity in the Haihe River Basin. *Journal of Geographical Sciences, 23*, 1080–1090.

GAFRD. (2020). General Authority for Fish Resources Development. In *Fish statistics year book-2020* (30th ed.). Ministry of Agriculture and Land Reclamation.

Goda, A. M., Mohamed, M. A., Hassaan, M. S., & Sharawy, Z. (2015). Bio economic features for aquaponic systems in Egypt. *Turkish Journal of Fisheries and Aquatic Sciences, 15*, 525–532. https://doi.org/10.4194/1303-2712-v15_2_40

Goddek, S., Joyce, A., Kotzen, B., & Burnell, G. M. (Eds.). (2019). *Aquaponics food production systems, combined aquaculture and hydroponic production technologies for the future*. Springer International Publishing. https://doi.org/10.1007/978-3-030-15943-6

Godfray, H. C. J., Beddington, J. R., & Crute, I. R. (2010). Food security: The challenge of feeding 9 billion people. *Science, 327*, 812–818.

Guyondet, T., Comeau, L. A., Bacher, C., Grant, J., Rosland, R., & Sonier, R. (2018). Climate change influences carrying capacity in a coastal embayment dedicated to shellfish aquaculture. *Estuaries and Coasts, 38*, 1593–1618. https://doi.org/10.1007/s12237-014-9899-x

Halwart, M. (2020). *Fish farming high on the global food system agenda in 2020* (Vol. 61, pp. II–III). FAO Aquaculture Newsletter.

Halwart, M., & Gupta, M. (2004). *Culture of fish in rice fields*. FAO/WorldFish. https://www.researchgate.net/publication/23550931_Culture_of_Fish_in_Rice_Fields

Handa, A., Min, H., Wang, X., Broch, O. J., Reitan, K. I., & Reinertsen, H. (2012). Incorporation of fish feed and growth of blue mussels (Mytilus edulis) in close proximity to salmon (Salmo salar) aquaculture: Implications for integrated multi-trophic aquaculture in Norwegian coastal waters. *Aquaculture, 356*, 328.

Hanjra, M. A., & Qureshi, M. E. (2010). Global water crisis and future food security in an era of climate change. *Food Policy, 35*, 365–377.

Hu, L., Ren, W., Tang, J., Li, N., Zhang, J., & Chen, X. (2013). The productivity of traditional rice-fish co-culture can be increased without increasing nitrogen loss to the environment. *Agriculture, Ecosystems & Environment, 177*, 28–34.

Ibrahim, L. A., Abu-Hashim, M., Shaghaleh, H., Elsadek, E., Hamad, A. A. A., & Alhaj Hamoud, Y. A. (2023). Comprehensive review of the multiple uses of water in aquaculture-integrated agriculture based on international and national experiences. *Water, 15*, 367. https://doi.org/10.3390/w15020367

Igbadun, H. E., Mahoo, H. F., Tarimo, A. K. P. R., & Salim, B. A. (2006). Crop water productivity of an irrigated maize crop in Mkoji subcatchment of the Great Ruaha River Basin, Tanzania. *Agricultural Water Management, 85*, 141–150.

Johnson, G., Buzby, K., Semmens, K., Holaskova, I., & Waterland, N. L. (2017). Evaluation of lettuce between spring water, hydroponic, and flow-through aquaponic systems. *International Journal of Vegetable Science, 23*(1), 456–470. https://doi.org/10.1080/19315260.2017.1319888

Kangmin, L. (2010). Extending integrated fish farming technologies to mariculture in China. Biotechnology, Vol. IX, Encyclopedia of life support systems. http://www.eolss.net/sample-chapters/c17/e6-58-08-10.pdf

Khairy, W. M., El-Ashmawy, N., & Ragab, E. (2020). Analyzing the evolution of environmental impacts due to fish farms expansion—A pilot study in The Northern Nile Delta. *International Journal of Engineering Research & Technology (IJERT), 9*(8) ISSN: 2278-0181. https://www.ijert.org/analyzing-the-evolution-of-environmental-impacts-due-to-fish-farms-expansion

Kibria, G., Haroon, Y. A. K., & Dayanthi, N. (2017). Climate change impacts on tropical and temperate fisheries, aquaculture, and seafood security and implicationsis—A review. *Livestock Research for Rural Development, 29*, 22.

Kilemo, D. B. (2022). The review of water use efficiency and water productivity metrics and their role in sustainable water resources management. *Open Access Library Journal, 9*, e7075. https://doi.org/10.4236/oalib.1107075

Krastanova, M., Sirakov, I., Ivanov, S., Yarkov, D., & Orozova, P. (2022). Aquaponic systems: Biological and technological parameters. *Biotechnology and Biotechnological Equipment, 36*(1), 305–316. https://doi.org/10.1080/13102818.2022.2074892

Kumar, M. K. (2021). Conceptual issues in water use efficiency and water productivity (Chapter 3). In *Current directions in water scarcity research* (Vol. 3, pp. 49–63). Elsevier. https://doi.org/10.1016/B978-0-323-91277-8.00005-8

Kumar, K., & Ayyappan, S. (1998). *Current practices in integrated aquaculture from India* (Working paper no.5). Central Institute of Freshwater Aquaculture. https://www.scribd.com/document/160656465/Integrated-Fish-Farming-Current-Practices#

Li, C., Zhang, B., Luo, P., Shi, H., Li, L., Gao, Y., Lee, C. T., Zhang, Z., & Wu, W. (2019). Performance of a pilot-scale aquaponics system using hydroponics and immobilized biofilm treatment for water quality control. *Journal of Cleaner Production, 208*, 274–284. https://doi.org/10.1016/j.jclepro.2018.10.170

Licamele, J. D. (2009). *Biomass production and nutrient dynamics in an aquaponics system* [Ph.D. thesis]. University of Arizona (AZ). http://hdl.handle.net/10150/193835

Little, D. C., & Edwards, P. (2004). *Integrated livestock-fish farming systems*. Food and Agriculture Organization of the United Nations (FAO). https://www.researchgate.net/publication/259999240

Liu, D., Tang, R., Xie, J., Tian, J., Shi, R., & Zhang, K. (2020). Valuation of ecosystem services of rice-fish coculture systems in Ruyuan County, China. *Ecosystem Service, 41*, 101054.

Long, P., Huang, H., Liao, X., Fu, Z., Zheng, H., Chen, A., & Chen, C. (2013). Mechanism and capacities of reducing ecological cost through rice-duck cultivation. *Journal of the Science of Food and Agriculture, 93*, 2881–2891.

Love, D. C., Fry, J. P., Genello, L., Elizabeth, S., Hill, E. S., Frederick, J. A., Li, X., & Semmens, K. (2014). An international survey of aquaponics practitioners. *PLoS One, 9*(7), e102662-10. https://doi.org/10.1371/journal.pone.0102662

Love, D. C., Fry, J. P., Li, X., Hill, E. S., Genello, L., Semmens, K., & Thompson, R. E. (2015). Commercial aquaponics production and profitability: Findings from an international survey. *Aquaculture, 435*, 67–74. https://doi.org/10.1016/j.aquaculture.2014.09.023

Malau, C. (2021). *A dictionary of Vurës, Vanuatu*, ANU Press.

Maulu, S., Hasimuna, O. J., Haambiya, L. H., Monde, C., Musuka, C. G., Makorwa, T. H., Munganga, B. P., Phiri, K. J., & Nsekanabo, J. D. (2021). Climate change effects on aquaculture production: Sustainability, implications, mitigation, and adaptations. *Frontiers in Sustainable Food Systems, 5*, 609097. https://doi.org/10.3389/fsufs.2021.609097

Melaku, S., & Natarajan, P. (2019). Status of integrated aquaculture—Agriculture systems in Africa. *International Journal of Fisheries and Aquatic Studies, 7*(4), 263–269.

Moayeri, M., Siadat, H., Pazira, E., Abbasi, F., Kaveh, F., & Oweis, T. Y. (2011). Assessment of maize water productivity in southern parts of the Karkheh river basin, Iran. *World Applied Sciences Journal, 13*, 1586–1594.

Mohanty, R. K., Verma, H. N., & Brahmanand, P. S. (2004). Performance evaluation of rice-fish integration system in rainfed medium and ecosystem. *Aquaculture, 23*, 125–135.

Molden, D. J. (1999). *Indicators for comparing performance of irrigated agricultural systems* (Research report 20). IWMI.

Molden, D., & Sakthivadivel, R. (1999). Water accounting to assess use and productivity of water. *Water Resources Development, 15*(1/2), 55–71.

Molden, J. D., Oweis, T., Steduto, P., & Kijne, J. (2007). Pathways to increase agricultural water productivity. In *Water for food water for life. A comprehensive assessment of water management in Agriculture*. EARTHSCAN. https://www.researchgate.net/publication/266382480

Molden, D. J., Oweis, T. J., Steduto, P., & Bindraban, P. S. (2010). Improving agricultural water productivity: Between optimism and caution. *Agricultural Water Management, 97*(4), 528–535. https://doi.org/10.1016/j.agwat.2009.03.023

Muendo, P. N., Stoorvogel, J. J., Verdegem, M. C. J., Mora-Vallejo, A., & Verreth, J. A. (2011). Ideotyping integrated aquaculture systems to balance soil nutrients. *Journal of Agriculture and Rural Development in the Tropics and Subtropics, 112*, 157–168.

Mulokozi, D. (2021). *Integrated agriculture and aquaculture systems (IAA) for enhanced food production and income diversification in Tanzania* [Ph.D. thesis], Department of Physical Geography, Stockholm University, Stockholm. https://stockholmuniversity.zoom.us/j/68951088257

Ngoan, L. D. (2018). Effects of climate change in aquaculture: Case study in Thua Thien Hue Province, Vietnam. *Biomedical Journal of Scientific & Technical Research, 10*, 2018. https://doi.org/10.26717/BJSTR.2018.10.001892

Odema, M., Adly, I., Wahba, A., & Ragai, H. (2017). Smart aquaponics system for industrial internet of things (IoT). Advances in intelligent systems and computing. In *Proceedings of the International Conference on Advanced Intelligent Systems and Informatics* (Vol. 639, pp. 844–854)

Ofori, J., Abban, E. K., Otoo, E., & Wakatsuki, T. (2005). Rice-fish culture: An option for smallholder Sawah rice farmers of the West African lowlands. *Ecological Engineering, 24*, 235–241.

Panigrahi, G. K., Panda, S., & Padhi, S. N. (2016). Aquaponics: An innovative approach of symbiotic farming. *International Journal of Bioassays, 5*(9), 4808. https://doi.org/10.21746/ijbio.2016.09.005

Pant, J., Demaine, H., & Edwards, P. (2004). Assessment of the aquaculture subsystem in integrated agriculture-aquaculture systems in Northeast Thailand. *Aquaculture Research, 35*, 289–298. https://doi.org/10.1111/j.1365-2109.2004.01014.x

Phong, L. T., Udo, H. M. J., van Mensvoort, M. E. F., Tri, L. Q., Bosma, R. H., Nhan, D. K., & van der Zijpp, A. J. (2008). Integrated agriculture aquaculture systems in the Mekong Delta, Vietnam: An analysis of recent trends. *Asian Journal of Agriculture and Development, 4*, 51–66.

Phong, L. T., van Dam, A. A., Udo, H. M. J., van Mensvoort, M. E. F., Tri, L. Q., Steenstra, F. A., & van der Zijpp, A. J. (2010). An agroecological evaluation of aquaculture integration into farming systems of the Mekong Delta. *Agriculture, Ecosystems & Environment, 138*, 232–241.

Popma, T., & Masser, M. (1999). *Tilapia: Life history and biology. Southern Regional Aquaculture Center* (SRAC Publication No. 283). United States Department of Agriculture.

Prinsloo, J. F., & Schoonbee, H. J. (1987). Investigations into the feasibility of duck-fish-vegetable integrated agriculture-aquaculture system for developing areas in South Africa. *Water SA, 13*(2), 109–118.

Purba, S. (1998). *The economics of rice fish production systems in North Sumatra, Indonesia: An empirical and model analysis. Farming systems and resource economics in the tropics* (Vol. 31). Wissenschafverlag Vauk Kiel KG.

Rakocy, J., Shultz, R. C., Bailey, D., & Eric, S. (2004). Aquaponic production of tilapia and basil: Comparing a batch and staggered cropping system. *Acta Horticulturae, 648*, 63–69. https://api.semanticscholar.org/CorpusID:45134183

Reiriz, M. J. F., Robinson, S. M. C., Cranford, P. J., & Labarta, U. (2013). Absorption efficiency of mussels Mytilus edulis and Mytilus galloprovincialis cultured under Integrated Multi-Trophic Aquaculture conditions in the Bay of Fundy (Canada) and Ria Areo Betanzos (Spain). *Aquaculture, 388*, 182–192.

Rutkayova, J., Vácha, F., Maršálek, M., Beneš, K., Civišová, H., & Horká, P. (2017). Fish stock losses due to extreme floods–findings from pondbased aquaculture in The Czech Republic. *Journal of Flood Risk Management, 11*, 351–359. https://doi.org/10.1111/jfr3.12332

Sae-Lim, P., Kause, A., Mulder, H. A., & Olesen, I. (2017). Breeding and genetics symposium: Climate change and selective breeding in aquaculture. *Journal of Animal Science, 95*, 1801–1812. https://doi.org/10.2527/jas2016.1066

Seggel, A., De Young, C., & Soto, D. (2016). *Climate change implications for fisheries and aquaculture: Summary of the findings of the intergovernmental panel on climate change fifth assessment report* (FAO Fisheries and Aquaculture Circular No. 1122). Rome, Italy.

Shaalan, M., El-Mahdy, M., Saleh, M., & El-Matbouli, M. (2017). Aquaculture in Egypt: Insights on the current trends and future perspectives for sustainable development. *Reviews in Fisheries Science & Aquaculture, 26*, 99. https://doi.org/10.1080/23308249.2017.1358696

Sharma, N., Acharya, S., Kumar, K., Singh, N., & Chaurasia, O. P. (2018). Hydroponics as an advanced technique for vegetable production: An overview. *Journal of Soil and Water Conservation, 17*, 364–371.

Soliman, F. N., & Yacout, A. D. (2016). Aquaculture in Egypt: Status, constraints and potentials. *Aquaculture International, 24*(5), 1201–1227. https://doi.org/10.1007/s10499-016-9989-9

Steduto, P., Hsiao, T. C., Fereres, E., & Raes, D. (2012). *Crop yield response to water* (FAO irrigation and drainage paper 66). Food and Agriculture Organization of the United Nations.

Sukanya, S., & Sabu, J. (2023). Climate change impacts on water resources: An overview. In *Visualization techniques for climate change with machine learning and artificial intelligence, Chapter 4* (pp. 55–76). Elsevier. https://doi.org/10.1016/B978-0-323-99714-0.00008-X

The General Authority for Fish Resources Development (GAFRD). (2022). Egyptian Marine Fisheries and Its Sustainability Report, national Institute of Oceanography and Fisheries, Egypt.

Tran, N., Crissman, C., Chijere, A., Chee, H. M., Jiau, T. S., & Valdivia, R. O. (2013). Ex-ante assessment of integrated aquaculture agriculture adoption and impact in Southern Malawi. WorldFish. Penang: Work. Pap. AAS-2013-03 CGIAR Res. Program Aquat. Agric. Sys. Malaysia. https://www.researchgate.net/publication/257965404_Exante_assessment_of_integrated_aquaculture-agriculture_adoption_and_impact_in_Southern_Malawi

Troell, M., Eide, A., Isaksen, J., Hermansen, Ø., & Crépin, A. S. (2017). Seafood from a changing Arctic. *Ambio, 46*, S368–S386.

Tugie, D., Abebe, A., & Endebu, M. (2017). Potential of integrated fish-poultry-vegetable farming system in mitigating nutritional insecurity at small scale farmer's level in East Wollega, Oromia, Ethiopia. *International Journal of Fisheries and Aquatic Studies, 5*(4), 377–382.

United Nations. (2017). *World population prospects: The 2017 revision, key findings and advance tables*. Population Division, United Nations.
United Nations. (2019). *World population prospects, report ST/ESA/SER.A/423*. Department of Economic and Social Affairs, Population Division, United Nations UN.
United Nations. (2020). UN world water development report: Water and climate change. https://www.unwater.org/water-facts/water-and-climate-change
Van Der Heijden, P. G. M. (2012). Water use at integrated aquaculture agriculture farms. Experiences with limited water resources in Egypt. *Global Aquaculture Advocate, 4*, 28–31.
Verdegem, M. C. J., & Bosma, R. H. (2009). Water withdrawal for brackish and inland aquaculture, and options to produce more fish in ponds with present water use. *Water Policy, 11*, 52–68.
Verdegem, M. C. J., Bosma, R. H., & Verreth, J. A. J. (2006). Reducing water use for animal production through aquaculture. *Water Resources Development, 22*, 101–113.
Villarroel, M., Junge, R., Komives, T., König, B., Plaza, I., Bittsánszky, A., & Joly, A. (2016). Survey of aquaponics in Europe. *Water, 8*(10), 468. https://www.researchgate.net/publication/309336645
Waktola, B. A., Devi, L. P., Sreenivasa, V., & Lakew, A. (2016). A study on the profitability of fish and horti-crop integrated farming at Nono District, West Shoa Zone, Ethiopia. *Greener Journal of Agricultural Sciences, 6*(2), 41–48.
Wally, A., & Akingbe, O. O. (2022). *An overview of the aquaculture industry in Egypt Country* (Report Number: EG2022-0003). USDA.
Zolnikov, T., Ramirez-Ortiz, D., Raymond, J., Chambers, D., Brears, R. C., & Cook, D. (2019). A scoping review on global climate change vulnerability, adaptation, and resiliency. *Spotlight on Climate Change Research*. https://doi.org/10.35831/sorccr/26072019trz
Zwart, S. J., & Bastiaanssen, W. G. M. (2004). Review of measured crop water productivity values for irrigated wheat, rice, cotton and maize. *Agricultural Water Management, 69*, 115–133.

Integrated Solar Power and Rainwater Harvesting System

Yuvraj Rathore, Tapash Nayak, Ramesh N, Shalabh Kumar Mishra, and Nawnit Kumar

Abstract Water scarcity is a growing concern around the world, with many regions facing droughts, water shortages, and poor water quality. At the same time, renewable energy sources, such as solar power and wind power, are gaining popularity as a clean and sustainable alternative to fossil fuels. In this context, the design of a device that can both conserve rainwater and harness solar energy can provide a solution to two pressing issues. This manuscript presents an automatic tracking solar plate which can be also used for water harvesting. The proposed design consists of solar panel rain sensor, motors, and turbine. During the sunlight, solar energy is converted into electrical energy, and during the rainy season, rain water is stored and also used in production of energy using turbine. This literature describes, analyze, and evaluate the performance and implications of the proposed device.

Keywords Automatic tracking · Rain sensor · Renewable energy · Solar panel

1 Introduction

In the contemporary landscape of environmental sustainability, two pressing challenges loom large: the burgeoning issue of water scarcity and the urgent need for renewable energy sources [International Renewable Energy Agency (IRENA) (2020), REN21 (2020), Owusu-Sekyere & Mireku (2019)]. Across the globe,

Y. Rathore · T. Nayak
Bharati Vidhyapeeth's College of Engineering, India, New Delhi

R. N
Atria Institute of Technology, Bengaluru, India

S. K. Mishra (✉)
Madan Mohan Malaviya University of Technology, Gorakhpur, U.P., India

N. Kumar
Sri Guru Tegh Bahadur Institute of Management & IT, New Delhi, India

J. C. Correia Leitão et al. (eds.), *Sustainable Water Management and Irrigation Systems*, Natural Resource Management and Policy 62,
https://doi.org/10.1007/978-3-031-93587-9_6

communities grapple with the far-reaching impacts of droughts, water shortages, and compromised water quality, while the imperative to transition away from fossil fuels toward cleaner, more sustainable energy alternatives gains increasing momentum [World Bank (2019), United Nations (2015)]. Recognizing the intricate interplay between these critical concerns, this chapter embarks on an exploration of a multifaceted solution that addresses both water conservation and renewable energy generation in tandem.

At its core, this chapter endeavors to introduce and elucidate the concept of an automatic tracking solar plate, a visionary device engineered to simultaneously harness solar energy and facilitate rainwater harvesting. In essence, it represents a convergence of technological innovation and ecological pragmatism, embodying a proactive approach to resource management in the face of mounting environmental challenges.

By delving into the intricate design, functionality, and potential impact of this integrated device, the chapter aims to shed light on its transformative potential. Through the seamless integration of solar panel rain sensors, motors, and turbines, this innovative solution capitalizes on the dual opportunities presented by sunlight and rainfall, offering a comprehensive approach to sustainable resource utilization.

Furthermore, the chapter endeavors to provide a comprehensive analysis of the operational dynamics of the proposed device, elucidating its functionality and efficacy in varying environmental conditions. Through meticulous examination and evaluation, the performance metrics and practical implications of this novel solution are scrutinized, offering valuable insights into its viability and applicability in real-world contexts.

Ultimately, this chapter seeks to contribute to the ongoing discourse on sustainable development by presenting a tangible innovation that embodies the principles of resource efficiency, environmental stewardship, and technological ingenuity. By bridging the realms of water conservation and renewable energy generation, the automatic tracking solar plate emerges as a promising tool in the collective endeavor to build a more resilient and sustainable future for generations to come.

2 Methodology

Water harvesting stands as a venerable practice ingrained in human civilization, embodying the art of capturing and utilizing runoff for various purposes [Fathollahzadeh & Esfahani (2018), Kazem & Al-Karaghouli (2017), Angelakis & Koutsoyiannis (2005)]. It encompasses a spectrum of techniques aimed at harnessing rainwater, a precious resource often underutilized in many regions. Of particular significance is rainwater harvesting, a method that entails the deliberate collection of runoff from impermeable surfaces such as rooftops or paved areas, as well as from natural ground surfaces [Gupta & Kumar (2018), Tiwari & Singh (2008),United Nations (2018)].

In countries like India, where the ebb and flow of life are intimately intertwined with the rhythms of monsoon rains, hydroelectric power emerges as a cornerstone of the energy landscape. Hydro technologies, harnessing the kinetic energy of flowing water, are revered for their status as a green energy source amidst the growing chorus for renewable alternatives to fossil fuels. Within this realm, small-scale hydroelectric systems hold promise, offering localized electricity generation independent of centralized grids and catering to the needs of individual households or communities.

The practice of rooftop harvesting embodies a fusion of practicality and sustainability, providing a cost-effective means of capturing rainwater and replenishing local groundwater reserves. By channeling rainwater from rooftops into storage tanks or cisterns, communities can mitigate the impacts of water scarcity and reduce their reliance on dwindling surface and groundwater sources.

In parallel, the advent of water markets underscores the growing pressures on water resources, driven by escalating demand and diminishing supplies. These markets, characterized by the commodification and trade of water rights, reflect the urgent need to address water scarcity through innovative solutions such as rainwater harvesting.

On the energy front, solar power emerges as a beacon of hope in the quest for clean, renewable energy sources. Solar panels, leveraging photovoltaic technology to convert sunlight into electricity, hold immense potential for meeting burgeoning energy demands while minimizing environmental impacts [Chandel et al. (2011), Qureshi & Hasanuzzaman (2017), Jain & Sharma (2020)]. Ensuring optimal alignment of solar panels with the sun's trajectory becomes paramount to maximizing energy production and enhancing the efficiency of solar installations [Al-Karaghouli & Kazmerski (2017), Abdulla & Al Marzooqi (2021)].

The convergence of water harvesting and renewable energy generation offers a synergistic solution to the dual challenges of water scarcity and energy insecurity, particularly in urban areas. By harnessing the power of the sun to both conserve water and generate electricity, communities can achieve a holistic approach to sustainable development, fostering resilience and self-sufficiency in the face of changing environmental realities.

3 Proposed Work

Rainwater harvesting systems have been employed for decades, yet their widespread adoption still lags behind conventional energy sources. However, recent advancements have ushered in a new era with the development of innovative devices integrating additional features aimed at reducing the costs associated with renewable energy. One such device has emerged, engineered to maximize solar energy collection while concurrently conserving rainwater. Originally conceptualized by students at CIT College Gubbi, the incorporation of cost-effective technologies has rendered these devices more accessible and appealing to both consumers and businesses alike.

Central to the functionality of these devices is energy storage, an indispensable component in the energy landscape of 2020. Our team has addressed the limitations of previous iterations of renewable energy systems, revolutionizing energy storage capabilities to facilitate a more efficient and user-friendly design. This includes the implementation of versatile batteries suitable for both industrial and domestic use, alongside pumped hydro storage and compressed air energy technologies, offering unparalleled potential to transform conventional storage methods.

Moreover, our devices leverage popular energy storage technologies that synergize with solar and rainwater harvesting systems. Notably, we have integrated a suite of widely utilized tools, including foldable canopy systems that safeguard devices during inclement weather and enhance reliability in adverse conditions such as high winds or heavy rainfall [Akbarzadeh et al. (2019), Khan & Islam (2017), Singh & Tiwari (2018)]. Additionally, grid stability is bolstered through the incorporation of smart grid technology, facilitating seamless integration of renewable energy sources, energy storage solutions, and electric vehicles, thereby enhancing grid efficiency and reliability.

Furthermore, energy efficiency emerges as a critical facet of the energy sector often overlooked. By implementing energy-efficient appliances, building insulation, and efficient lighting solutions, substantial reductions in energy consumption and carbon emissions can be achieved, resulting in significant cost savings for consumers and businesses alike.

In conclusion, the energy sector is undergoing a profound transformation, spurred by the imperative to address urban energy challenges, particularly in densely populated areas like factories, multiplexes, and hospitals. Innovations such as trolley systems for rainwater collection between skyscrapers and solar concentrator plants underscore the ingenuity driving this transformation. Through meticulous optimization approaches like mixed-integer linear programming, the feasibility of deploying solar thermal power plants for electricity generation can be assessed, facilitating expanded grid coverage and meeting growing electricity demands while reducing reliance on fossil fuels and freshwater resources. This research offers a sustainable pathway towards energy generation, simultaneously addressing economic, social, and environmental imperatives, ultimately paving the way for a brighter, more sustainable future.

3.1 System Architecture

Solar panel 36 W, 18 V solar panel will be used on the house terrace to generate electricity from Sunrays.

- Battery: The battery will be used to store the charge, for this charge. The circuit will be built which charges the 12 V, 1 A battery.

- The ARM controller LPC2148 series for the control process will be used for the whole system; it is one of the advanced controllers for fast processes.
- The rainwater harvesting device includes several components, such as a water level sensor to indicate when the rain gutter is full and sends a signal to the controller. It also utilizes a rainfall sensor to detect when it starts raining and a DC generator to produce 12 V DC output voltage when the fan rotates from the high-pressure water in the gutter.
- L293D motor driver IC L293D is a typical motor driver IC which allows DC motor to drive in either direction.
- LCD: On LCD, we will display the sensor output, i.e., rainy season condition, and level sensor condition to show the water gutter's full capacity.

4 Project Concept

In this project, we will utilize solar panels installed on rooftops to harness solar energy during periods of no rainfall. The solar panels will convert sunlight into electricity, which will be stored in a battery for later use [Tiwari & Sodha (2007), Sharma & Sharma (2018), Ahmadi & Rahmani (2017), Tong et al. (2017), Goel & Kumar (2020)]. A charging circuit will facilitate the transfer of electricity from the solar panels to the battery, ensuring efficient storage. Subsequently, the electricity stored in the battery will be supplied to a controller circuit, which governs the operation of the entire system. The illustrative representation of the proposed design is shown in Fig. 1.

To detect rainfall, we will incorporate a rain sensor into our project. When rain is detected, the rain sensor will send a signal to the controller circuit. In response, the controller will activate the rain gutter mechanism, aligning it to capture rainwater from the rooftop. Once the rain gutter reaches a predetermined level of water accumulation, a level sensor will transmit a signal to the controller. The controller will then signal an electronic valve to open, allowing water to pass through the valve.

As water flows through the valve, it will exert pressure on a turbine mechanism. This high-pressure water will drive the turbine to rotate, thereby generating electricity. The electricity produced can be utilized to power various household appliances or stored in a battery for future use. The voltage of the generated electricity typically ranges up to 12 V, contingent upon the intensity of the water pressure applied to the turbine.

The turbine assembly comprises either a movable magnet and fixed coil or a fixed magnet and movable coil. As the magnet or coil rotates within the turbine assembly, electricity is generated through electromagnetic induction, thereby facilitating the conversion of hydraulic energy into electrical energy.

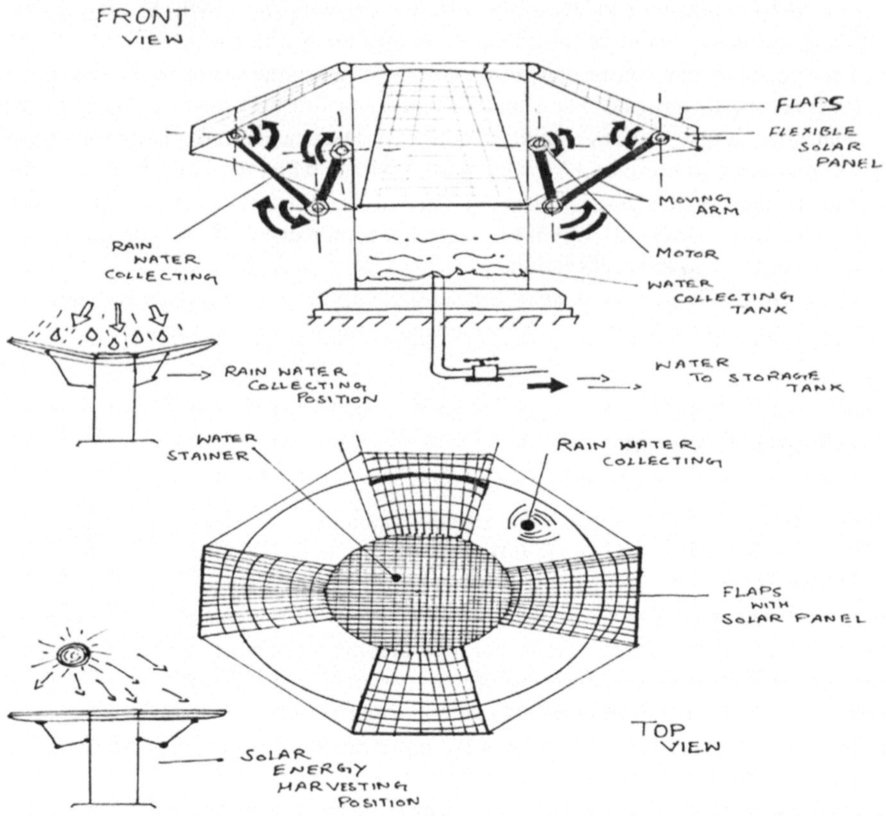

Fig. 1 Illustrative representation of the proposed design

5 Devices Used

The final hardware of the "Integrated Solar Power and Rainwater Harvesting System" is manufactured as illustrated in Fig. 2. This device consists of several essential components which are briefly described in the following subsections.

5.1 Dc Gear Motor

A DC gear motor is a type of electric motor that incorporates a gearbox to achieve a desired output speed and torque. It consists of two main components: a DC motor and a gearbox. The DC motor converts electrical energy into mechanical energy through the interaction of magnetic fields, causing the motor shaft to rotate.

DC gear motors are widely used in various applications where precise control of speed and torque is required, such as robotics, automation, automotive systems, and

Fig. 2 Integrated solar power and rainwater harvesting system

industrial machinery. They offer several advantages, including high efficiency, compact size, and reliable performance. Additionally, they can be customized to meet specific application requirements by selecting the appropriate motor and gearbox specifications, such as gear ratio, motor speed, and torque rating.

5.2 Arduino UNO

The Arduino Uno is a popular microcontroller board that serves as the cornerstone of the Arduino platform. Developed by the Italian company Arduino LLC, the Uno is designed to provide an accessible and versatile platform for hobbyists, educators, and professionals alike to prototype and develop electronic projects [Wang & Wang (2020)].

At its core, the Arduino Uno features the Atmega328P microcontroller, clocked at 16 MHz, which provides ample processing power for a wide range of applications. The board includes a variety of input and output pins, including digital input/output pins, analog input pins, and specialized pins for communication protocols such as UART, SPI, and I2C.

5.3 Rain Sensor

A rain sensor is an electronic device designed to detect the presence of rainfall. It functions as a switch, toggling its state based on the presence or absence of rain. The fundamental principle behind its operation is that, during rainfall, the switch remains closed. This closure indicates the detection of rain.

Typically, a rain sensor module or board comprises nickel-coated plates that operate based on the principle of resistance. These plates react to the moisture in the environment, changing their resistance accordingly. As rain falls and the moisture level increases, the resistance between the plates alters, triggering a response from the sensor.

5.4 Solar Panel

A solar panel, also referred to as a solar electric panel, photovoltaic (PV) module, or PV panel, constitutes an assembly of photovoltaic solar cells systematically arranged within a rectangular frame. These solar cells possess the capability to convert sunlight into electrical energy, specifically in the form of direct current (DC) electricity [Jäger-Waldau (2018), Sharma & Chandel (2016), Parida et al. (2011), Li et al. (2019), Sovacool (2019)].

When sunlight interacts with the solar cells, it triggers a photovoltaic effect, causing the cells to generate electric current. This electrical energy can then be harnessed for various applications, such as powering electronic devices, charging batteries, or supplying electricity to residential or commercial buildings [Huld et al. (2010), Alam & Lee (2019), Kalogirou (2004)].

Multiple solar panels are often grouped together to form a larger system known as a photovoltaic system or solar array. These arrays can be installed on rooftops, open fields, or other suitable locations to maximize exposure to sunlight.

6 Conclusions and Future Scope

In conclusion, the manuscript presents a promising solution to address the pressing issues of water scarcity and the transition to renewable energy sources. The automatic tracking solar plate, designed to conserve rainwater while harnessing solar energy, offers a holistic approach to sustainability. By integrating solar panel rain sensors, motors, and turbines, the device demonstrates the potential to maximize energy generation during sunlight hours while efficiently harvesting rainwater during the rainy season. Through comprehensive analysis and evaluation, the performance and implications of the proposed device have been scrutinized, highlighting its potential to contribute to water conservation and renewable energy generation efforts.

Moving forward, there are several avenues for further exploration and development of the proposed device. Firstly, refinement of the design and optimization of components could enhance the efficiency and reliability of the system, ensuring consistent performance under varying environmental conditions. Additionally, research into innovative materials and technologies could lead to advancements in rainwater harvesting and solar energy conversion, further improving the device's effectiveness and scalability.

Furthermore, the integration of smart technologies, such as IoT (Internet of Things) sensors and data analytics, could enable remote monitoring and optimization of the device, enhancing its usability and adaptability in diverse settings. Collaborative efforts with stakeholders from academia, industry, and government sectors could facilitate real-world testing and implementation of the device in communities facing water and energy challenges.

Overall, the automatic tracking solar plate represents a promising innovation with significant potential to address the dual challenges of water scarcity and renewable energy generation. Continued research, innovation, and collaboration will be essential in realizing the full benefits of this technology and fostering sustainable development on a global scale.

References

Abdulla, F. A., & Al Marzooqi, M. (2021). An overview of solar-powered desalination technologies. *Desalination, 502*, 114970.

Ahmadi, M. H., & Rahmani, R. (2017). A review on recent advances in solar collectors and thermal energy storage systems. *Renewable and Sustainable Energy Reviews, 70*, 1072–1088.

Akbarzadeh, A., Hossainpour, S., & Hossainpour, S. (2019). Rainwater harvesting and solar energy utilization: A review. *Renewable and Sustainable Energy Reviews, 109*, 510–522.

Alam, M. M., & Lee, J. T. (2019). Technological development of photovoltaic solar cells: A review. *Journal of Industrial and Engineering Chemistry, 80*, 100–114.

Al-Karaghouli, A., & Kazmerski, L. L. (2017). Energy consumption and water production cost of conventional and renewable-energy-powered desalination processes. *Renewable and Sustainable Energy Reviews, 73*, 738–752.

Angelakis, A. N., & Koutsoyiannis, D. (2005). Urban water cycle, globalization, and climate change: The case of Athens. *Water Resources Management, 19*(1), 13–27.

Chandel, S. S., Agarwal, V., & Mathur, J. (2011). Performance analysis of a low capacity solar still. *Solar Energy, 85*(5), 998–1004.

Fathollahzadeh, H., & Esfahani, J. A. (2018). A review on solar water pumping systems: Types and optimal designs. *Renewable and Sustainable Energy Reviews, 82*, 1326–1342.

Goel, S., & Kumar, A. (2020). Design and development of an automatic solar tracking system. *International Journal of Renewable Energy Research, 10*(2), 820–829.

Gupta, A., & Kumar, A. (2018). Rainwater harvesting: A sustainable solution to water scarcity. *International Journal of Environmental Science and Technology, 15*(4), 849–862.

Huld, T. A., Dunlop, E. D., & Ossenbrink, H. A. (2010). Photovoltaic solar electricity: Status and future prospects. *Solar Energy, 85*(8), 1580–1608.

International Renewable Energy Agency (IRENA). (2020). *Renewable power generation costs in 2019*. IRENA.
Jäger-Waldau, A. (2018). *PV status report 2018*. European Commission, Joint Research Centre.
Jain, D., & Sharma, R. (2020). A review on recent advances in solar still technologies. *Renewable and Sustainable Energy Reviews, 123*, 109756.
Kalogirou, S. A. (2004). Solar thermal collectors and applications. *Progress in Energy and Combustion Science, 30*(3), 231–295.
Kazem, H. A., & Al-Karaghouli, A. (2017). Renewable energy desalination: An emerging solution to close the water gap in the Middle East and North Africa. *Renewable and Sustainable Energy Reviews, 71*, 163–173.
Khan, M. N. I., & Islam, M. M. (2017). Rainwater harvesting and renewable energy for sustainable water and energy resources management in Bangladesh. *Journal of Cleaner Production, 142*, 4281–4293.
Li, Y., Wang, R., & Ma, J. (2019). Development and application of an automatic solar tracking system for maximizing photovoltaic energy production: A review. *Renewable and Sustainable Energy Reviews, 112*, 417–434.
Owusu-Sekyere, E., & Mireku, J. (2019). Renewable energy in agriculture: A narrative review. *Renewable and Sustainable Energy Reviews, 105*, 23–36.
Parida, B., Iniyan, S., & Goic, R. (2011). A review of solar photovoltaic technologies. *Renewable and Sustainable Energy Reviews, 15*(3), 1625–1636.
Qureshi, A. W., & Hasanuzzaman, M. (2017). Present status and future scope of solar energy in Bangladesh. *Renewable and Sustainable Energy Reviews, 76*, 60–70.
REN21. (2020). Renewables 2020 global status report.
Sharma, S. K., & Chandel, S. S. (2016). Review of solar photovoltaic water pumping system technology for irrigation and community drinking water supplies. *Renewable and Sustainable Energy Reviews, 58*, 153–164.
Sharma, V., & Sharma, R. (2018). A review on solar energy use in industries. *Renewable and Sustainable Energy Reviews, 82*, 2973–2997.
Singh, S., & Tiwari, G. N. (2018). Review of water harvesting system, performance and applications. *Renewable and Sustainable Energy Reviews, 91*, 893–913.
Sovacool, B. K. (2019). The political economy of solar photovoltaics. *Energy Research & Social Science, 50*, 228–237.
Tiwari, G. N., & Singh, H. N. (2008). Water distillation methods using solar energy. *Solar Energy, 82*(5), 435–449.
Tiwari, G. N., & Sodha, M. S. (2007). *Solar energy fundamentals, design, modeling and applications*. Alpha Science International Ltd.
Tong, X., Lu, L., & Yang, H. (2017). Concentrated solar power generation using high-efficiency thermoelectric conversion. *Renewable and Sustainable Energy Reviews, 69*, 361–373.
United Nations. (2015). *Transforming our world: The 2030 agenda for sustainable development*. United Nations.
United Nations. (2018). *World water development report 2018*. Nature-Based Solutions for Water.
Wang, L., & Wang, F. (2020). Design and implementation of a solar tracking system based on Arduino and LabVIEW. *Journal of Physics: Conference Series, 1666*(1), 012030.
Wang, X., Gao, W., Zhang, Y., & Deng, Z. (2018). Development status and future trends of solar energy in China. *Renewable and Sustainable Energy Reviews, 81*, 2434–2441.
World Bank. (2019). *Quality unknown: The invisible water crisis*. World Bank.

Case Studies

Exploring Sustainable Practices of Micro-irrigation in Farming: A Case Study of SLECI in Morocco

Jan Siering, Bassou Bouazzama, Sabri Ahmed, Nadya Wahid, Hicham El Yousfi, Jelena Zascerinska, and Regina Krause

Abstract The effects of the scarcity of water resources and the obligation to improve the efficiency of water use in agriculture require the development of localized irrigation technology. The research aims at exploring sustainable practices of micro-irrigation technology SLECI (Self-regulating, Low Energy, Clay based Irrigation) in farming based on the analysis of the case study carried out in Morocco. The exploratory research was used in this work. The study focused on SLECI technology implementation in Morocco. The analysis of SLECI implementation in Morocco allows concluding that SLECI serves as a sustainable practice of micro-irrigation in farming. From the social and environmental perspectives of sustainability, SLECI showed positive results. The economic costs of SLECI use in Morocco, in comparison to conventional drip irrigation system, appeared to be higher.

J. Siering · R. Krause (✉)
Hochschule Wismar, University of Applied Sciences: Technology, Business and Design, Wismar, Germany
e-mail: jan.siering@hs-wismar.de; regina.krause@hs-wismar.de

B. Bouazzama
National Institute of Agronomic Research, Beni Mellal, Morocco

S. Ahmed
National Institute of Agronomic Research, Errachidia, Morocco

N. Wahid
University Sultane Moulay Slimane, Beni Mellal, Morocco
e-mail: n.wahid@usms.ma

H. El Yousfi
University Mohammed V of Rabat, Rabat, Morocco

J. Zascerinska
University of Wismar, Wismar, Mecklenburg-Vorpommern, Germany

J. C. Correia Leitão et al. (eds.), *Sustainable Water Management and Irrigation Systems*, Natural Resource Management and Policy 62,
https://doi.org/10.1007/978-3-031-93587-9_7

Keywords Agriculture · Micro-irrigation · Morocco · SLECI (Self-regulating, Low Energy, Clay-based Irrigation) technology

1 Introduction

Agriculture, a vital sector of economy, is a dynamic and inherently complex entity. Its fundamental role in human survival, coupled with the diverse challenges it faces, highlights the crucial importance of sustainable agricultural practices. When it refers to farming, sustainable agriculture means the practice of farming based on the principles of respecting ecology and saving natural resources (EPRS, 2012).

In agriculture and in farming, in particular, water, being a natural resource, is appearing scarce in the Mediterranean countries including Morocco due to global warming, natural disasters such as droughts and floods. In the last 5 years, Morocco has seen the loss of almost 6.5 billion m^3 of water, including 700 million m^3 in 2022, although the latter was the driest year since 1945, adding that 300 million m^3 of water have also been lost since September 2022 (Agricolae, 2023). Morocco is ranked 22nd among the countries most exposed to the risks of water insecurity (Pereira et al., 2023).

Agriculture is under great spatiotemporal, inter-, and intra-annual rainfall irregularity, and more than 70% of the mobilizable surface resources are spread over three watersheds. To mobilize these resources, the State has built more than 100 dams with a total storage capacity of around 16 billion m^3 and created a dense network of wells and boreholes providing an annual volume of 2.8 billion m^3 which remain dependent on climatic conditions.

Over the past 10 years, the policy of economical use of irrigation water has become a central issue in many irrigated regions of the Mediterranean (Mélanie, 2007; Molle & Sanchis-Ibor, 2019). Since they are faced with the problem of aging irrigation infrastructure, shortage of water needed for irrigation, climate change (drought), and population growth (food security). Thus, the Mediterranean sustainable development strategy invests in the management of water demand through the improvement of agronomic and economic efficiency (Plan Bleu, 2008; Fernandez & Mouliérac, 2010). This encourages irrigators to adjust their agricultural practices by adopting water-saving irrigation technology to adapt to soil, crop, and climate conditions. Thus, support tools facilitating long-term and short-term investment decisions at the farm level help irrigators to improve economic efficiency and performance environmental impact of their farms. Agriculture, vital for human survival, is also influenced by fluctuating factors such as climate, diseases, and market dynamics; its sustainability among other issues relies on access to adequate financing and strategic investment decisions and requires a solid and tailored financial strategy.

Effects of the scarcity of water resources and the obligation to improve the efficiency of water use in agriculture require the development of localized irrigation technology (Wahid et al., 2022). Therefore, innovative irrigation systems such as

SLECI (Self-regulating, Low Energy, Clay based Irrigation) technology must be developed and tested in experimental fields. Micro-irrigation SLECI technology was developed by Prof. Dr. Harald Hansmann at the Institute for Polymer and Produktionstechnologies (IPT) and patented by the Wismar University of Applied Sciences (HSW) in Germany (patent number: DE 102019005311.7) (Malchev et al., 2022). It should be pointed that the concept of the use of clay is known since the ancient era and then used by the Romans for a long time when the clay pots had been used for irrigation. SLECI technology is developed according to a new concept which combines clay pot irrigation and subsurface drip irrigation (Dührkoop, 2020). SLECI, therefore, requires verification in testing on the crop field to validate its sustainability.

The previously published research related to SLECI was devoted to the analysis of SLECI effectiveness. A pilot installation was tested at the Fruit Growing Institute—Plovdiv (FGI), Bulgaria (Malchev et al., 2022). The pilot allowed for a preliminary assumption that, on the one hand, better nutrient in comparison to the non-irrigation control was provided, and, on the other hand, the better water regime (no overwatering) compared to the drip irrigation was ensured. SLECI was acknowledged to be a self-regulating subsurface irrigation technique that uses the actual suction force of the surrounding soil for regulation of the water release of the system (Malchev et al., 2022). SLECI was found to be a very simple in concept, production, and installation and therefore fulfils all demands of irrigation systems in rural environments (Malchev et al., 2022).

However, SLECI sustainability remains under-explored.

It should be noted that sustainability is conventionally evaluated from three perspectives (Ahrens & Zaščerinska, 2012):

- Economy
- Society
- Environment

The aim of this work is to explore sustainable practices of micro-irrigation technology SLECI in farming based on the analysis of the case study carried out in Morocco.

2 Methodology

The research methodology was enabled by the research question: Is the use of the micro-irrigation technology SLECI sustainable in farming?

As the present research aims at gaining understanding (Edgar & Manz, 2017), the exploratory research was used in this work. The emphasis of the exploratory research is on perspective and relative importance (Edgar & Manz, 2017). The exploratory research does not imply a well-established methodology as the exploratory research deals with a not well investigated phenomenon. For the experimental part of the explorative research, the case study was applied (Yin, 2003).

The purpose of the case study was to analyze the economic and financial aspects of SLECI implementation in Morocco underpinning the implications for researchers and farmers.

The methodology of the present case study proceeded

- From SLECI installation in Morocco in Phase 1
- Through collection of data relevant of the analysis of economic and financial aspects of SLECI in Phase 2
- Via data analysis and interpretation in Phase 3
- To the elaboration of implications for researchers and investors in Phase 4

The case study started in 2022. The case study has not been finished yet as plant developments can be observed in the second season only.

The case study aimed at generalization of theoretical propositions (analytical generalization) and not to enumerate frequencies (statistical generalization) (Yin, 2003). Consequently, the focus is on the SLECI technology implementation in Morocco and not on the SLECI users. The collected data were interpreted by researchers who participated in the current research (Ahrens et al., 2018).

3 Results

3.1 SLECI Production

Water in agriculture, when water is scarce, has to be used effectively.

The use of porous materials as well as the exploitation of natural interactions of the soil is a solution for keeping water secure and saved in agriculture.

Irrigation of plants, in particular, in dry areas of the world, poses a special challenge to irrigation systems that need to be integrated. This work deals with the functionality of soil in the context of irrigation technologies, specifically the transport of water by using porous materials as emitters like SLECI.

3.1.1 Soil Function

Soil functions fall into two categories:

1. Ecological soil functions: They include all naturally occurring interactions of the soil. In this context, soil is an important link between the outer spheres of the earth below the atmosphere. These include the lithosphere, biosphere, pedosphere, and atmosphere. In addition, various vital material cycles take place, such as the carbon cycle or the water cycle.
2. Soil functions used by humans: The use of soil by humans as a source of raw materials, building ground, or cultivated soil (agricultural use) can impair its ecological functionality.

However, the transformation of the soil on earth is due to the satisfaction of material needs and (basic) needs designed for food.

The humus content of the soil determines the fertility for plants by providing important minerals and nutrients. The climate, but also anthropogenic activities, can influence the content.

3.1.2 The Soil and the Crop Production

The importance of soil for crop-based agriculture is fundamental, as it serves as a source of nutrients, an agent of vital matter transformation processes, a root anchoring medium, and protection.

The properties and structure of today's soils are and have been caused by ongoing matter transformation and matter transfer processes. The physical, chemical, and biogenic weathering leads to a downward displacement of various substances. This results in the formation of typical horizontal soil layers, which differ in their characteristics such as color or pH value. These are also called soil horizons.

Plant growth is characterized by interactions between the plant and the surrounding soil. Most plants develop in situ and are thus sessile living organisms. Thus, plants must adapt to local environmental conditions. Growth is a dynamic process influenced by soil structure, chemical compounds, and water availability. This also means that changes in environmental conditions cause plant feedback. For example, plants respond to drought by closing their stomata to minimize transpiration, i.e., the evaporation of water.

3.1.3 The Water Transport in the Soil

Water transport in the soil is a fundamental factor for agricultural irrigation of plants. In nature, water cycles take place that consider the changes of state of water during its transport through the individual spheres. In agriculture, nutrients and water are selectively provided to crops, often during periods when plant growth/reproduction would be virtually impossible due to a dry climate or nutrient-poor soil composition.

After precipitation, part of the water penetrates the soil and percolates at a certain rate. Water transport is determined by soil structure and composition, volume changes caused by swelling, biological activities, and changes in voids. Soil is also capable of storing water and holding it against gravity (adhesive water). This is achieved through adhesion and the capillary forces that act in this context.

The roots of plants can absorb quantities of water in areas of adhesive water. Water can be transported within the plant by diffusion and osmosis via the root hairs. Stagnant water prevents adequate uptake of oxygen and nutrients. On the other hand, the soil will hardly release water to the plant if the water content is too low. When the so-called permanent wilting point is exceeded, the plant will continuously stop its activities and wilt.

3.1.4 Irrigation Methods

Common irrigation methods are designed to provide the plant with the amount of water, possibly with soluble fertilizer, that it can absorb. It must be taken into account that portions of the supplied water evaporate and seep away, which cannot be absorbed by the plant. Accordingly, optimal pre-irrigation calculations must take place to determine the proper amount of irrigation water.

A variety of irrigation systems with advantages and disadvantages determine the market. They can be divided into aboveground and belowground systems. The following lists a selection of aboveground systems:

Surface Irrigation Description: Irrigation takes place above the earth's surface. Often, the water falls on the arable land to be irrigated in a targeted or untargeted manner. Practical examples are as follows:

(a) Spray sprinkler:
Nozzles are used to distribute a quantity of water, usually on a round surface. The range is small and its accuracy depends on the weather.
(b) Mobile irrigation vehicles:
Water is distributed in a moving manner by means of nozzles attached to a nozzle cart. The expensive purchase is offset by high flexibility and accuracy.
(c) Drip irrigation:
Small amounts of water can be delivered "drip" to the plant location via emitters or boreholes. The water-saving effect in arid regions is offset by the labor involved and the high material costs.
(d) Gravity irrigation or flood irrigation:
It is the oldest and still remains the most dominant in the world today, irrigation by furrow, by wedging, and by flat-bottomed basin.
Problematic in the application of aboveground irrigation systems is the loss of water through evaporation. As the air temperature increases, so does the evaporation rate of water. This depends on the air pressure and the moisture content in the air.

Therefore, in particularly arid regions, subsurface irrigation systems are of interest:

Subsurface Irrigation Description: The irrigation lines are laid underground. Water distribution is usually carried out directly in the root zone. Practical examples are as follows:

(a) Drip irrigation:
Emitters or boreholes installed on the pipes release water selectively. The preinstallation effort is higher than for aboveground drip irrigation, but less water is needed. In the literature, drip irrigation is often included among the micro-irrigation systems.
(b) Low-pressure micro-irrigation:
The subsurface irrigation is carried out through tiny outlets of a larger surface than drippers. In part, the suction effect of the surrounding soil is exploited,

depending on the system. Here, too, the labor and material costs are high. The water can be effectively distributed into the root area. Known are soaker/nano hoses and the use of porous materials such as ceramics.

3.1.5 Low-Pressure Micro-irrigation Systems

An irrigation system that uses the natural suction pressure of the surrounding soil to achieve water distribution requires only a small amount of upstream pressure in the hose system. The irrigation water in the pipes is separated from the soil by a porous medium or membrane. Since the soil consists of a large number of particles and pores of different sizes, a so-called osmotic pressure is created in which the water attempts to achieve pressure equalization by passing through the porous layer of the solid medium/membrane into the soil.

3.1.6 The Use of Clay Materials as Porous Emitters: The SLECI System

Fired clay belongs to the ceramics and has a large specific surface area. The firing temperature profile determines, among other things, the size and shape of the material's pore system. In order to be able to use the clay as a porous emitter for water, it is fired below the densifying sintering process. The size of the pores is in the micro-meter and also nano-meter range. The shape and the interconnecting voids influence the water transport in the material.

Porous clay is suitable as a semi-permeable connecting membrane between the irrigation water and the surrounding soil. Several millennia ago, this property was already used in the cultivation of plants by filling large clay pots with water and inserting them into the soil.

Wismar University of Applied Sciences is currently enhancing the SLECI system as part of the MED-WET project co-funded by the European Union to improve irrigation efficiency in the Mediterranean region. In this system, cylindrical clay bodies are connected to irrigation pipes that are buried in the soil directly in the root zone. The clay material has a distinct pore network that has the ability to store water. A clay body filled with water reacts to the activity of the surrounding soil, which also consists of a large number of pores. Due to decreasing soil moisture, the clay bodies release the water to the soil. The capillary action in the clay body itself then leads to a suction pressure in the irrigation lines, through which the water must flow from the water reservoir. The plant plays a decisive role in this process, as it draws the water from the soil. When the plant's demand for water increases, the clay body releases all the more water as the moisture in the soil decreases.

The ecological interactions with the soil result in a water-saving effect, as losses due to infiltration and evaporation are reduced. In addition, no additional energy is required as the system is self-sufficient. However, the system requires special requirements to ensure that its performance is not limited in the long term, such as preventing foreign matter from blocking the pores. The installation costs and material and labor requirements of subsurface irrigation systems are higher compared to aboveground systems. However, the increasing regional water shortage leads to an increasing interest in these technologies.

3.2 SLECI Installation in Morocco

Four sites were selected to compare the new SLECI technique with an ordinary micro-irrigation. In the four sites, the conventional micro-irrigation equipment was installed a few weeks before the installation of the SLECI-based irrigation in the period of 27 November 2022–5 December 2022. The installation of the SLECI elements was planned for four test fields with different tree fruits:

- Citrus
- Dates
- Olives
- Vine grapes

3.2.1 Criteria for Choosing Experimental Sites

In order to test the performance of the two irrigation systems in more contrasting conditions in the Moroccan context, three more important criteria were considered when choosing the sites:

- The type of climate is an important criterion since it determines the climatic demand and therefore the water needs of the crops to be satisfied by the irrigation technique.
- The soil is also an important criterion that influences the three-dimensional water transfers, the wetting rate, and the volume of soil moistened by the water distributor.
- The nature of the crop is also important by its cycle, its root system, and the importance of its water needs during the cycle. The origin and quality of the irrigation water may be a condition for the risk of clogging.

Consideration of these criteria resulted in the choice of four sites characterized in Table 1.

Table 1 Summary of the sites hosting the comparative trials of the irrigation systems (by Bouazzama and Sabri)

	Site 1	Site 2	Site 3	Site 4
Climate[a]	Semiarid	Arid pre-Saharan	Semiarid	Temperate Mediterranean
Soil type	Loamy clay	Loamy	Clay-loam	Sandy clay
Water origin	Dam	Well	Dam + well	Well
Crop practiced	Citrus	Date palm	Olive tree	Grape vine

[a] Classification of Köppen-Geiger

Further, the test platforms installed at the four sites are described. In the description of the citrus site, the installation of the SLECI systems will be presented in detail. In order to avoid repetitions in the text, only the specificities of other sited that differ from the citrus site will be indicated. It is worth pointing that the work by Malchev et al. (2022) does not contain a detailed description of SLECI installation.

3.2.2 Site 1: CITRUS—INRA-Field

3.2.2.1 Field Information

- In the trial field of orchard housing, the orange trees of the "Maroc late" variety are planted at 7 m * 7 m spacing.
- The part equipped by SLECI contains 44 trees divided into 12 rows and 4 columns.
- The part equipped by conventional drip contains 39 trees divided into 13 rows and 3 columns.
- The pumping station is located 700 m from the orchard. The water is taken from a geo-membrane basin. After being pumped, the water passes through a sand filter, disc filter (124 μm), and screen filters.
- To ensure the water permanently, the water is then routed to 4 tanks of 4500 m^3 each placed on the roof of the pumping station as shown in Fig. 1. The tanks are equipped with sensors of low and high levels to automatically activate the pump.
- At the entrance to the orchard, 4 tanks of 4500 m^3 each are placed on the ground surface to store water intended for conventional drip irrigation. To do this, a solar pump with a filtration system (screen filters and disc filter) is installed to inject the water at a pressure of 1 bar. The overview of the citrus trial installation is given in Fig. 2.

3.2.2.2 Installation of the Conventional Drip Irrigation

- Four subplots equipped with conventional drip with two 16 mm diameter high-density polyethylene (HDPE) pipes per tree line. The ramps are spaced 2 m apart. Each lateral or, in other words, dripper's pipe contains integrated self-regulating drippers over the pressure range of 0.8–3 bar and with a nominal flow rate of 3.7 L/h and a spacing of 80 cm.
- The ramp doors are PVC PN 6 pipes with a diameter of 50 mm. At the end of each ramp door, purge valves for the evacuation of impurities are placed.
- At the entrance to each plot, there is a meter flow, a 2.5 bar pressure gauge, and a handwheel valve.

Fig. 1 Storage basin and raised tanks. (Bouazzama's Own Elaboration)

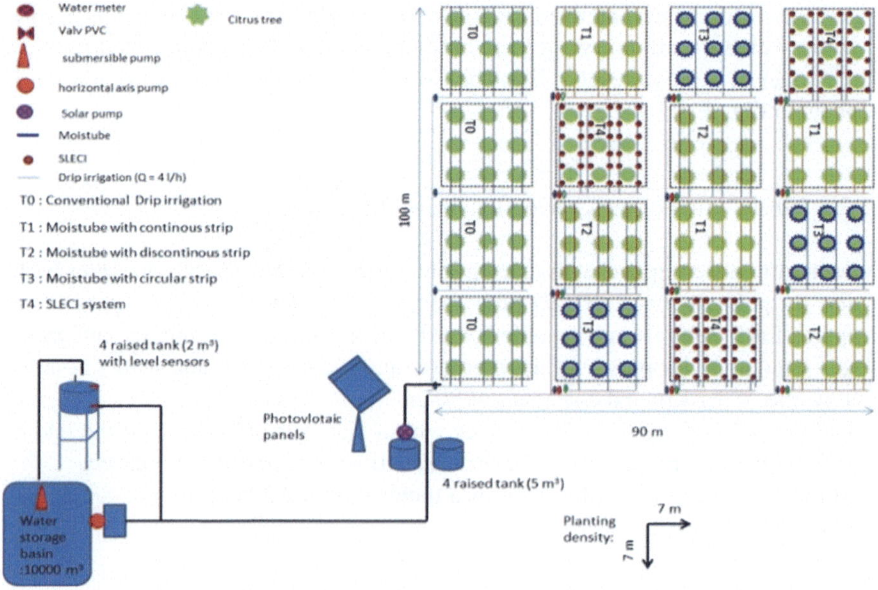

Fig. 2 Citrus trial setup. (Bouazzama's Own Elaboration)

3.2.2.3 Installation of the SLECI System

3.2.2.3.1 Installation of the VCA

The Vertical Control Assembly (VCA) is responsible to reduce the hydraulic pressure to 0.2 bar since the pressure at the entrance to the orchard is 0.7 bar. The Float Valve is fastened at the top of a stand as demonstrated in Fig. 3.

The distance between the SLECI level to the Float Valve is 2.0 m. The inlet of the Float Valve is connected with the smart water meter and a filter system (50 and 1 μm). The tanks are connected via PVC DN50 pipes and HDPE pipes with this filter system. The outlet of the Float Valve is similar linked with HDPE and PVC DN50 pipes.

3.2.2.3.2 Positioning of the Clay Tubes

To irrigate the trees, DN16 HDPE tubes are connected on each side of a row. The SLECI elements are linked via 6 × 4 mm tubes with the DN16 tubes as branch lines as reflected in Fig. 4.

Irrigation of the citrus trees is planned from two sides of the tree by burying one SLECI element into the soil from each side.

The holes in the soil are drilled by a hand auger in a depth of 60 cm. The distance to the trunk of the trees is about 50 cm estimated by the installers' agricultural experience. The hole must be in that position; the roots can effectively suck water out of

Fig. 3 Installation of the vertical control assembly (VCA) (citrus field). (Siering's Own Elaboration)

Fig. 4 Connection of SLECI branch lines to DN16 tubes (citrus field). (Siering's Own Elaboration)

the soil. The hole must offer enough space to position the SLECI clay cylinder in the soil. The small tubes of the SLECI elements must be adjusted to a length according to the position in the soil and the DN16 tube laying on the soil surface. A cutter and small straightway connectors are necessary to extend the length.

3.2.2.3.3 Connection Options

The connection of the 6 × 4 mm tubes (SLECI elements) to the DN16 tube is made with straightway connectors (option A) and T-barbs (option B) as depicted in Fig. 5.

For option A, the DN16 tube must be prepared with a small hole drilling by a spiky object. The straightway connector is pushed in the hole with its thick nozzle. For option B, the DN16 tube must be cut into two pieces and pushed on the DN16 nozzles of the connector. The T-barb has a 4 mm nozzle connecting with the 6 × 4 mm tube.

3.2.2.3.4 Filling the Boreholes

The holes have to be filled with fine sand. Here ≤2 mm grain size sand is used. First the 60 cm hole is filled to a distance of 10–15 cm from the soil surface as illustrated in Fig. 6.

After the successful test run of the SLECI system, the SLECI element was placed into the hole, and the residual empty part of the hole was filled to the surface. The soil surface is covered with surrounding soil.

Fig. 5 Connection of the SLECI branch lines to the DN16 tubes (left: option A—straightway connector; right: option B—Tbarb) (citrus field). (Siering's Own Elaboration)

Fig. 6 Filling the drilled holes with sand (citrus field). (Siering's Own Elaboration)

3.2.2.4 Test Run of the Irrigation Equipment

- The conventional drip system.
- The tightness test was carried out after the start-up of the solar pump and the opening of all the valves to ensure that there are no losses at the level of the junc-

tions and start at the joint between the sub-main line and lateral. Also, a visual control of the operation of all drippers. Similarly, pressure measurements were carried out at the end of lines as proposed in Figs. 7 and 8.

- SLECI system.
- After having connected the SLECI elements via tubes to the VCA, the test runs can start. For the test run, the valves of the pipes/tubes should be open. The system is monitored to:

Fig. 7 Pressure measurement. (Bouazzama's Own Elaboration)

Fig. 8 Flow measurement. (Bouazzama's Own Elaboration)

- Find possible leakages.
- Ensure the working process of the Float Valve.
- Check the SLECI elements for their conductivity.
- Find possible pressure drops.
- Check the water meter on its flow rate.

The SLECI elements are working if water drops are visible as pointed in Fig. 9. If no water is released, the trouble shooting, described in the manual, must be conducted. A few leakages were found especially at connection locals. Replacement of parts solved these problems. All the installed SLECI elements worked reliably.

3.2.3 Site 2: Date Palm–Errachidia INRA Center

3.2.3.1 Field Information

- About 30 trees divided into 6 rows of 5 trees each. Three lines are reserved for SLECI system while three others are reserved for ordinary micro-irrigation.
- The rows of date palms irrigated by SLECI are scattered between rows of alternative irrigation technology. However, they are supplied by a DN30 line, so the installation of a Vertical Control Assembly Station (VCA) is sufficient as indicated in Fig. 10.
- The trees are 6 years old and belong to the *Majhoul* variety, one of the most cultivated varieties in Moroccan palm groves.
- The water source is a storage basin fed from wells.

3.2.3.2 Installation of the Conventional Drip Irrigation

- The head station is made up of a horizontal axis pump (Q = 5 m^3/h with HMT = 20 mce) which draws water from the pool. The water passes through a filtration station consisting of a sand filter and a 124 μm disc filter.

Fig. 9 SLECI elements with a good conductivity; water drops flow out (citrus field). (Siering's Own Elaboration)

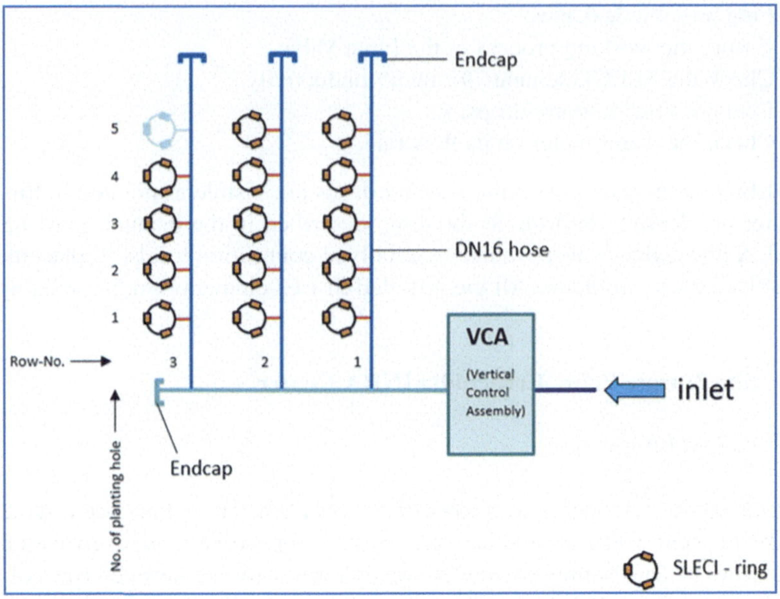

Fig. 10 Irrigation Plan—Date Field. (Siering's Own Elaboration)

- The ramp holder is in PVC Dn 50 mm. Each line is equipped with two Dn 16 mm PEDH ramps. Each tree is fed by two self-regulating branch drippers with a nominal flow rate of 25 L/h.
- At the entrance to the sector, a Dn 32 flow meter with a nominal flow rate of 6 m^3/h is installed to record the volumes of water per irrigation.

3.2.3.3 Installation of the SLECI System

The installation of the VCA was done in a similar way as on the citrus field with some modifications as revealed in Fig. 11.

The clay bodies were connected to DN6 size hoses and placed in a circle around the date palm tree at a radius of about 1.5 m as specified in Figs. 12 and 13. Three XXL-type SLECI elements were installed to serve each tree. The connecting device between the ramps and the clay elements are similar to the citrus orchard.

3.2.4 Site 3: Olive Tree–Private Field (Beni Mellal)

3.2.4.1 Field Information

- 240 olive plants divided into 4 rows of 60 trees each. Two lines for SLECI system and two others for the conventional system as shown in Fig. 15.

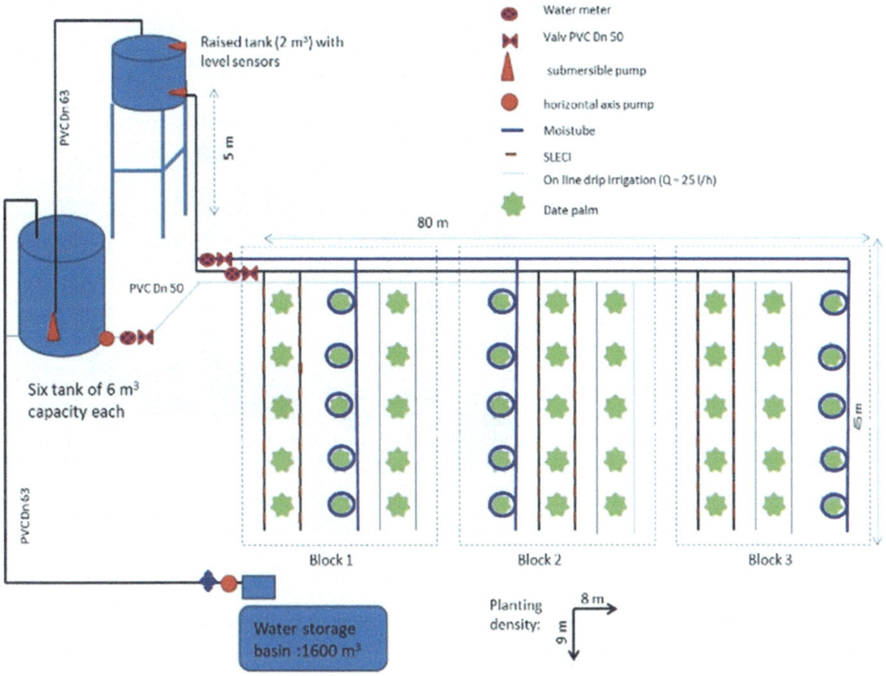

Fig. 11 Schematic field map—Date palm—Errachidia site. (Bouazzama's and Sabri's Own Elaboration)

Fig. 12 Positioning of the clay bodies as a ring installation (date field). (Siering's Own Elaboration)

Fig. 13 Vertical control system (VCA) of the date field. (Siering's Own Elaboration)

The trees are 4 years old, of the *Arbosana* variety, planted at a spacing of 3 m between rows and 1.35 m between shrubs in the same row.

- The olive trees had a height of about 1.5 m.
- Water comes for the entire orchard from a geomembrane storage basin which is supplied with water from the dam, if available, or currently from several wells.

3.2.4.2 Conventional Micro-irrigation Equipment

- The two lines of olive trees, isolated to serve as a control for comparison, are still linked to the general irrigation installation of the orchard. Upstream of the two lines, a Dn 63 meter was added in addition to a shut-off valve.

3.2.4.3 Installation of the SLECI System

The installation of the VCA was not necessary, because already approx. 0.2 bar upstream pressure should be reached by the installed water tank of 2000 L at 2 m height. The water meter was installed horizontally at the outlet of the tank using reducers as illustrated in Fig. 14.

A preinstalled sand filter and a 300 μm filter were already located at the inlet of the water tank. Regarding infiltration, 1 and 50 μm filters were additionally installed after the 300 μm filter.

The shrubs of the olive tree are 4 years old as disclosed in Fig. 15.

A single SLECI element was reserved for each foot to satisfy its water requirements as displayed in Fig. 16.

Fig. 14 Water tank in 2-m height (olive field). (Siering's Own Elaboration)

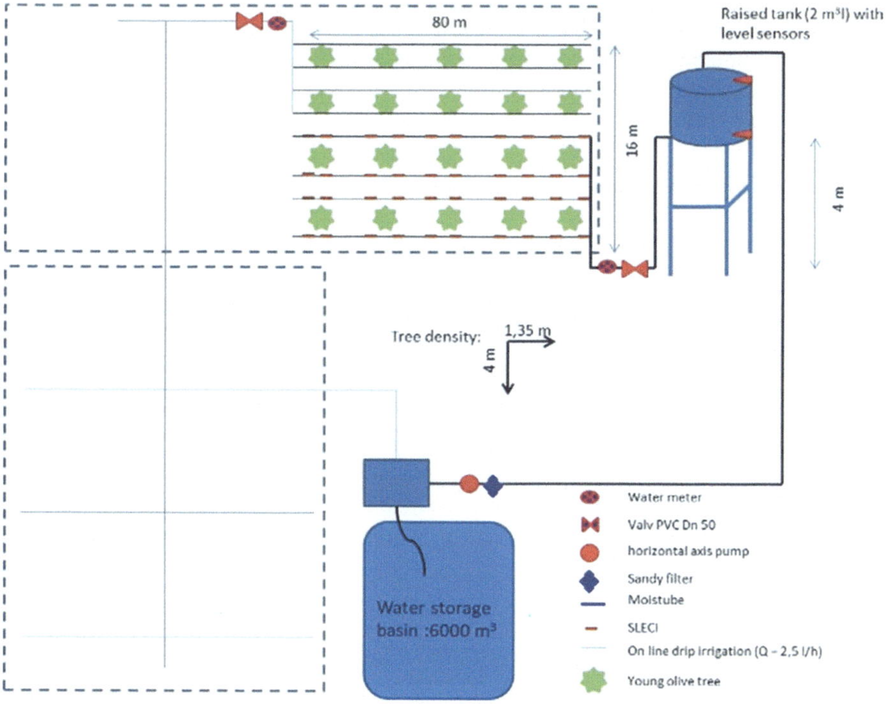

Fig. 15 Schematic field map—young olive tree. (Bouazzama's Own Elaboration)

Fig. 16 Olive tree rows. (Siering's Own Elaboration)

The clay bodies were positioned at a distance of approx. 10–15 cm from the shoot axis as marked in Fig. 17.

They were connected to the DN16 main lines on one side along the rows via straightway connectors. The boreholes were created with a depth of approx. 60 cm and a diameter of approx. 8 cm.

3.2.5 Site 4: Vine Grapes—Private Field (Demnat Region)

3.2.5.1 Field Information

- 120 trees in 6 rows each contain 20 plants of the "Victoria" variety. The vineyard was planted in April 2022. The spacing between rows is 4 m while the vines are spaced 2 m apart on the same row.
- Hilly terrain with slope, also along the rows with an average drop of 0.5–0.8 m between successive planting lines.
- Water source: the water comes from a geomembrane storage basin installed on the natural ground at a height difference of 17 m compared to the plot where the vineyard is planted. The basin is fed by a solar pump from a well located 2 km away near a river.

Fig. 17 Positioning of the clay bodies (olive field). (Siering's Own Elaboration)

3.2.5.2 Installation of Conventional Drip Irrigation

The slope of the land between the geomembrane storage basin allows the water to rich the vineyard at a pressure of 1.15 bar as reflected in Fig. 18.

This advantage made it possible to avoid the installation of a pump to pressurize the water. A PVC pipe, Dn 63 mm is used to convey the water. A 124 µm disc filter is used to remove any impurities from the water coming from the geomembrane pool. Downstream, a DN 50 m and a DN 50 PVC shut-off valve were installed. The ramp holder is in PVC Dn 50 mm. Each of the three vineyard lines is equipped with a Dn 16 HDPE pipe on which are inserted PC Junior self-regulating drippers with a nominal flow rate of 8 L/h under a pressure ranging from 0,8 to 3 bar. The vines being young, one dripper per vine is sufficient.

3.2.5.3 Installation of the SLECI System

To ensure the water supply of SLECI elements at a pressure of 2 m, 2 tanks of 2000 L each at ground level on the natural ground at a high position of 2 m in relation to the vineyard as mirrored in Fig. 19 were installed.

A filtration system composed of 1 and 50 µm filters is installed upstream of the two tanks. A float inside the tanks was installed to stop and open depending on the water level in the tanks. Such design allowed avoiding the installation of VCA. A water meter was installed at the outlet of the two tanks on a 50 mm PVC pipe as reproduced in Fig. 20.

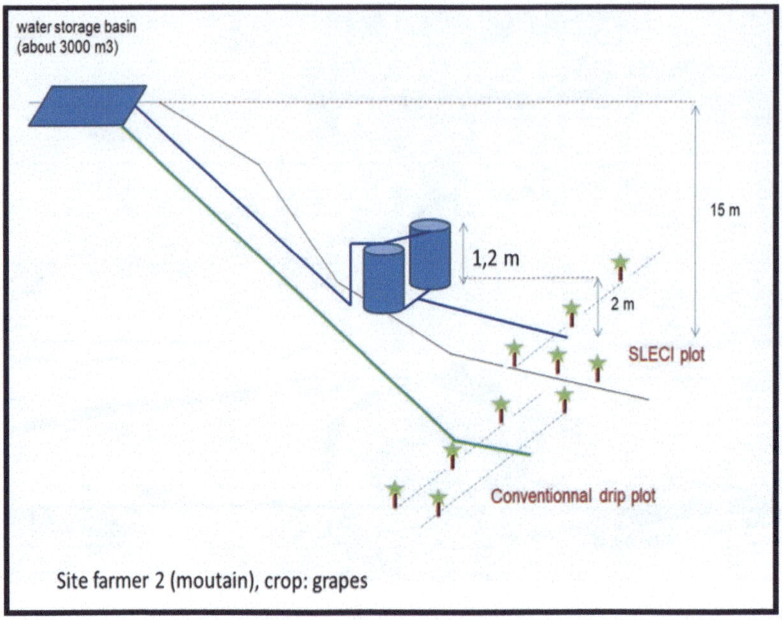

Fig. 18 Position of storage basin and tanks in relation to the plot. (Bouazzama's Own Elaboration)

Fig. 19 2000 L tanks (vine grape field) (Siering's Own Elaboration)

Fig. 20 Left: installed smart water meter; right: installed filter system (Vine Grape Field). (Siering's Own Elaboration)

Fig. 21 Positioning of the clay bodies (vine grape field). (Siering's Own Elaboration)

A single M-type SLECI element was assigned to each foot to ensure its water requirements as disclosed in Fig. 21.

The connection of SLECI elements and its location in relation to the plant are the same as in case of the olive trees. Due to the presence of rocks, holes of 30–40 cm were made instead of 60 cm initially planned.

The leak tests carried out at each site after the installation as well as the average flow rates of the drippers and the pressures measured at the end of the ramps show that the installation was successful. Monitoring of growth parameters, yields, and assessment of clogging risks are planned for the comparison of the two irrigation systems.

3.3 Challenges of Changing Irrigation Technologies and Financial Mechanisms

According to the general sensitivity analysis in the Mediterranean region, macro-environmental and sectoral policy analysis reveals that, in the context of technological change, the irrigated agriculture sector presents two major challenges on the environmental and economic levels and thus benefits from significant public support (Richefort, 2009; Wahid, 2023b). It must in fact contribute, on the one hand, to saving water resources and, on the other hand, to increasing the farm productivity. The public incentive aims to improve the cost-benefit ratio of the adoption of new irrigation techniques by farmers via subsidizing farmers with equipment and training and advisory support for irrigators.

In addition, technical change on irrigated areas appears quite slow and remains unfinished. The reasons which motivated some farmers to improve their irrigation system and others not to change equipment and practices deserve further investigation. Impact of public support is to be investigated since a significant proportion of farmers have equipped themselves alone, without having to resort to subsidies. The role of advice and training interventions by support structures in the dynamics of equipping farmers can also be explored. For Public Authorities, the question of a best strategy to stimulate technological innovation in the irrigation sector finally arises.

A wide range of decision support tools had been developed at the level of farms and irrigation systems. The analysis of the selection of new technologies (the case of SLECI) by a social system using decision support tools is the subject of this section.

3.3.1 Decision Support Tools for Reasoned Irrigation

An analysis of the variability of water consumption appears relevant to highlight the role of irrigation equipment on water demand, taking into account the diversity of equipment between farms. Reasonable irrigation in the Mediterranean regions requires frequent decisions which must adjust to the variability of the water needs of the crop depending on the climate, the growth of the plant, and technical hazards. This complexity can encourage irrigators to use, as a precaution, more water than actual needs. To help the farmer or decision-maker, several simple complementary software programs have recently been developed. These include, in particular, two irrigation decision support tools:

- OSIRI (Chopart et al., 2007a).
- FIVE-Core (Chopart et al., 2007b).

To use them, information on climate and useful soil water reserves must be as local as possible. Two more specific software programs supply these abovementioned tools with information (Chopart, 2004) on

1. The climate: ECLAIR.
2. The water reservoir in the soil RACINE.

As such, and although not essential to the functioning of OSIRI and FIVE-CoRe, they can be considered as part of the range of tools contributing to decision support in irrigation but are presented more succinctly.

3.3.2 Tool to Help Analyze the Economic, Social, and Environmental Benefits of Irrigation

Cost-benefit analysis (CBA) is a decision-making tool integrating both a quantitative aspect of costs and a qualitative aspect of the project effectiveness. There are two types of CBA:

- Financial CBA.
- Economic CBA.

A financial CBA is carried out from the perspective of people involved in the project. The financial analysis constitutes a simplification of the CBA method. Its principle is based on taking into account the costs and benefits realized during the project.

An economic CBA is carried out from the broader perspective of the impact on society. This difference will imply, for the calculation of costs and benefits, the taking into account of all costs and benefits, including externalities, which are sometimes difficult to convert into accounting profit. This latter approach involves defining a method of financial evaluation of nonfinancial benefits, such as environmental impact.

The sustainable economic and financial performance of investments in irrigation infrastructure ultimately comes from the additional income earned by the farmer or the additional net national economic benefits resulting from that investment. Economically motivated farmers invest in irrigation improvement only if the expected net present value of their income is positive (Merriam & Freeman, 2007; Chopart et al., 2007a, b). The valuation resulting from a reallocation of water to other users also influences the economic efficiency of water.

Economically motivated farmers only invest in improving irrigation through innovative technology in a project situation if the expected net present value of their income is positive (Merriam & Freeman, 2007; Chopart et al., 2007a, b). The recovery resulting from a reallocation of water to other users also influences the economic efficiency of water. The recovery resulting from a reallocation of water to other users also influences the economic efficiency of water.

Characteristics of adoption context for new technology irrigators are based on the key factors in the diffusion of the water-saving benefit of the new technology. Four key factors are identified:

- The price of irrigation water.
- The price of crop yield.
- The transfer of information.
- The trust of irrigators in their institutions.

Firstly, the economic context linked to the cost of irrigation water seems to be a determining factor in explaining individual adoption decisions as well as the aggregate rate of adoption of water-saving irrigation technologies. However, the weak evolution of the price of irrigation water in certain regions of the southern Mediterranean suggests that this factor has only a limited impact on the rate of technological adoption (Molle & Sanchis-Ibor, 2019). Thus, the adoption of new water-saving technology becomes more real and responsible when the cost of water pricing increases while the yield benefit for farmers decreases. Furthermore, a sharp increase in the price of water seems contrary to the objective of saving money while ensuring the sustainability of farms.

On the other hand, the nature of the market in which irrigators participate probably plays a role in their desire to innovate in terms of irrigation. In the Mediterranean region, the price of culture and the uncertainties linked to the culture market can have two major impacts:

1. The first concerns crop rotation choices. Irrigators may thus be tempted to modify their crop rotation in an attempt to increase their financial margins.
2. The second concerns technological choices. Some irrigators may therefore postpone their investments in irrigation equipment and wait to see how the market outlook evolves. Other irrigators reassured by the state subsidy to the market can equip themselves more easily.

The social context linked to the transfer of information influences the rate of adoption of innovative water-saving irrigation technologies. When the number of irrigators who have already adopted is low, the adoption decision constitutes an innovative action. In this case, access to information is generally formal. When the number of irrigators is already high, the transfer of information is mainly due to word of mouth spread by irrigators who have already adopted (the innovators). In particular, the influence of certain irrigators can prove decisive and encourage others to behave like them. Furthermore, it seems that the more formal the access to information is (Chamber of Agriculture), the faster irrigators adopt it. Conversely, the more the empirical access to information (word of mouth), the slower the diffusion. Finally, the process of imitation inherent in the diffusion of each innovation can be more or less marked depending on the characteristics of the new technology.

Finally, the institutional context linked to the trust of irrigators in their institutions can modify technological choices. It seems logical to consider that the diffusion of new irrigation technologies is beneficial for all water stakeholders. The willingness of irrigators to equip themselves with water-saving irrigation technologies is therefore similar to their desire to cooperate with institutions. The degree of cooperation may depend, in particular, on the subsidy rates for each material and the confidence of irrigators in the institutions which supervise them. This confidence is illustrated by the financing choices (recourse or not to subsidies) as well as the choices of suppliers and installers of irrigation technology equipment made by irrigators. This factor is directly linked to the characteristics of the technology market and the advisory activities for prescribing new technologies carried out by institutions with irrigators (e.g., Chamber of Agriculture).

Characteristics of Irrigation Technologies The diffusion of water-saving irrigation technologies is strongly linked to the characteristics of the innovated technologies, namely, adaptability, complexity, and the evolution of the quality-price ratio of new, more water-saving irrigation technologies. The adaptability of water-saving irrigation technologies to the local and collective context, as well as their complexity, can modify technological decisions. Indeed, certain types of irrigation technology seem better suited than others to certain types of crops, to certain types of soil, as well as to the know-how of farmers.

On the other hand, the evolution of the quality-price ratio of water-saving irrigation technologies seems to be an important parameter when choosing equipment for irrigators. This report depends, in particular, on the evolution of irrigators' practices. When technology changes, there may be a significant learning cost to achieve value for money with complex new technology.

3.4 Costs and Benefits of the SLECI Technology

Innovative irrigation systems, e.g., SLECI technology, have to be examined before a positive investment decision is justified in terms of the following:

- Economic profitability.
- The benefit results' contribution to the improvement of the well-being of farmers.

Costs and benefit analysis (CBA) is a decision support tool that can facilitate discussion between stakeholders on evaluation of financial and economic efficiency. It provides a structured framework for presenting all the elements of the decision, thus promoting the transparency of the decision-making process.

The implementation of the cost and benefit analysis process goes through six steps.

- Step 1: Definition of the project.
  The SLECI technology is the project situation at the level of three experimental pilot sites in Morocco Farmer 1 (Olivier), Farmer 2 (Vine) and INRA 1 (Citrus).

The referential situation is drip irrigation. This localized irrigation is a cheaper type of irrigation adapted by farmers.

- Step 2: Identification of the relevant impacts of the project.
  Following a thorough and in-depth bibliographic synthesis, we have developed a questionnaire with socioeconomic and agronomic indicators for estimating the financial costs of equipment, operation and maintenance, and water consumption water of SLECI irrigation compared to drip irrigation as a referential situation. There are renewal costs directly induced by the depreciation of irrigation equipment. The benefit estimation is linked to sales revenue and water and labor saving using the irrigation systems. These benefits will have repercussions on the social and environmental benefits.
- Step 3: Drawing a timeline for expected costs and benefits.
  These lifetimes conditions determine the cost of water, when comparisons between networks are implemented. We are based on the lifetime of the irrigation water distribution network to the plot ($\approx$10 years), while considering that the water pumping station is the same for both SLECI project and drip reference situations.
- Step 4: Monetize costs and benefits.
  The initial identification costs and relevant impacts of the crop were used to calculate the costs and benefits of SLECI irrigation system compared to drip referential situation. Data were obtained from crop season in 2022/2023. The cost of installing of the SLECI irrigation system compared to drip reference situation in Morocco is estimated for 1 ha. During the project, we have visited and conducted a survey from farmers who were supported with irrigation system by the project. The cost estimation was done on the discount investment costs using national price change indices. We identify by inventory and express in current euros the amounts of investments that they agree to finance, and then we calculate the operating and maintenance costs.
  We inventoried the costs for equipment, infrastructure, and building which are used for each pilot site. Annual repair and maintenance costs would be added to the initial investment for the irrigation system. The fixed costs occur regardless of farming and will generally be the depreciation and interest. The deprecation is calculated by the straight line method. Operating costs (social cost) are related to the daily activities of farming and vary depending on the working time per irrigation system, frequency of irrigation, amount of water applied per irrigation system, and consumption of fuel. The labor cost is the same for both systems of irrigation. Environmental costs are measured by the cost of used water and lost water.
  The benefits of irrigation system are the crop revenue, increased yields using the irrigation system, water, and operating cost savings.
- Step 5: Discounting of costs and benefits.
  The results of the survey were firstly processed; all the inputs and outputs are converted into monetary values. In this study, techno-economic data were processed with EXCEL to produce conventional indicators at pilot site.

The cost-benefit analysis indicators are based on gross margin, net present value, the benefit-cost ratio, and economic efficiency.

– Step 6: Calculation and results of CBA performance indicators.

3.4.1 Results and Analysis of Economic, Social, and Environment Costs

The installation of natural material of clay SLECI elements is very easy. The initial investment of equipment and fixed costs of SLECI irrigation system, installed in four pilot sites with different crops in Morocco, are a little higher than the installation of drip irrigation as shared in Table 2.

The difference of economic cost between the two irrigation systems SLECI vs. Drip is 23% for equipment cost and 3% for fixed cost. According to the effect of the scale economy, the initial cost of installation calculated on 1 ha decreases by increasing the quantity of its production. Moreover, the economic cost of SLECI irrigation system can be saved in the social and environmental costs.

On a social cost scale, the SLECI technology is advantageous and recommended especially for regions where labor for water exploitation is critical. SLECI is a self-regulating underground nano-irrigation technology according to the water needs of the crop, the climate, and the quality of the soil. It is a technique where the flow of irrigation water is continuous. Therefore, it does not require high water exploitation labor cost (Citrus: = 40.6 €/ha, Olivier: 28.7 €/ha, Vine: 54.8 €/ha) compared to the cost of labor required for drip irrigation system (Citrus: = 1037 €/ha, Olivier: 921.4 €/ha, Vine: 654.5 €/ha) as discussed in Table 3. Hence, the SLECI social benefit is 996.4 €/ha, Olivier: 892.7 €/ha, Vine: 599.6 €/ha.

Table 2 Economic costs of two compared system irrigation in Morocco (Wahid, 2023a)

Nr		Citrus	Olive	Vine
1	**Initial investment of equipment cost (€/ha)**			
1.1	SLECI system irrigation	1318.8	1318.8	1318.8
1.2	Drip irrigation system	1086.6	1086.6	1086.6
2	**Fixed cost (€/ha)**			
2.1	SLECI system irrigation	131.9	131.9	131.9
2.2	Drip irrigation system	108.7	108.7	108.7

Table 3 Social cost linked to operating cost of compared systems irrigation in Morocco (Wahid, 2023a)

	Citrus	Olive	Vine
Operating cost of non-harvested labor (€/ ha)			
SLECI irrigation system	40.6	28.7	54.8
Drip irrigation system	1037.0	921.4	654.5
Additional gain (€/ha)	996.4	892.7	599.6

Table 4 Economic costs of two compared system irrigation in Morocco (Wahid, 2023a)

	Citrus	Olive	Vine
Cost of water applied (€/ha)			
SLECI irrigation system	86.6	27.1	54.1
Drip irrigation system	585.7	270.5	202.9
Additional gain (€/ha)	499.1	243.5	148.8
Water saving (€/ ha)	1629.4	180.0	110.0

Besides economic and social costs, the SLEC irrigation system can also help the environment. The environmental cost, linked to used water cost, is observed very low in case of SLECI (Citrus: = 40.6 €/ha, Olivier: 27.1 €/ha, Vine: 54.1 €/ha) if compared to drip irrigation (Citrus: = 405.8 €/ha, Olivier: 270.5 €/ha, Vine: 202.9 €/ha) as illuminated in Table 4.

The average of environmental benefit from the use of SLECI irrigation system is near to 297 €/ha. This environmental benefit is related to water saving benefit of 640 €/ha (Citrus: 1629.4 €/ha, Olivier: 180 €/ha, and Vine 110 €/ha) as was proposed in Table 4. The water savings estimated by SLECI provide nearly 80% compared to drip as an improvement in hydraulic irrigation efficiency on the plot (Elmajetni et al., 2023). SLECI irrigation is beneficial for plants as well. Providing water with the SLECI methods means that the water reaches the roots more directly. If more water goes to the plant and less to evaporation, then it can help plants grow more quickly. More water goes in local roots' plant less weed grow. Therefore, it will have less treatment with herbicide and insecticide chemicals. In addition, saving in the water by SLECI irrigation compared to drip irrigation promotes the release of more and more water for domestic, commercial, and industrial uses. Reducing water exploitation helps reduce energy use and provides great financial benefits to users. It also benefits the environment by cutting down on the release of carbon dioxide, which has been a major contributor to global warming. In addition, reducing fuel use decreases other pollutants released when coal, oil, and other fuels are burned to create energy. This eco-friendly feature of SLECI irrigation will become increasingly important as the world's weather becomes more extreme. Lengthy droughts are becoming more common in some Mediterranean areas. Water-stingy irrigation systems can help ensure that farmers can continue to keep their plants healthy.

3.4.2 Cost and Benefit Analysis

The financial evaluation and economic efficiency indicators are given in Table 5.

We observed that the cash flow is negative. This means that the cash outflows costs are greater than the cash inflows benefit in the first year of SLECI installation. Moreover, the economic efficiency of SLECI system irrigation of different crop testified in pilot sites in Morocco is between 0.2 (Citrus) and 0.04 (Vine) as shown

Table 5 Cost and benefit analysis of two compared system irrigation in Morocco (Wahid, 2023a)

	Citrus	Olive	Vine
Gross margin (t/ha)	−3692.1	−2975.7	−1767.4
Net present value	−4829.8	−3925.4	−3564.7
Benefit-cost ratio	−3.2	−3.1	−1.0
Economic efficiency of SLECI system irrigation	0.2	0.1	0.04
Strategic value of irrigation water (€/m^3)	0.03	0.02	0.01

in Table 5. Strategic value of irrigation water (€/m^3) varied from 0.03 (€/m^3) (Citrus) to 0.01 (€/m^3) (Vine). This efficiency indicator can be corrected by reconsidering the cropping pattern adaptability under SLECI irrigation system.

3.5 Financing Methods for Farmers and Investment Choice Indicators as well as Financial Plan

Given the demands of contemporary agriculture focused on sustainability and innovation, e.g., SLECI, various financing methods have emerged globally. Whether through traditional bank loans, agricultural credits, subsidies, partnerships with private investors, or crowdfunding, farmers today have a broad range of options. However, obtaining financing is just the first step of the challenge. Farmers must also decipher investment opportunities and risks using profitability indicators, evaluating risks, and optimizing resource efficiency. Success in agriculture does not only depend on access to funds but also on how these resources are strategically allocated, managed, and implemented. Armed with in-depth knowledge of financing sources and relevant evaluation tools, farmers can design robust strategies to ensure the growth and resilience of their businesses. In this section, different financing methods available for farmers, indicators that guide investment choices, and components of an effective financing plan are explored.

3.5.1 Analysis of Financing Methods in Agriculture

Agriculture, as a fundamental pillar of the economy and food security, requires substantial investments to ensure its sustainability and growth. Financing is, therefore, a major issue for farmers. There are various financial mechanisms available to farmers:

(a) **Traditional bank loans.**

Financial and credit institutions play a central role in agricultural financing. These entities offer loans, often conditioned by the collateralization of tangible assets such as agricultural land or equipment. The terms of these loans, whether in interest rates (fixed or variable) or repayment duration, are generally determined based on the borrower's creditworthiness and the nature of the investment.

(b) **Agricultural cooperatives.**

Cooperatives represent an organizational model based on resource pooling. By bringing several farmers under one entity, cooperatives can mobilize more substantial financial resources. This allows them to offer their members financial services tailored to their needs, especially access to capital, often under more favorable conditions than the traditional market.

(c) **Subsidies and government assistance.**

Recognizing agriculture's vital role, many governments allocate specific subsidies and financial aids to support farmers. These programs can take the form of direct subsidies, tax exemptions, or preferential rate loans. They generally aim to promote sustainable agricultural practices, support innovation, or ensure food security.

(d) **Microfinance institutions.**

Faced with challenges in accessing traditional financing, especially for small farmers, microfinance institutions have emerged as a viable alternative. By offering small loans, often without collateral, they allow otherwise ineligible farmers to finance their projects. Although the interest rates might be higher than those of traditional banks, the flexibility and accessibility of microcredits make them an attractive option for many.

(e) **Crowdfunding and peer-to-peer lending.**

The advent of digital has disrupted the financial landscape. Crowdfunding or peer-to-peer lending platforms allow farmers to directly solicit financing from the general public. By describing their projects on these platforms, they can attract interested investors or donors. This financing mode, although relatively new, is gaining popularity due to its transparency and the closeness it establishes between project leaders and financiers.

(f) **Self-financing.**

Self-financing, fundamentally rooted in an agricultural enterprise's ability to fund its operations and investments using its own internal resources, mainly undistributed profits, is a pivotal financial strategy within the agricultural realm (Myers, 2001). This approach enables farmers to evade borrowing costs and constraints frequently tied to external financing, such as agricultural loans (Fazzari & Petersen, 1993). Self-financing grants farmers increased autonomy and better control over their investment decisions, such as procuring new equipment or expanding cultivated lands. However, while it offers independence, relying solely on self-financing can limit growth opportunities. In the agricultural sector, where investment needs might be substantial for initiatives like irrigation or adopting new technologies, self-financing might be insufficient if the accumulated profits are constrained (Scholtens & van Wensveen, 2003).

3.5.2 Indicators for Investment Choices: An Agronomic Perspective

Agriculture, as an inherently dynamic sector, requires prudent investment decisions to optimize productivity while minimizing risks. Investment decisions are influenced by various indicators serving as benchmarks for farmers. Here is an exploration based on academic sources of these indicators:

(a) **Return on investment (ROI).**
ROI, as elucidated by Kaufmann and Carter (2009), is a barometer to evaluate the financial relevance of an investment. In agriculture, this might mean assessing whether investing in high-quality seed varieties or technologically advanced equipment is likely to generate superior returns in terms of production or efficiency.

(b) **Market demand.**
The evolving dynamics of the agricultural market, shaped by the projection of product demand, dictate decisions to invest in certain crops or livestock. Farmers who understand this demand, both current and future, and adapt their production accordingly align better with market preferences (O'Sullivan et al., 2011). This strategic approach advantageously positions them to maximize the profitability of their investments.

(c) **Technological advances.**
Technology has profoundly transformed agriculture, introducing both more efficient and co-responsible practices. Piesse and Thirtle (2010) suggest that innovations, whether in precision agriculture or biotechnology, have the potential to revolutionize seeding, farming, and harvesting techniques. These technological advances, besides offering a competitive edge, can lead to yield increases while reducing operational costs.

(d) **Soil and climate analysis.**
Suitability of soil and climate is critical for the success of each crop. The choice of crops tailored to these specific conditions can greatly influence quality and yield (Jones et al., 2009). Consequently, a thorough analysis of soil and weather conditions is essential, providing farmers with invaluable insights, guiding them towards wiser and optimized investment choices.

(e) **Risk assessment.**
The agricultural field is naturally exposed to various hazards, whether climatic or related to threats such as pests and diseases. Hardaker et al. (2015) emphasize the critical importance of rigorous risk assessment. This approach not only enables the development of proactive strategies to anticipate and mitigate these risks but also ensures efficient resource management while maximizing return on investment.

3.5.3 Development of a Comprehensive Financing Plan

The agricultural sector is characterized by its inherent complexity, due to its sensitivity to climate variations, technological advancements, and market fluctuations. In the face of this complexity, having a solid financing plan is imperative. Here is a detailed analysis of the crucial steps in developing such a plan:

(a) **Define clear objectives.**
Any agricultural enterprise, whether small or large scale, has distinct aspirations. These ambitions can have a quantitative nature, like an increase in production, or qualitative, such as improving product quality. Defining clear and precise objectives is critical as it guides financial and operational decisions,

ensuring optimal resource use to achieve set objectives. An effective financial plan stems from a clear understanding of agricultural enterprise's ambitions. Whether aspiring for expansion, adopting new technologies, or diversifying products, it is essential to set concrete milestones (Barry et al., 2000).

(b) **Assess current financial health.**

This step transcends mere accounting. It involves a holistic evaluation encompassing not only assets and liabilities but also other vital parameters like the debt ratio, liquidity, and profitability. An in-depth analysis of these elements provides an accurate snapshot of the enterprise's current financial health, forming the foundation on which any investment strategy rests (Harwood et al., 1999).

(c) **Identify financial needs.**

With clearly defined objectives and a solid financial assessment, it becomes easier to pinpoint specific financial needs. These can encompass aspects like acquiring new equipment, hiring additional personnel, or integrating new agricultural technologies. It is essential to have a detailed view of these financial needs, distinguishing immediate requirements from long-term projected investments (Bierlen & Barry, 2000). Such an approach is vital to prevent unexpected expenses and ensure resource availability at the right time.

(d) **Choose appropriate financing methods.**

Financing mechanisms abound, from traditional bank loans to private investments, not forgetting government subsidies. Each mechanism has its own advantages and limitations. The choice of financial instrument should be judiciously calibrated based on the identified needs, financing cost, flexibility offered, and the risk profile and ambitions of the agricultural entity (Patrick, 1992).

(e) **Create a repayment strategy.**

Any financing, whether through borrowing or investment, entails repayment obligations. It is crucial to establish a solid repayment strategy to ensure these obligations are met without affecting the agricultural enterprise's daily activities. Generating adequate cash flows is vital (Moss, 1997). This might require optimizing operations or adjusting the budget based on projected revenues.

(f) **Monitor and adjust.**

The agricultural sector is characterized by its constant dynamism: the emergence of new technologies, regulatory changes, and market fluctuations. In this ever-changing context (Boehlje & Eidman, 1984), it is imperative that the financial plan is not only effective but also agile enough to adapt to evolutions. This demands a regular revision of the plan to incorporate contextual changes, thus ensuring its relevance and long-term effectiveness.

4 Discussion

Climate change and its associated impacts are the subject of climate conferences held around the world and are increasingly determining global politics. Availability of water means a local existential question not only of human life. In increasingly dry areas, irrigation of crops according to the phrase "as before" is no longer possible. Despite the competence in the application of conventional irrigation technologies, a rethinking is certainly necessary. Alternative water-saving technologies are interesting despite their higher price or more complex retrofitting. This can be explained by a simple problematic situation. In a region characterized by increasing water shortages, farmers have less irrigation water available from wells or water basins. In order to be able to irrigate as before, water deliveries by tanker trucks may initially be necessary. However, this supply of water is expensive. This will make water-saving technologies such as micro-irrigation affordable in the future.

Negative cash flows do not necessarily indicate that the SLECI irrigation project is loss-making in the medium or long term. It is necessary to allow time for the plants to react to the environmental conditions. This means that the yield of the plant will decrease during the first years of adaptation to regain its optimal yield (Malchev et al., 2022). Thus, the conversion to SLECI irrigation could accompany yield drops of 50% in the first year, 20% in the second year, followed by an increase of 10% in the third year, and 20% in the fourth year. If SLECI irrigation is well conducted, it is predicted that optimum yields will reach a 40% increase compared to the reference situation from the fifth, which will stabilize.

Various financing mechanisms available to farmers, the key indicators guiding investment decisions, and the essential components for devising a robust financial plan have been discussed in this work. The explored various financing methods offer a range of options to meet the unique needs of each farmer, while the investment indicators provide essential tools to guide their decisions in an informed manner. However, financial planning does not stop simply at fund acquisition. It encompasses a strategic vision that integrates ongoing assessment, adjustment, and adaptability in the face of sectoral changes. In essence, the future prosperity of agriculture lies not just in the ability to access financial resources but in the capacity to use them wisely, anticipating challenges and seizing opportunities. In an era where agriculture intersects with environmental, economic, and social challenges, a well-thought-out and effectively implemented financial strategy will undoubtedly be one of the pillars of the sector's sustainable resilience and growth.

Development actions should be adapted to the specificities of different territories (Nejjari & Elyousfi, 2023). This might help ensure the economic and social development of the country through better development of the resource, e.g., water, and more rigorous management of its scarcity for current and future generations (Nejjari & Elyousfi, 2023).

5 Conclusions

The analysis of the SLECI implementation in Morocco allows concluding that SLECI serves as a sustainable practice of micro-irrigation in farming. From the social and environmental perspectives of sustainability, SLECI showed positive results. The economic costs of the SLECI use in Morocco, in comparison to conventional drip irrigation system, appeared to be higher.

The limitation of the research is the involvement of only one country, namely, Morocco, in the Mediterranean area in the case study.

Future work intends to engage more countries from different regions in the study of SLECI implementation for micro-irrigation in agriculture.

The research question for future work is: What are economic costs of the use of SLECI for micro-irrigation in farming in other countries?

Implications for researchers imply to adapt the number of clay bodies of the SLECI–irrigation system to the needs of a particular tree in irrigation:

- If the tree suffers from the water scarcity, then to increase the number of clay bodies.
- If the tree receives more water than needed, then to decrease the number of clay bodies.

Implications for investors lay the path to a balancing approach when considering economic and financial aspects of SLECI implementation and maintenance as micro-irrigation technology by small holder farmers.

Acknowledgments This research carried out within the project MED-WET "Improving MEDiterranean irrigation and Water supply for smallholder farmers by providing Efficient, low-cost and nature-based Technologies and practices" (Project ID 1646) funded by Partnership for Research and Innovation in the Mediterranean Area (PRIMA) program.

References

Agricolae. (2023). *Drought, Morocco presents the water highway: By September 400 million cubic meters along 66 km to irrigate 176 thousand hectares*. Free Press. https://agricolae.eu/drought-morocco-presents-the-water-highway-by-september-400-million-cubic-meters-along-66-km-to-irrigate-176-thousand-hectares/

Ahrens, A., & Zaščerinska, J. (2012). Sustainable development in engineering education: A pedagogical approach. In S. Vaitekunas (Ed.), *ATEE Spring University 2012 Conference 20 Years of Sustainable Development: Learning from Each Other* (pp. 115–130). Klaipeda University.

Ahrens, A., Purvinis, O., Zaščerinska, J., Micevičienė, D., & Tautkus, A. (2018). *Burstiness management for smart, sustainable and inclusive growth: Emerging research and opportunities*. IGI Global. https://doi.org/10.4018/978-1-5225-5442-4

Barry, P. J., Ellinger, P. N., Hopkin, J. A., & Baker, C. B. (2000). *Financial management in agriculture*. Interstate Publishers.

Bierlen, R., & Barry, P. J. (2000). Financial structure, technological change, and farm size. *Journal of Agricultural and Applied Economics, 32*(3), 549–560.

Boehlje, M., & Eidman, V. R. (1984). *Farm management*. John Wiley & Sons.

Chopart, J. L. (2004). *RACINE: logiciel de traitement de données racinaires à partir de comptages sur profils de sol. Notice d'utilisation*. Note Cirad Réunion.

Chopart, J.-L., Mézino, M., Aure, F., Le Mézo, L., Mété, M., & Vauclin, M. (2007a). OSIRI: A simple decision-making tool for monitoring irrigation of small farms in heterogeneous environments. *Agricultural Water Management, 87*, 128–138.

Chopart, J. L., Mézino, M., Le Mézo, L., & Fusillier, J. L. (2007b). FIVE-CoRe: A simple model for farm irrigation volume estimates according to constraints and requirements. Application to sugar cane in Reunion (France). In *Proceedings ISSCT, Vol. 26, XXXVI ISSCT Congress*, Durban, South Africa, Abstracts Book (pp. 98–99).

Dührkoop, A. (2020). Saving water and energy with an intelligent subsurface irrigation system. *The International Journal for Rural Development Rural 21*. https://www.rural21.com/english/current-issue/detail/article/saving-water-and-energy-with-an-intelligent-subsurface-irrigation-system.html

Edgar, T. W., & Manz, D. O. (2017). Chapter 3—Starting your research. In T. W. Edgar & D. O. Manz (Eds.), *Research methods for cyber security. Syngress* (pp. 63–92). Elsevier. https://doi.org/10.1016/B978-0-12-805349-2.00003-0

Elmajetni, L., Wahid, N., Zascerinska, N., Sering, Y., Hansmann, H., Sabri, A., & Bouazzama, B. (2023). Hydraulic characteristics of SLECI and moistube irrigation system as influenced by pressure operating, water quality and soil texture? In *10th International Micro Irrigation Conference on the Topic of Micro-Irrigation in the Era of Technological Innovation and Digital Transformation*, Dakhla, Morocco, 25–27 January 2023. http://10imic.ma

EPRS (European Parliamentary Research Services). (2012). *Sustainable agriculture*. Free access: https://epthinktank.eu/2012/09/04/sustainable-agriculture/#:~:text=under%20Shutterstock%20licence-,Sustainable%20agriculture%20has%20been%20defined%20as%20an%20integrated%20system%20of,life%20for%20farmers%20and%20society

Fazzari, S. M., & Petersen, B. C. (1993). Working capital and fixed investment: New evidence on financing constraints. *The Rand Journal of Economics, 24*(3), 328–342.

Fernandez, S., & Mouliérac, A. (2010). Évaluation économique de la gestion de la demande en eau en Méditerranée. Sophia Antipolis: *Plan Bleu*, Centre d'activités régionales, PNUE, Environnement et Développement en Méditerranée, 44 p.

Hardaker, J. B., Huirne, R. B., & Anderson, J. R. (2015). Risk in agriculture. *Encyclopedia of Agriculture and Food Systems, 5*, 312–328.

Harwood, J., Heifner, R., Coble, K., Perry, J., & Somwaru, A. (1999). *Managing risk in farming: Concepts, research, and analysis*. USDA Economic Research Service.

Jones, P. G., Thornton, P. K., & Diaz, W. (2009). PGIS and P3DM modeling to support climate change adaptation. *PLA Notes, 60*, 23–28.

Kaufmann, B., & Carter, M. R. (2009). Risk and the limits of microfinance. *Journal of Agrarian Change, 9*(3), 365–389.

Malchev, S., Kornov, G., & Hansmann, H. (2022). Innovative clay-based micro-irrigation system "SLECI" (self-regulating, low energy, clay-based irrigation): Preliminary results from cherry orchard trials. *Journal of Agricultural, Food and Environmental Sciences, 76*(5), 45–55. https://journals.ukim.mk/index.php/jafes/article/view/1892

Mélanie, T. (2007). Politiques environnementales et agriculture. L'agriculture nouvelle défis (édition), pp 207–212.

Merriam, J. L., & Freeman, B. J. (2007). Operational cost benefits study of flexible on-farm irrigation supply systems. *Journal of Irrigation and Drainage Engineering, 133*, 12–16.

Molle, F., & Sanchis-Ibor, C. (2019). Irrigation policies in the Mediterranean: Trends and challenges. In *Irrigation in the Mediterranean. Technologies, institutions and policies, Chapter 10*. Springer. https://doi.org/10.1007/978-3-030-03698-0_10

Moss, C. B. (1997). Return, risk, and aggregation. *Journal of Agricultural and Resource Economics, 22*(2), 245–261.

Myers, S. C. (2001). Capital structure. *The Journal of Economic Perspectives, 15*(2), 81–102.

Nejjari, S., & Elyousfi, H. (2023). The blue economy in the new development model: The case of Morocco. In *2023 International Interdisciplinary PhD Workshop (IIPhDW)*, Wismar, Germany, 2023 (pp. 1–6). https://doi.org/10.1109/IIPhDW54739.2023.10124417

O'Sullivan, D., Hutchinson, W. G., & Matthews, A. (2011). The influence of specification bias upon agri-food demand elasticities. *European Review of Agricultural Economics, 38*(3), 363–384.

Patrick, G. (1992). Risk in agriculture: Decision framework. *Journal of Agribusiness, 10*(2), 71–82.

Pereira, D., Leitao, J. C. C., Gaspar, P. D., Fael, C., Falorca, I., Khairy, W., Wahid, N., El Yousfi, H., Bouazzama, B., Siering, J., Hansmann, H., Zascerinska, J., Camilleri, S., Busuttil, F., Borg, M., Mizzi, J., Micallef, R., & Cutajar, J. (2023). Exploring irrigation and water supply

technologies for smallholder farmers in the Mediterranean region. *Sustainability, 15*, 6875. https://doi.org/10.3390/su15086875
Piesse, J., & Thirtle, C. (2010). Agricultural R&D, technology and productivity. *Philosophical Transactions of the Royal Society B, 365*(1554), 3035–3047.
Plan Bleu. (2008). *Les Perspectives du Plan Bleu sur le développement durable en Méditerranée*. Plan Bleu.
Richefort, L. (2009). *Processus de sélection des technologies d'irrigation par les agriculteurs: entre interactions sociales et choix rationnels. Economies et finances*. Université de la Réunion.
Scholtens, B., & van Wensveen, D. (2003). The theory of financial intermediation: An essay on what it does (not) explain. *SUERF Studies, 2003*(1), 1–26.
Wahid, N. (2023a). *Cost-benefit analysis blueprint for water efficient irrigation in selected Mediterranean countries*. MED-WET project, T3.2 report.
Wahid, N. (2023b). Analyse macro-environnementale de l'eau d'irrigation en Méditerranée: Vers une efficience économe en eau. Éditions Universitaires Européennes, 57 p. ISBN 978-620-3-45104-7
Wahid, N., Bouazzama, B., Yousfi, H. E., & Zascerinska, J. (2022). A histrocial overview of evolution of the irrigated agriculture sector in Morocco. *Journal of Regional Economic and Social Development, 14*, 109–118. https://doi.org/10.17770/jresd2022vol14.6896
Yin, R. K. (2003). *Case study research, design and methods* (Vol. 5, 3rd ed.). Sage.

Impact of Climate Change on Sustainable Water Management: A Study on Urban Bangladesh

A. K. M. Mahmudul Haque, Imran Hossain, Md. Sohel Rana, Md. Asfaq Salehin, and Md. Shamim Hossen

Abstract Climate change has profound effects on agriculture, public health, the economy, and sustainable development, making it difficult for Bangladesh to manage its water resources. Adaptive solutions are necessary to strengthen resilience and decrease risks, as highlighted by economic evaluations that emphasize the large economic losses associated with climate-related occurrences. In addition to stressing the value of integrated approaches to water sanitation, hygiene promotion, and disease prevention, health assessments also highlight the connections between water, climate, and public health. Agricultural research highlights the significance of sustainable agricultural practices and adaptive measures by elucidating how vulnerable rural communities are to variations in temperature extremes, precipitation patterns, and water availability. More effective management of water resources, increased water efficiency, and greater risk resistance against climate change are all made possible by technical advancements. This chapter highlights the necessity of taking coordinated action to mitigate the effects of climate change on Bangladesh's water resources through an interdisciplinary lens. It also highlights the significance of investing in research and development, stakeholder engagement, and integrated approaches to promote sustainable water management and guarantee water security for all.

Keywords Climate change · Water management · Sustainability · Water quality · Governing agencies · Infrastructure · Urban Bangladesh

---

A. K. M. Mahmudul Haque
Department of Political Science, University of Rajshahi, Rajshahi, Bangladesh
e-mail: akmmahmudul@ru.ac.bd

I. Hossain (✉) · M. Asfaq Salehin
Department of Political Science, Varendra University, Rajshahi, Bangladesh
e-mail: imran.pol@vu.edu.bd; imranbd.ru@gmail.com; asfaq@vu.edu.bd

J. C. Correia Leitão et al. (eds.), *Sustainable Water Management and Irrigation Systems*, Natural Resource Management and Policy 62,
https://doi.org/10.1007/978-3-031-93587-9_8

1 Introduction

Global water resources are under severe threat from climate change, which will affect both their quantity and quality (Kundzewicz et al., 2018). According to the Intergovernmental Panel on Climate Change (IPCC), increasing temperatures cause modifications to runoff dynamics, the melting of glaciers, and changes to patterns of precipitation, all of which have an impact on freshwater availability (IPCC, 2021; Hussain et al., 2021). Due to its location and dense population, Bangladesh, a nation in a deltaic region, is particularly vulnerable to these changes, exacerbating preexisting vulnerabilities (Parven et al., 2022). The nation faces two challenges at once: saline intrusion in coastal areas due to sea-level rise and regular floods caused by large rainfall events, which impact water supply and quality (Rabbani et al., 2018). Furthermore, under climate change scenarios, extreme weather events like cyclones get stronger, putting additional strain on infrastructure and water resources (Rahman et al., 2023).

The way water is managed in cities is significantly impacted by these changes in climate. Bangladesh's metropolitan areas are experiencing a growing demand for water resources as a result of unplanned urbanization, rapid population growth, and inadequate infrastructure (Hossen et al., 2024; Rahaman et al., 2023; Hossain et al., 2024e). These problems are made worse by climate change, which increases the likelihood of contamination, water scarcity, and infrastructure damage in urban areas (Rahman et al., 2023). Additionally, the IPCC emphasizes how altered precipitation patterns brought about by climate change, along with higher rates of evapotranspiration, affect groundwater recharge, thus taxing urban water supplies (IPCC, 2021). To ensure water security and sustainability in urban settings, it is imperative to develop efficient management techniques that take into account the unique effects of climate change on Bangladesh's urban water resources.

In urban settings, sustainable water management is critical to addressing climate change difficulties and ensuring the resilience of water systems. Sustainable water management methods include integrated water resource management, pollution control, and effective use of water, according to the United Nations World Water Development Report (Hossain et al., 2024b; Makanda et al., 2022). Sustainable management methods are essential for preserving water quality, guaranteeing fair access, and reducing the effects of climate change in metropolitan areas where there is a high demand for water and a limited supply of resources (Hossain et al., 2024a; Pradhanang & Jahan, 2021). Furthermore, biodiversity and ecosystem conservation

M. Sohel Rana
Department of Political Science, University of Barisal, Barishal, Bangladesh

Department of Political Science, Texas Tech University, Lubbock, TX, USA
e-mail: msrana@bu.ac.bd

M. Shamim Hossen
Department of Population Science and Human Resource Development, University of Rajshahi, Rajshahi, Bangladesh

are critical to the general welfare of urban inhabitants, and these goals are further advanced by sustainable water management (Vannevel & Goethals, 2020).

There are several advantages to using sustainable water management techniques in cities. Cities can lessen their reliance on outside water supplies and ease pressure on freshwater ecosystems by supporting water conservation techniques like rainwater collection and water recycling (Hossain et al., 2023c; Makanda et al., 2022; Pokhrel et al., 2022). Moreover, sustainable water management practices improve the effectiveness of water distribution networks, lowering leakage losses and facilitating urban dwellers' access to dependable and safe water sources (Pokhrel et al., 2022; Hossain et al., 2023a, c). Lowering operating expenses and strengthening water infrastructure resistance to hazards associated with climate change also help cities become more economically sustainable (World Bank, 2020).

Furthermore, sustainable water management practices, such as guaranteeing that every member of urban communities has access to clean water and sanitation services, promote social fairness. This is especially crucial for vulnerable groups living in peri-urban and informal settlements, where access to water resources may be restricted and water-related illnesses are common (WHO, 2021). Cities may address the needs of excluded groups and foster social cohesion and resilience by giving priority to inclusive and participatory approaches to water management (Hossen et al., 2024; Hardoy et al., 2019).

Understanding the particular difficulties and dynamics of water management in Bangladesh's urban setting is crucial for placing this study in its proper context. Bangladesh, one of the world's most populous nations, is confronted with a multitude of water-related problems, which are made worse by the effects of climate change and fast urbanization (Ashrafuzzaman et al., 2023). The demand for water resources in urban areas is rising due to the fact that about 35% of the population lives there, and urbanization is expected to expand (Hossain et al., 2024e; World Bank, 2020).

Bangladeshi urban regions confront complex issues related to water management because of environmental degradation, insufficient infrastructure, and population increase (Hossen et al., 2024; Rahaman et al., 2023). Existing water supply and sanitation infrastructure are strained by the growth of informal settlements and peri-urban areas, which results in unequal access to services. Water management in Bangladesh's urban environment is made more difficult by problems with institutional capability, financing, and governance (Rahman et al., 2023; Hossain et al., 2023c). Ineffective coordination and planning for the development and administration of water infrastructure are impeded by weak regulatory frameworks and disjointed governance structures (Hossain et al., 2023b; World Bank, 2020). Finally, there are major obstacles to fully addressing urban water concerns due to the limited financial resources available for investment in water-related initiatives.

Bangladesh's urban context offers a difficult and complex environment for managing water resources, made worse by the interplay of concerns about governance, socioeconomic inequality, fast urbanization, and climate change vulnerabilities (Hossen et al., 2024). It is imperative to comprehend these contextual subtleties in order to formulate efficacious strategies aimed at augmenting the resilience and sustainability of urban water systems in Bangladesh. This study's objective is to thoroughly evaluate how climate change is affecting Bangladesh's water resources, with a focus on how it may affect public health, agricultural production, economic growth, and technical advancements. This study intends to provide insights into the potential and difficulties of sustainable water management in the face of environmental change and climate variability by combining current literature and empirical data.

2 Climate Change and Water Resources in Bangladesh

Bangladesh is particularly vulnerable to the effects of climate change because of its location and socioeconomic status, which are reflected in a variety of observable patterns and implications (IPCC, 2021). Temperature rise is one of the most notable developments; throughout the past few decades, Bangladesh has seen an increase in mean annual temperatures (Mullick et al., 2019). The country's many areas see variations in the amount, timing, and distribution of rainfall due to changes in precipitation patterns brought about by this warming trend (Mullick et al., 2019).

One of the most obvious effects of climate change in Bangladesh is the rise in the frequency and intensity of extreme weather events, such as storms, cyclones, and heavy rainfall (Hossain et al., 2024c; Islam et al., 2021). These occurrences have long-term effects on the management of water resources, in addition to causing immediate harm to human lives, livelihoods, and infrastructure (Bell et al., 2018). Strong rainfall events, for example, might result in floods in river basins and low-lying areas, affecting water supply, polluting water sources, and damaging water infrastructure (Kabir & Hossen, 2019).

Bangladesh is disproportionately affected by sea-level rise, another major effect of climate change, because of its vast coastline and level topography (Alam et al., 2018). Groundwater supplies get salinized in coastal locations due to their growing susceptibility to saltwater intrusion, making them unfit for agricultural or drinking use (Sangadi et al., 2022). This phenomenon creates hurdles for agricultural productivity and food security, as well as aggravating water constraints in coastal zones.

In addition, variations in river flow patterns brought about by climate-change-induced glacial melt in the Himalayas impact Bangladesh's water supply downstream (UNEP, 2018). Water-related issues in the nation can be made worse by variations in the time and volume of river flows, which can affect ecosystem dynamics, hydropower production, and irrigation systems (WHO, 2021). All things considered, the trends of rising temperatures, altered precipitation patterns, more frequent extreme events, rising sea levels, and glacial melt highlight how urgently effective adaptation and mitigation measures are needed to address the effects of climate change on Bangladesh's water resources.

Bangladesh faces numerous difficulties made worse by its geographic and socioeconomic circumstances, and its water resources are particularly vulnerable to the effects of climate change (Rahman et al., 2019). The increased frequency and severity of extreme weather events, such as storms, cyclones, and heavy rainfall, which frequently cause floods in low-lying areas and river basins, is one of the main causes of concern (Hossain et al., 2024c). In addition to causing disruptions to the availability of water, these floods contaminate water sources, hence increasing the risk of waterborne infections and endangering public health (World Bank, 2020).

Socioeconomic issues like poverty, unplanned urbanization, and fast population expansion make Bangladesh's water resources even more vulnerable to climate change (Hossain et al., 2024c; WHO, 2021). The nation's ability to adapt to and lessen the effects of climate change on water supplies is hampered by inadequate infrastructure, shoddy governance frameworks, and a lack of financial resources (Hossain et al., 2024e). As a result, in order to address the vulnerability of water resources to climate change, integrated and comprehensive approaches are needed, and Bangladesh's water management policies must take into account the intricate interactions between environmental, social, and economic aspects (Hossain et al., 2024a; IPCC, 2021).

Bangladesh's urban water management system faces unique difficulties, which are made worse by the effects of climate change and the country's fast urbanization. High population densities, poor infrastructure, and restricted access to clean drinking water and sanitary facilities are characteristics of the nation's metropolitan areas (Hossain et al., 2024e; Rahman et al., 2019). The escalating occurrence and severity of extreme weather phenomena, like storms, cyclones, and heavy rainfall, which frequently cause flooding and waterlogging in urban areas, exacerbate these difficulties (World Bank, 2020).

The outdated and insufficient infrastructure for sanitation and water delivery is one of the main issues facing urban water management in Bangladesh (Haque et al., 2020). Water distribution systems in many cities are antiquated and ineffective, which causes leaks and contaminates water sources, resulting in losses (Haque et al., 2020). Inadequate drainage systems also make flooding during monsoon seasons worse, damaging infrastructure and property.

In addition, the strain on water resources and urban infrastructure is increased by unplanned development and fast urbanization (UN-Habitat, 2020). The absence of essential water and sanitation facilities in informal settlements and peri-urban areas can result in environmental deterioration and health risks. In addition, the expansion of urban areas into wetlands and other bodies of water diminishes the ability of natural water storage and increases the danger of flooding and water scarcity.

Urban water management systems are put under additional stress due to alterations in precipitation patterns and an increase in the frequency of extreme weather events caused by climate change. Additional concerns for coastal communities include saltwater intrusion into groundwater supplies and sea-level rise, which impact water availability and quality (WHO, 2021). To tackle these obstacles, comprehensive strategies are needed that give top priority to resilient infrastructure development, sustainable water management techniques, and efficient governance frameworks. Building climate-resilient and sustainable urban water systems in Bangladesh also requires encouraging stakeholder and community participation in decision-making processes (IPCC, 2021).

3 Sustainable Water Management Practices

Sustainable water management principles integrate social, economic, and environmental factors in a comprehensive manner to handle water-related issues (Mishra et al., 2021; Hossain et al., 2023c). The optimal use of water resources, with an emphasis on reducing water waste and optimizing water consumption across several sectors, is one of the core principles (Namany et al., 2019; Hossain et al., 2024d). This entails putting into effect techniques and technologies that improve water conservation, such as demand management plans, leak detection methods, and water-efficient irrigation systems (Gupta et al., 2020; Hossain et al., 2023a).

The preservation and restoration of ecosystems that offer essential functions for water supply, management, and purification is another fundamental idea (Mishra et al., 2021). Water quality, biodiversity, and ecosystem resilience depend on the health of aquatic ecosystems, such as rivers, wetlands, and forests (Pelletier et al., 2020). Planning and decision-making procedures for water resources must incorporate the ideas of ecosystem-based approaches in order to practice sustainable water management.

In addition, a fundamental tenet of sustainable water management is the advancement of social justice and inclusivity, guaranteeing that every member of the public has fair access to sanitary facilities and clean water (UN-Habitat, 2020; Hossain et al., 2023c). This entails considering the needs of vulnerable and marginalized groups in water planning and management procedures, including women, children, the elderly, and individuals with disabilities (WHO, 2021). Building ownership, trust, and responsibility in water management projects also requires encouraging community participation and stakeholder engagement.

Furthermore, integrated water resource management (IWRM) strategies that take into account the interdependence of water systems and the various ways that water is used in various sectors are crucial to sustainable water management (Katusiime & Schütt, 2020; Skoulikaris et al., 2018). In order to maximize water allocation, reduce conflict, and improve water security, IWRM encourages coordination and collaboration among diverse stakeholders, including government agencies, local communities, civil society organizations, and the commercial sector (Katusiime & Schütt, 2020).

All things considered, the fundamentals of sustainable water management offer a framework for directing strategies, policies, and initiatives to solve water-related issues in a way that is socially, ecologically, and financially feasible. Stakeholders may develop resilient and sustainable water systems that satisfy the requirements of present and future generations while preserving the well-being of ecosystems and people by adhering to these principles.

Adopting sustainable water management methods in Bangladesh is fraught with difficulties, but it also offers a great deal of promise for resolving water-related problems in the face of rapidly increasing urbanization and climate change. The nation's finite water supplies and the rising demand brought on by industrialization and population expansion provide a significant obstacle (Hasnat et al., 2018).

Particularly in metropolitan areas, this increased demand puts stress on the water infrastructure already in place and exacerbates the problem of water shortages. Furthermore, inadequate wastewater management and sanitation facilities exacerbate water pollution and worsen the state of the water (Howladar et al., 2021).

Sustainable water management initiatives are further complicated by Bangladesh's susceptibility to the effects of climate change, including floods, cyclones, and sea-level rise (Hadi, 2019). Significant health and socioeconomic hazards are presented by climate-related catastrophes, which frequently disrupt water supply systems, damage infrastructure, and increase waterborne infections (World Bank, 2020). Furthermore, ensuring equitable access to water and sanitation services is made more difficult by the socioeconomic disadvantages that are common in Bangladesh, such as gender inequality, poor access to education, and poverty (WHO, 2021).

4 Assessing the Impact of Climate Change on Urban Water Management in Bangladesh

4.1 Analysis of Climate Change Data and Trends

Researchers typically use secondary sources, such as academic papers, government publications, and scientific reports, to analyze data and trends related to climate change in urban Bangladesh. A thorough summary of past climate data, projected future climate, and trends in temperature, precipitation, and extreme weather occurrences may be found in these sources. The Intergovernmental Panel on Climate Change (IPCC), for example, conducts studies that provide insightful information on regional and global climate trends, including those that affect Bangladesh (IPCC, 2021). Furthermore, localized information on climatic variability and change is provided by national-level climate assessments, such as those carried out by the Ministry of Environment, Forests, and Climate Change and the Bangladesh Meteorological Department (Rashid, 2023).

When analyzing data related to climate change, one common approach is to use statistical techniques like trend analysis and time series modeling to look at long-term patterns in important climatic variables like temperature and precipitation (Xie et al., 2018). Meteorological stations located throughout Bangladesh provide historical climate data, which is frequently used to evaluate trends in temperature and rainfall patterns throughout time. According to Hosseinzadehtalaei et al. (2020), this approach aids in the identification of important changes in climate parameters, such as rising temperatures, altered rainfall patterns, and fluctuations in the frequency and severity of extreme weather events.

Additionally, in order to comprehend probable future climatic scenarios and their effects on Bangladesh's urban water management, experts examine forecasts from climate models. Based on various scenarios of greenhouse gas emissions, climate models, such as those created by the IPCC and regional climate modeling centers, simulate the conditions of the future climate (Alizadeh, 2022). Under various future scenarios, these forecasts offer important information on possible changes in temperature, precipitation, sea-level rise, and other climate variables. Researchers can evaluate the range of potential climate consequences and uncertainties by examining the results of climate models; this information helps guide policy decisions and adaptation plans (World Bank, 2020).

The examination of climate change data and patterns for urban Bangladesh is dependent on many secondary sources, such as official publications, scientific papers, and climate model estimates. Researchers may learn a great deal about the causes and effects of climate change in urban settings by methodically examining past climate data and projected climate scenarios. In order to improve climate resilience and sustainable water management in Bangladesh's urban centers, this study provides critical support for decision-making processes and the development of adaptation strategies (UNEP, 2018).

4.2 Identification of Key Factors Impacting Urban Water Management

Increased urbanization and population expansion are major factors influencing urban water management. In recent decades, Bangladesh's urban population has grown significantly, placing a strain on the country's infrastructure and increasing demand for water services (Hossain et al., 2024a; Hasnat et al., 2018). The demand for water resources increases as cities grow and urban areas become denser; thus, efficient management techniques are required to guarantee sufficient supply and delivery.

Furthermore, Bangladesh faces major issues in managing urban water resources due to insufficient infrastructure and the restricted availability of safe drinking water and sanitation services. There are dangers to public health, contamination, and water scarcity in many metropolitan areas due to inadequate wastewater treatment facilities, sanitation infrastructure, and water delivery networks (WHO, 2021). For sustainable urban water management, filling in these physical gaps and enhancing service delivery are top considerations.

Furthermore, Bangladesh's urban water management is beset with greater difficulties as a result of climate change. Urban water resources and infrastructure are seriously at risk from the effects of climate change, which include rising sea levels, more frequent and intense extreme weather events, and altered precipitation patterns (IPCC, 2021). Storms, cyclones, and floods have the potential to taint water sources, destroy distribution networks, and cause waterborne illnesses, in addition to degrading the environment (Nichols et al., 2018).

Urban water management techniques are significantly shaped by institutional and governance considerations, in addition to these challenges. Effective water governance and infrastructure development are hampered by weak regulatory frameworks, disjointed governance structures, and a lack of funding (Haque et al., 2020; Hossain et al., 2023b). In order to improve the resilience and sustainability of Bangladesh's urban water systems, it is imperative that institutional capacity be strengthened, stakeholder collaboration be encouraged, and governance mechanisms be improved (World Bank, 2020).

In general, a thorough examination of population dynamics, infrastructure requirements, the effects of climate change, and governance issues is necessary to identify the critical elements influencing urban water management in Bangladesh. Policymakers and water managers can create focused interventions and strategies to solve the complex issues of urban water management and guarantee water security and sustainability in Bangladesh's rapidly expanding cities by knowing these aspects and how they interact.

4.3 Assessment of the Current State of Urban Water Management in Bangladesh

Urban water management relies heavily on infrastructure, which includes stormwater drainage networks, wastewater treatment plants, and water delivery systems. The availability and functionality of these infrastructures greatly impact the accessibility, quality, and dependability of water in urban regions (Hoque et al., 2019). Secondary sources highlight inequalities in service delivery and infrastructure deficiencies, providing information on the availability and state of water and sanitation infrastructure in various Bangladeshi towns and areas (Rahman et al., 2019).

Furthermore, urban water management practices are significantly shaped by institutional arrangements and governance procedures. Secondary sources shed light on the institutional structures, policies, and regulatory frameworks controlling Bangladesh's water governance (UN-Habitat, 2020). These resources provide evaluations on the efficiency of water governance mechanisms with regard to accountability, transparency, stakeholder participation, and decision-making procedures.

Another component of urban water management that needs to be evaluated is service delivery. Secondary sources offer information on coverage rates, water quality criteria, and service dependability related to urban areas' access to safe drinking water and sanitation services (WHO, 2021). Additionally, these sources shed light on the difficulties underprivileged people, such as slum dwellers and informal settlement residents, have in obtaining water and sanitation services (World Bank, 2020).

Additionally, examining vulnerabilities and adaptation strategies is necessary to evaluate how resilient urban water management is to the effects of climate change. For Bangladesh's urban water systems, secondary sources include information on vulnerability assessments, adaptation plans, and climate change estimates (IPCC, 2021). These resources shed light on the dangers posed by climate-related natural disasters like hurricanes, floods, and sea-level rise, as well as the steps that must be taken to improve the resilience of water infrastructure and service delivery.

Evaluating Bangladesh's urban water management situation now necessitates a multifaceted examination of the country's governance, service delivery, infrastructure, and climate resilience. Through the utilization of secondary sources, scholars can develop an all-encompassing understanding of the obstacles and prospects confronting urban water management in Bangladesh, as well as pinpoint tactics for enhancement and sustainable growth.

4.4 Effects on the Economy

Bangladesh's water supplies are seriously threatened by climate change, which has serious economic ramifications for the country. The availability, quality, and distribution of water are significantly impacted by temperature and precipitation changes, as well as an increase in the frequency and severity of extreme weather events (IPCC, 2021). These changes in hydrological systems have a profound impact on livelihoods and economic growth across a number of economic sectors, including infrastructure development, manufacturing, and agriculture.

In Bangladesh, one of the main economic effects of climate change on water supplies is seen in the agriculture industry. A large percentage of the workforce is employed in agriculture, which also significantly boosts gross domestic product (GDP) (Miah et al., 2020). Agriculture continues to be an important part of the nation's economy. However, altered precipitation patterns and a rise in the frequency of floods and droughts impair agricultural output, resulting in decreased crop productivity, yield losses, and jeopardized food security (Muluneh, 2021). Poverty and economic inequality are made worse by these effects, which directly affect farmers' incomes, rural lives, and national food security.

Furthermore, Bangladesh's industrial development and infrastructure are impacted by variations in water quality and availability brought on by climate change. Water is essential to the industrial sector's energy generation facilities, cooling systems, and manufacturing processes. Temperature and water supply fluctuations can cause industrial operations to be disrupted, which can result in lower competitiveness, higher costs, and delays in output (World Bank, 2020). Furthermore, industries including manufacturing, leather goods, and textiles are negatively impacted by low water quality brought on by pollution and salinization.

Furthermore, infrastructure development and urban water management are just two areas where the economic effects of climate change on water supplies go beyond those of industry and agriculture. The demands on water supply, sanitation, and infrastructure services are rising as a result of Bangladesh's fast urbanization and population expansion (UN-Habitat, 2020). However, the resilience of infrastructure is seriously threatened by flooding, sea-level rise, and water scarcity brought on by climate change, which can result in damages, interruptions, and higher maintenance costs (UNEP, 2018). These effects have wider economic repercussions, such as decreased productivity, higher healthcare costs, and declining property prices, in addition to affecting urban inhabitants' access to water and sanitation (WHO, 2021).

In conclusion, Bangladesh's water supplies are greatly impacted by climate change, which has serious financial ramifications for a number of industry sectors. Integrated strategies that give priority to resilience building, sustainable water management, and climate adaptation are needed to address these issues. Bangladesh has the potential to reduce the financial effects of climate change on its water resources and maintain sustainable development in the face of environmental obstacles by making investments in climate-resilient infrastructure, encouraging water-efficient practices, and strengthening its adaptive capacity.

4.5 Dynamics of Populations

Bangladesh's water resources are significantly impacted by climate change, which also interacts with population dynamics to increase vulnerabilities and challenges. Bangladesh's water supplies are particularly under pressure due to the country's highly populated and quickly expanding population, which is made worse by changes brought on by climate change. The nation's high population density, low-lying terrain, and susceptibility to natural disasters amplify the effects of climate change on the availability, distribution, and quality of water (IPCC, 2021). The rivalry for scarce water resources is intensified by rapid population increase, especially in metropolitan areas where the demand for sanitation and water services is rising quickly (UN-Habitat, 2020).

There are various ways in which population dynamics and the consequences of climate change on Bangladesh's water resources interact. Firstly, altered precipitation patterns and a rise in the frequency of extreme weather events—like cyclones and floods—have a direct effect on infrastructure and population centers, causing instability, loss of livelihood, and increased susceptibility (Rahman et al., 2019). These issues are made worse by rapid urbanization since peri-urban and informal settlements frequently lack access to clean water and proper sanitation, increasing health hazards and socioeconomic inequality (WHO, 2021).

Additionally, as the population grows, more demands are made on infrastructure and water supplies, increasing the amount of water needed for industrial, agricultural, and residential uses. Water security and livelihoods are further threatened by this increased demand, which intensifies competition for water resources, especially during times of drought and water scarcity (Haque et al., 2020). The quantity and quality of water resources are further jeopardized by population growth's role in pollution, environmental deterioration, and unsustainable water use.

Additionally, the socioeconomic effects of climate change on water resources are amplified by demographic trends, such as migration from rural to urban regions and population concentration in susceptible coastal areas (UNEP, 2018). Migration from rural to urban areas places pressure on already-limited resources by adding to the strain on urban water infrastructure and services (World Bank, 2020). In a similar vein, dense populations along the shore increase the likelihood of erosion, flooding, and saltwater intrusion, endangering freshwater supplies and livelihoods even more (WHO, 2021).

Population dynamics and the effects of climate change on Bangladesh's water resources are linked, making the nation's vulnerabilities and difficulties greater. Integrated solutions that take into account the intricate relationships between population dynamics, water management techniques, and climate change are needed to address these issues. Bangladesh can lessen the effects of climate change on its water resources and increase its ability to withstand future environmental challenges by tackling socioeconomic inequality, investing in resilient infrastructure, and encouraging sustainable water management practices.

4.6 Effects on Health

The effects of climate change on Bangladesh's water supplies have a substantial impact on public health, aggravating preexisting conditions and posing new threats to the populace. Water-related health concerns include food insecurity, hunger, and waterborne illnesses. These are caused by changes in precipitation patterns, rising temperatures, and an increase in the frequency of extreme weather events (IPCC, 2021). Vulnerable groups, especially those residing in impoverished and disadvantaged communities, are disproportionately impacted when these effects combine with socioeconomic variables.

A major health effect of climate change on Bangladesh's water supplies is an elevated risk of waterborne illnesses, such as cholera, typhoid fever, and diarrheal illnesses. Waterborne disease epidemics are caused by flooding and water contamination brought on by climate change, which jeopardizes the safety of drinking water sources (Rahman et al., 2019). Children, the elderly, and people with weakened immune systems are among the vulnerable groups most at risk of acquiring these illnesses, which raises the rates of morbidity and death (WHO, 2021).

Food insecurity and hunger are further exacerbated by altered precipitation patterns and flooding incidents, especially in rural areas where agriculture is the main source of income. Floods have the potential to cause shortages and price surges by destroying crops, contaminating the food supply, and upsetting food distribution networks (Rahman et al., 2024). This makes hunger and food insecurity worse, especially for vulnerable groups like women and children (Haque et al., 2020). Malnutrition also impairs immune systems, making people more vulnerable to infections carried by water and other health risks.

Furthermore, the effects of climate change on water supplies worsen mental health issues, especially in areas where flooding and relocation occur frequently. Affected populations may experience trauma, stress, and anxiety as a result of flooding, which may lead to mental health issues and psychosocial distress (World

Bank, 2020). These effects on mental health may have long-term repercussions on people's well-being, social cohesiveness, and economic output.

The effects of climate change on Bangladesh's water resources have a substantial influence on public health, making food insecurity, mental health issues, and waterborne illnesses worse. Integrated approaches that incorporate public health interventions, water management plans, socioeconomic development projects, and climate adaptation measures are needed to address these health implications (UNEP, 2018). Bangladesh can address the health hazards associated with climate change and safeguard the welfare of its populace in the face of environmental difficulties by making investments in robust water infrastructure, expanding access to clean drinking water and sanitation, and fortifying its health systems.

4.7 Production of Agriculture

Bangladesh's water resources are significantly impacted by climate change, especially when it comes to agricultural production, which is the main driver of the country's economy and provides a living for millions of people. Crop yields, agricultural production, and water availability are all directly impacted by temperature variations, precipitation patterns, and extreme weather. These effects compound food insecurity, provide obstacles for farmers, and jeopardize the sustainability of the nation's agricultural sector when combined with socioeconomic concerns.

Production in agriculture is one of the main effects of climate change on Bangladesh's water resources. Crop growth cycles, planting and harvesting dates, and the availability of water for irrigation are all impacted by changes in precipitation patterns, which include variations in the timing, intensity, and distribution of rainfall (Rahman et al., 2019). Particularly in rain-fed agricultural areas, an increase in the frequency of droughts and dry spells causes crop failure, water stress, and decreased soil moisture. On the other hand, severe precipitation and flooding can submerge fields, ruin crops, and remove rich topsoil, resulting in large yield losses and financial difficulties for farmers (Haque et al., 2020).

Furthermore, crops experience increased heat and water stress due to rising temperatures brought on by climate change, which hinders agricultural growth and development. Elevated temperatures hasten the rate of evapotranspiration, which raises the need for water for irrigation and decreases the efficiency of crop water usage (UN-Habitat, 2020). This leads to lower crop quality, lower yields, and lower agricultural productivity—especially for crops that are susceptible to heat, such as wheat and rice (World Bank, 2020). Furthermore, in agricultural systems that depend on livestock for revenue and food security, temperature extremes can cause heat stress in animals, which can have an impact on animal health and productivity.

Moreover, the availability and quality of irrigation water—which is essential for maintaining Bangladesh's agricultural output—are similarly impacted by the effects of climate change on water resources. Irrigation systems face difficulties because of changes in river flows, groundwater levels, and water salinity brought on by sea-level rise and saltwater intrusion, especially in coastal regions (WHO, 2021). Elevated salt in irrigation water can cause crop production reduction, soil erosion, and heightened susceptibility to pests and diseases (UNEP, 2018). These effects exacerbate socioeconomic inequality and food insecurity by disproportionately affecting vulnerable communities and smallholder farmers.

Bangladesh's agricultural output is greatly impacted by the effects of climate change on water resources, which presents issues for food security, rural livelihoods, and the country's economic growth. Integrated strategies that incorporate agricultural innovations, sustainable water management techniques, and climate adaptation measures are needed to address these issues (IPCC, 2021). Bangladesh can lessen the effects of climate change on agricultural production and guarantee food security for its expanding population by making investments in climate-resilient agriculture, enhancing the management of water resources, and assisting smallholder farmers in their adaptation efforts.

5 Sustainable Strategies for Managing Water Resources in the Face of Climate Change

Water resources are facing serious problems from climate change everywhere, including Bangladesh, where the effects are more severe because of the nation's physical and socioeconomic vulnerabilities. As a result of these difficulties, the necessity for technologically driven, sustainable solutions that both adapt to and lessen the effects of climate change on water supplies is becoming increasingly apparent. These technological advancements are essential for strengthening water resource management, increasing water efficiency, and building resilience to climate-change-related risks.

Adopting cutting-edge water management technologies, such as satellite images, geographic information systems (GIS), and remote sensing, to efficiently monitor and manage water resources is one technologically viable alternative. Real-time data on precipitation, river flows, groundwater levels, and soil moisture are provided by these technologies, facilitating prompt and well-informed solutions to water-related issues. Remote sensing technologies, for instance, may monitor changes in vegetation cover and land usage, which aids in identifying regions susceptible to erosion, drought, and water scarcity.

Moreover, water scarcity can be effectively addressed and water availability can be improved through the application of cutting-edge water harvesting and conservation technology (Zhang et al., 2020). For example, rainwater harvesting systems reduce dependency on surface and groundwater sources by collecting and storing rainwater for use in industry, agriculture, and home settings (Abdulla et al., 2021). Similarly, processed wastewater can be safely reused for nonpotable uses like irrigation, industrial processes, and urban landscaping, thanks to wastewater treatment and recycling technologies (WHO, 2021). These innovations support circular water management techniques, lessen pollution, and ease the strain on water resources.

Smart water infrastructure and management technologies can also be implemented to maximize water distribution, reduce water loss, and enhance water efficiency (IPCC, 2021). Water allocation and conservation can be made more accurate through real-time monitoring of water usage, leak detection, and demand management, made possible with the use of smart meters, sensors, and automation technologies (Sánchez et al., 2020). The resilience and adaptive capacity of water systems are increased by integrated water management platforms, which are driven by artificial intelligence and predictive analytics and provide decision-support tools for water planners and managers (UN-Habitat, 2020).

Furthermore, in order to promote sustainable water management, nature-based solutions, like ecosystem restoration and green infrastructure, work in tandem with technical advancements (Souliotis & Voulvoulis, 2022). Urban wetlands, green roofs, and permeable pavements all contribute to better water quality, reduced floods, and increased biodiversity in cities (Stefanakis, 2019). In a similar vein, efforts to restore watersheds and afforest land safeguard natural water sources, control water flows, and improve ecosystem services—all of which support long-term water security and climate resilience.

Technologically sound, sustainable solutions are essential for mitigating the effects of climate change on Bangladesh's and other countries' water supplies. Bangladesh can strengthen its resilience to climate-related threats, improve water security, and advance sustainable development by utilizing cutting-edge technology, including remote sensing, water conservation, smart infrastructure, and nature-based solutions. However, robust institutional backing, multistakeholder cooperation, and investments in infrastructure development and capacity building are necessary for the successful implementation of these solutions. Bangladesh can effectively handle the difficulties posed by climate change and secure a sustainable water future for its population and ecosystems by embracing technological innovation and using integrated methods to manage water resources.

6 Conclusion

This study clarifies the complex effects of climate change on Bangladesh's water resources by analyzing its consequences for different industries, demographics, and ecosystems. It is clear from a study on the technological, agricultural, health, and economic aspects that climate change would significantly affect Bangladesh's water supply, distribution, and quality. The results emphasize how critical it is to work together to address these issues and advance sustainable water management techniques in the face of environmental uncertainty.

The financial effects of climate change on water resources highlight how crucial it is to comprehend how hydrological extremes, climate variability, and economic performance are all intertwined. The economic effects of climate-related events are extensive and necessitate proactive steps to reduce risks and increase resilience. These include infrastructure development, urban water management, agriculture, and industry.

The necessity for integrated methods that give top priority to public health treatments, water sanitation, and hygiene promotion is further highlighted by the health effects of climate change on water resources. Food instability, mental health issues, and waterborne illnesses highlight the interdependence of water, health, and climate, necessitating comprehensive approaches to protecting human welfare and guaranteeing fair access to clean water and sanitation.

The effects of climate change on agricultural output also draw attention to how vulnerable rural people are and how crucial it is to implement adaptive policies that improve water efficiency, encourage climate-resilient crops, and assist smallholder farmers. Technologically enabled sustainable solutions present promising paths toward bettering water security, managing water resources, and increasing resilience to risks associated with climate change; however, their successful execution necessitates substantial institutional backing, stakeholder involvement, and research and development expenditures.

Overall, this study highlights the necessity of making coordinated efforts to mitigate the effects of climate change on Bangladesh's water resources. Bangladesh can effectively tackle the problems posed by climate change and secure water for its people and ecosystems by adopting sustainable water management practices, utilizing cutting-edge technologies, and promoting cooperation across many sectors and stakeholders. However, in order to create resilience and advance sustainable development in the face of environmental change, realizing this vision will require political will, financial investment, and coordinated actions at local, national, and international levels.

References

Abdulla, F., Abdulla, C., & Eslamian, S. (2021). Concept and technology of rainwater harvesting. In *Handbook of water harvesting and conservation: Basic concepts and fundamentals* (pp. 1–16). Wiley.

Alam, E., Momtaz, S., Bhuiyan, H. U., & Baby, S. N. (2018). Climate change impacts on the coastal zones of Bangladesh: Perspectives on tropical cyclones, sea level rise, and social vulnerability. *Bangladesh I: Climate change impacts, mitigation and adaptation in developing countries* (pp. 145–166).

Alizadeh, O. (2022). Advances and challenges in climate modeling. *Climatic Change, 170*(1), 18.

Ashrafuzzaman, M., Gomes, C., & Guerra, J. (2023). The changing climate is changing safe drinking water, impacting health: A case in the southwestern coastal region of Bangladesh (SWCRB). *Climate, 11*(7), 146.

Bell, J. E., Brown, C. L., Conlon, K., Herring, S., Kunkel, K. E., Lawrimore, J., et al. (2018). Changes in extreme events and the potential impacts on human health. *Journal of the Air & Waste Management Association, 68*(4), 265–287.

Gupta, A. D., Pandey, P., Feijóo, A., Yaseen, Z. M., & Bokde, N. D. (2020). Smart water technology for efficient water resource management: A review. *Energies, 13*(23), 6268.

Hadi, T. (2019). An analysis of water policies and strategies of Bangladesh in the context of climate change. *Asia-Pacific Journal of Rural Development, 29*(1), 111–123.

Haque, S. S., Yanez-Pagans, M., Arias-Granada, Y., & Joseph, G. (2020). Water and sanitation in Dhaka slums: Access, quality, and informality in service provision. *Water International, 45*(7–8), 791–811.

Hardoy, J., Gencer, E., & Winograd, M. (2019). Participatory planning for climate resilient and inclusive urban development in Dosquebradas, Santa Ana and Santa Tomé. *Environment and Urbanization, 31*(1), 33–52.

Hasnat, G. T., Kabir, M. A., & Hossain, M. A. (2018). Major environmental issues and problems of South Asia, particularly Bangladesh. In *Handbook of environmental materials management* (pp. 1–40). Springer.

Hoque, S. F., Hope, R., Arif, S. T., Akhter, T., Naz, M., & Salehin, M. (2019). A social-ecological analysis of drinking water risks in coastal Bangladesh. *Science of the Total Environment, 679*, 23–34.

Hossain, I., Haque, A. M., & Ullah, S. A. (2023a). Assessment of domestic water usage and wastage in urban Bangladesh: A study of Rajshahi City corporation. *The Journal of Indonesia Sustainable Development Planning, 4*(2), 109–121.

Hossain, I., Haque, A. K. M., & Ullah, S. M. (2023b). Role of city governance institutions in ensuring sustainable water: A study on Gazipur city of Bangladesh. *SSRN Electric Journal.* https://doi.org/10.2139/ssrn.4480760

Hossain, I., Ullah, S. A., & Haque, A. M. (2023c). Water and sanitation services at the local government level in Bangladesh: An analysis of SDG 6 implementation status and way forward. *Asia Social Issues, 17*(3), e265358.

Hossain, I., Haque, A. M., Kılıç, Z., Ullah, S. A., Azrour, M., & Mabrouki, J. (2024a). Navigating the challenges of urban water sustainability in Bangladesh: Towards a sustainable approach for urban governance. In *Smart internet of things for environment and healthcare* (pp. 1–25). Springer Nature Switzerland.

Hossain, I., Haque, A. M., Ullah, S. A., Azrour, M., Mabrouki, J., & Kılıç, Z. (2024b). Sustainable water management at City Corporation Level in Bangladesh: A comparative analysis between SCC and BCC. In *Sustainable and green technologies for water and environmental management* (pp. 215–237). Springer Nature Switzerland.

Hossain, I., Mahmudul Haque, A. K. M., & Akram Ullah, S. M. (2024c). Socio-economic dimensions of climate change in urban Bangladesh: A focus on the initiatives of local governing agencies. In *Climate crisis, social responses and sustainability: Socio-ecological study on global perspectives* (pp. 293–316). Springer Nature Switzerland.

Hossain, I., Haque, A. M., & Ullah, S. A. (2024d). Assessing sustainable waste management practices in Rajshahi City Corporation: An analysis for local government enhancement using IoT, AI, and Android technology. *Environmental Science and Pollution Research*, 1–19.

Hossain, I., Haque, A. M., & Ullah, S. A. (2024e). Policy evaluation and performance assessment for sustainable urbanization: A study of selected city corporations in Bangladesh. *Frontiers in Sustainable Cities, 6*, 1377310.

Hosseinzadehtalaei, P., Tabari, H., & Willems, P. (2020). Climate change impact on short-duration extreme precipitation and intensity–duration–frequency curves over Europe. *Journal of Hydrology, 590*, 125249.

Hossen, M. S., Haque, A. M., Hossain, I., Haque, M. N., & Hossain, M. K. (2024). Towards comprehensive urban sustainability: Navigating predominant urban challenges and assessing their severity differential in Bangladeshi city corporations. *Urbanization, Sustainability and Society, 1*(1), 1–17.

Howladar, M. F., Chakma, E., Koley, N. J., Islam, S., Al Numanbakth, M. A., Ahmed, Z., et al. (2021). The water quality and pollution sources assessment of Surma river, Bangladesh using, hydrochemical, multivariate statistical and water quality index methods. *Groundwater for Sustainable Development, 12*, 100523.

Hussain, S., Aslam, M. U., Javed, M., Zahra, M., Ejaz, H., Kubra, K. A., & Mushtaq, I. (2021). Impact of climatic changes and global warming on water availability. *Anthropogenic Pollution, 5*(2), 57–66.

IPCC. (2021). Climate change 2021: The physical science basis. Contribution of working group I to the sixth assessment report of the intergovernmental panel on climate change.

Islam, M. N., van Amstel, A., Islam, M. N., Tamanna, S., van Amstel, A., Noman, M., ... & Ghosh, A. (2021). Climate change impact and comprehensive disaster management approach in Bangladesh: A review. *Bangladesh II: Climate change impacts, mitigation and adaptation in developing countries* (pp. 1–39).

Kabir, M. H., & Hossen, M. N. (2019). Impacts of flood and its possible solution in Bangladesh. *Disaster Advances, 12*(10), 48–57.

Katusiime, J., & Schütt, B. (2020). Integrated water resources management approaches to improve water resources governance. *Water, 12*(12), 3424.

Kundzewicz, Z. W., Krysanova, V., Benestad, R. E., Hov, Ø., Piniewski, M., & Otto, I. M. (2018). Uncertainty in climate change impacts on water resources. *Environmental Science & Policy, 79*, 1–8.

Makanda, K., Nzama, S., & Kanyerere, T. (2022). Assessing the role of water resources protection practice for sustainable water resources management: A review. *Water, 14*(19), 3153.

Miah, M. D., Hasan, R., & Uddin, H. (2020). Agricultural development and the rural economy: The case of Bangladesh. In *Bangladesh's economic and social progress: From a basket case to a development model* (pp. 237–266). Springer.

Mishra, B. K., Kumar, P., Saraswat, C., Chakraborty, S., & Gautam, A. (2021). Water security in a changing environment: Concept, challenges and solutions. *Water, 13*(4), 490.

Mullick, M. R. A., Nur, R. M., Alam, M. J., & Islam, K. A. (2019). Observed trends in temperature and rainfall in Bangladesh using pre-whitening approach. *Global and Planetary Change, 172*, 104–113.

Muluneh, M. G. (2021). Impact of climate change on biodiversity and food security: A global perspective—A review article. *Agriculture & Food Security, 10*(1), 1–25.

Namany, S., Al-Ansari, T., & Govindan, R. (2019). Sustainable energy, water and food nexus systems: A focused review of decision-making tools for efficient resource management and governance. *Journal of Cleaner Production, 225*, 610–626.

Nichols, G., Lake, I., & Heaviside, C. (2018). Climate change and water-related infectious diseases. *Atmosphere, 9*(10), 385.

Parven, A., Pal, I., Witayangkurn, A., Pramanik, M., Nagai, M., Miyazaki, H., & Wuthisakkaroon, C. (2022). Impacts of disaster and land-use change on food security and adaptation: Evidence from the delta community in Bangladesh. *International Journal of Disaster Risk Reduction, 78*, 103119.

Pelletier, M. C., Ebersole, J., Mulvaney, K., Rashleigh, B., Gutierrez, M. N., Chintala, M., et al. (2020). Resilience of aquatic systems: Review and management implications. *Aquatic Sciences, 82*, 1–25.

Pokhrel, S. R., Chhipi-Shrestha, G., Hewage, K., & Sadiq, R. (2022). Sustainable, resilient, and reliable urban water systems: Making the case for a "one water" approach. *Environmental Reviews, 30*(1), 10–29.

Pradhanang, S. M., & Jahan, K. (2021). Urban water security for sustainable cities in the context of climate change. *Water, Climate Change, and Sustainability*, 213–224.

Rabbani, G., Munira, S., & Saif, S. (2018). Coastal community adaptation to climate change-induced salinity intrusion in Bangladesh. In *Agricultural economics-current issues*. IntechOpen.

Rahaman, M. A., Kalam, A., & Al-Mamun, M. (2023). Unplanned urbanization and health risks of Dhaka City in Bangladesh: Uncovering the associations between urban environment and public health. *Frontiers in Public Health, 11*, 1269362.

Rahman, M. M., Ahmad, S., Mahmud, A. S., Hassan-uz-Zaman, M., Nahian, M. A., Ahmed, A., et al. (2019). Health consequences of climate change in Bangladesh: An overview of the evidence, knowledge gaps and challenges. *Wiley Interdisciplinary Reviews: Climate Change, 10*(5), e601.

Rahman, M. M., Chowdhury, M. M. I., Al Amran, M. I. U., Malik, K., Abubakar, I. R., Aina, Y. A., & Rahman, S. M. (2024). Impacts of climate change on food system security and sustainability in Bangladesh. *Journal of Water and Climate Change, 15*(5), 2162–2187.

Rahman, M. C., Rahaman, M. S., Biswas, J. C., Rahman, N. M. F., Islam, M. A., Sarkar, M. A. R., et al. (2023). Climate change and risk scenario in Bangladesh. *Asia-Pacific Journal of Regional Science, 7*(2), 381–404.

Rashid, M. (2023). *Bangladeshi laws, policies, and plans to mitigate and adapt to climate change*. Springer.

Sánchez, A. S., Oliveira-Esquerre, K. P., dos Reis Nogueira, I. B., de Jong, P., & Filho, A. A. (2020). Water loss management through smart water systems. In *Smart village technology: Concepts and developments* (pp. 233–266). Springer.

Sangadi, P., Kuppan, C., & Ravinathan, P. (2022). Effect of hydro-geochemical processes and salt-water intrusion on groundwater quality and irrigational suitability assessed by geo-statistical techniques in coastal region of eastern Andhra Pradesh, India. *Marine Pollution Bulletin, 175*, 113390.

Skoulikaris, C., Filali-Meknassi, Y., Aureli, A., Amani, A., & Jiménez-Cisneros, B. E. (2018). Information-communication technologies as an integrated water resources management (IWRM) tool for sustainable development. In *Achievements and challenges of integrated river basin management* (pp. 179–181).

Souliotis, I., & Voulvoulis, N. (2022). Operationalising nature-based solutions for the design of water management interventions. *Nature-Based Solutions, 2*, 100015.

Stefanakis, A. I. (2019). The role of constructed wetlands as green infrastructure for sustainable urban water management. *Sustainability, 11*(24), 6981.

UNEP (2018). Annual Report 2018. https://www.unep.org/resources/annual-report-2018

UN-Habitat (2020). Annual Report 2020. https://unhabitat.org/sites/default/files/2021/05/annual_progress_report_2020_final.pdf

Vannevel, R., & Goethals, P. L. (2020). Identifying ecosystem key factors to support sustainable water management. *Sustainability, 12*(3), 1148.

World Bank (2020). Climate Risk Country Profile: Bangladesh. https://climateknowledgeportal.worldbank.org/sites/default/files/country-profiles/15502-WB_Bangladesh%20Country%20Profile-WEB.pdf

WHO (2021). World Health Statistics: Monitoring Health for the SDGs. https://iris.who.int/bitstream/handle/10665/342703/9789240027053-eng.pdf

Xie, Y., Zhang, Y., Lan, H., Mao, L., Zeng, S., & Chen, Y. (2018). Investigating long-term trends of climate change and their spatial variations caused by regional and local environments through data mining. *Journal of Geographical Sciences, 28*, 802–818.

Zhang, D., Sial, M. S., Ahmad, N., Filipe, A. J., Thu, P. A., Zia-Ud-Din, M., & Caleiro, A. B. (2020). Water scarcity and sustainability in an emerging economy: A management perspective for future. *Sustainability, 13*(1), 144.

Sustainable Water Management Practices Using Micro-irrigation Technology: A Case Study in Portugal and Future Research Directions

Artur Frazão, Pedro Dinis Gaspar, Cristina Fael, Isabel Falorca, Dina Pereira, and João Carlos Correia Leitão

Abstract Faced with an escalating scarcity of fresh water, the responsible management of water resources has become fundamentally important, especially in the agricultural sector, which is a major consumer. This study evaluates the sustainability and economic viability of a new micro-irrigation system, focusing on the self-regulating low-energy clay-tube irrigation system (SLECIS), and compares its performance with traditional drip irrigation systems (DISs) in Mediterranean climates, particularly in Portuguese cherry and peach orchards. The study's methodology includes a comparative analysis of the irrigation season, weather information (average air temperatures, average air humidity, soil water content, and evapotranspiration), and empirical data (water consumption and water flows) to assess water use efficiency, crop yield, and the long-term environmental impacts of each system

A. Frazão
University of Beira Interior, C-MAST—Center for Mechanical and Aerospace Sciences and Technologies, Covilhã, Portugal
e-mail: artur.frazao@ubi.pt

P. D. Gaspar (✉)
Dept of Electromechanical Engg, University of Beira Interior, COVILHÃ, Portugal
e-mail: dinis@ubi.pt

C. Fael · I. Falorca
University of Beira Interior, C-MADE—Centre of Materials and Building Technologies, Covilhã, Portugal
e-mail: cmsf@ubi.pt; ifalorca@ubi.pt

D. Pereira
CEG-IST, University of Lisbon, Lisboa, Portugal
e-mail: dina@ubi.pt

J. C. Correia Leitão
Department of Management and Economics, University of Beira Interior, Covilhã, Portugal
e-mail: jleitao@ubi.pt

J. C. Correia Leitão et al. (eds.), *Sustainable Water Management and Irrigation Systems*, Natural Resource Management and Policy 62,
https://doi.org/10.1007/978-3-031-93587-9_9

as regards water consumption and management. The results demonstrate that SLECIS, with its adaptive response to soil moisture levels, has higher water conservation rates than DIS. The study concludes that SLECIS may represent a strategic advancement towards sustainable agriculture, offering a viable solution to the mounting pressures of water scarcity and the growing demand for agricultural output. Further research is recommended to explore the scalability of SLECIS across various crops and more diverse climatic conditions.

Keywords Drip irrigation · Micro-irrigation systems · SLECI · Water efficiency

1 Introduction

Over the last few decades, the agricultural sector has been forced to adopt measures, regulations, and new technologies focusing on efficiency and productivity, thereby compromising the inherent resilience and adaptability naturally found in this economic sector, as well as its biological diversity in plants and wildlife (Outhwaite et al., 2022). The world population is growing as never before (Sadigov, 2022), and besides socio-economic consequences, countries where these effects are more visible are usually characterized by a generally low average income, multiple food safety problems, declining soil fertility due to over-exploitation, and increasing environmental degradation (Maja & Ayano, 2021; Balogh & Jámbor, 2020). Rudel (2020) states that this problem requires research on the sustainable intensification and industrialization of the agricultural sector, which is an ongoing process. Theoretically, this would increase agricultural production without damaging the surrounding ecosystems further, but in practice, various problems are encountered, such as a difficult trade-off between agricultural development and environmental protection and raising producers' awareness of this sustainable intensification and its economic and environmental benefits. This, coupled with the intensive monoculture preferred by large-scale producers, which has contributed to the global food supply for decades, has significantly impacted environmental sustainability. Even though crop diversity has been identified as one of the most important factors in stabilizing grain production globally, long-term studies also show that it stabilizes grain production at the field level (Li et al., 2021). The consequences of this widespread monoculture among producers include the homogenization of rural landscapes and increased erosion processes (Brychta & Podhrázská, 2020), soil contamination and reduced fertility (Purwanto & Alam, 2020), reduction in fauna and flora due to extensive fertilizer use (Raven & Wagner, 2021), and the negative effects of chemical pesticides on the health of soils and the public (Bisht & Chauhan, 2020).

Added to these problems on biodiversity among crops are other sectoral challenges related to an exponential decrease in the labour available (Hill et al., 2021), a rising consumer demand for transparent products and eco-friendly crops, and an increasing desire for high-quality agricultural products (Maheshwari, 2024).

Therefore, there is an urgent need to change the agricultural production paradigm to minimize environmental impacts and promote local fauna and flora without compromising crop productivity. According to Stein-Bachinger et al. (2022), a more heterogenic landscape (promoting semi-natural habitats and diversified croplands), a sustainable farming system (promoting crop choice considering rotation, animal husbandry, and soil management), and more sustainable crop management (considering the type and number of fertilizers and the frequency of use and toxicity of pesticides) will naturally interact, increasing the biodiversity of crops and, consequently, their yield.

For six decades, the Common Agricultural Policy (CAP) has been pivotal in ensuring food safety and fostering agricultural productivity in the European Union (EU) (European Commission, 2022a, b). Originally designed to address post-World War II food shortages and market instability, the CAP has continued to adapt to changing economic, social, and environmental contexts. Transitioning towards a model that supports a barrier-free market with common standards and high-quality products reflects the EU's broader goals of economic integration and consumer protection. In 2021, the CAP strategic plan was formally adopted by the European Council for the 2023–2027 period (European Commission, 2022a, b). This reviewed document still focuses on specific objectives linked to the common EU goals of social, environmental, and economic sustainability in the agricultural sector. The strategic plan aims to contribute to climate change mitigation and adaptation, promote sustainable energy development and efficient management of natural resources, and halt and reverse the loss of biodiversity while enhancing ecosystem services. There is also an alignment with the European Green Deal's multiple objectives and strategies by targeting pesticide use and nutrient loss in the member states through tight control by the European Commission and committing to reduce these parameters by 50% by 2030. This will include reducing nutrient runoff into water bodies; encouraging the expansion of organic farming (member states are encouraged to increase organic farming coverage by 25% by 2030); promoting high-diversity landscapes to mitigate nutrient runoff and boost local biodiversity (member states are expected to bring at least 10% of agricultural areas under such conditions); implementing good environmental and agricultural condition (GAEC) standards, such as GAEC 4, which requires the establishment of buffer strips alongside streams to prevent pollution from fertilizers and pesticides, and GAEC 7, which mandates crop rotation to improve the health of European agricultural soil; and reducing the use of chemical inputs (member states must meet these GAEC standards in order to receive financial support and incentives).

The symbiotic relationship between agriculture and the environment is crucial for sustainability and rural development. According to Eurostat (2023), there were 9.1 million farms in the EU in 2020, utilizing 1.55 million hectares of available land for agricultural production, meaning that almost two-fifths (37.8%) of the EU's total land area was being farmed. García-Tejero et al. (2014) emphasize the importance of responsible land and water management to preserve natural habitats and embrace environmental sustainability in the agricultural sector. However, improper practices can degrade soil, water, and biodiversity. Sustainable agriculture should prioritize responsible soil and water management, implementing techniques to meet current and future needs. Preserving resources without compromising environmental identities is essential, emphasizing a rational application of technology and economic sustainability. Even though FAO (2023) was issued 9 years later, it still agrees with this prioritizing and resource allocation. Some key policies to achieve the major objective of global environmental sustainability in the agricultural sector were suggested. Well-established practices widely recognized as potentially harmful to the surrounding environment have been reformed. More recently, Pereira et al. (2023) demonstrated the importance of adopting low-cost smart irrigation systems to mitigate the effects of climate change conditions. This can be achieved by increasing investment in research and development by improving technologies and practices, redirecting agricultural support, and leveraging trade for sustainable development. Aligning policies and incentives with long-term objectives and rolling back measures that distort markets and trade are also among the suggested measures.

Figure 1 illustrates the increasing focus on this symbiosis in the European Commission's policies since 1957.

When delving into the study and assessment of irrigation problems, it is essential to first understand the type of climate in question. In this case study, we will examine the Mediterranean climate, and we already have information available regarding

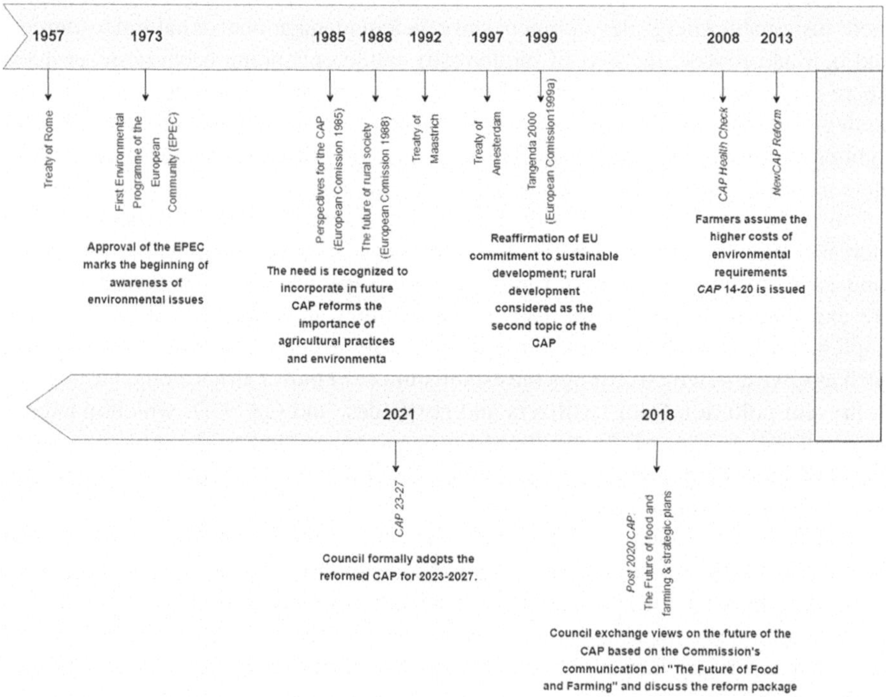

Fig. 1 Timeline of European policies regarding agricultural and sustainability problems in member states. (Adapted from García-Tejero et al., 2014; European Council, n.d.)

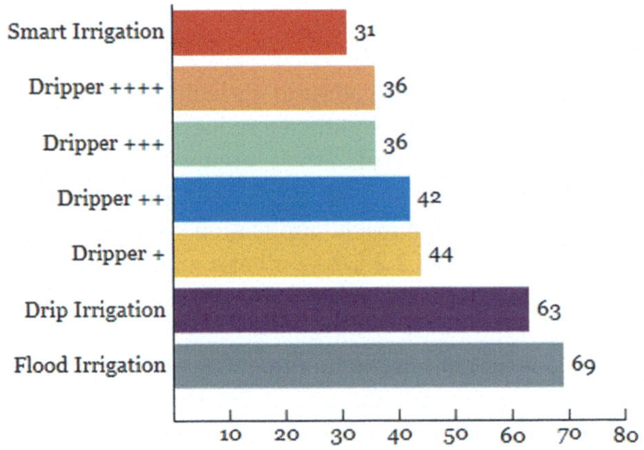

Fig. 2 Effect of irrigation technification on the water footprint of an apple. (Adapted from Blueberries Consulting Magazine, 2019)

water consumption for apple trees, for instance, as shown in Fig. 2, with water consumption in liters per fruit. This kind of climate can pose challenges for agricultural irrigation due to the limited availability of water for any crop or way of life. Additionally, climate change (Girona et al., 2010) and irregularities in rainfall patterns across different seasons (FAO, 2009) make this Mediterranean climate highly unpredictable. Therefore, water consumption for most crops in this climate becomes equally unpredictable.

The current chapter is structured as follows. In the second section, irrigation techniques and technologies of reference are briefly described. In the third section, self-regulating low-energy clay-tube irrigation (SLECI) technology and the characteristics of irrigation systems are presented. In the fourth section, field tests and the installation process are described, taking as reference the experimental fields organized in Portugal, applied to cherry and peach trees. The fifth section contains the main results and the corresponding discussion. The last section presents the conclusions based on the evidence obtained, as well as suggestions for future studies.

2 Irrigation Techniques and Technologies

Currently, the main technologies used in Portuguese agriculture include conventional drip irrigation, deficit irrigation, sub-surface irrigation (Galego et al., 2021), and sprinkler or centre-pivot irrigation.

2.1 Drip Irrigation

The widespread adoption of drip irrigation in agriculture has revolutionized water conservation in the sector, achieving significant reductions in water consumption by approximately 25–35% (Darouich et al., 2014). Figure 3 illustrates its application in vineyards using a rigid hose with an external dripper. Drip irrigation offers adaptability to varied terrains, precise nutrient application without clogging risks, and a low operational flow rate (around 2.2 L/h)—all at a relatively low installation cost. However, converting from other irrigation methods can be costly (Shock, 2006). Various drip irrigation typologies are available according to producers' specific needs, with design considerations including terrain topography, recommended water pressures, irrigation area, soil textures, and filtration capacity to prevent clogging. Effective planning involves ensuring uniform water distribution along the tape by employing weighted selection, determining appropriate irrigation lengths, and performing periodic cleaning for consistent results.

2.2 Deficit Irrigation

Due to water scarcity at certain times of the year, producers can choose to apply less water than the recommended amount for a specific crop during those periods or in spells of drier weather. This decision should consider the plant's developmental cycles to minimize the impact of this deficit irrigation on overall production and harvest (González-Altozano & Castel, 2003; De la Rosa et al. 2016; Pérez-Pérez et al., 2014).

Romero et al. (2022) provide a review focusing on the state of the art of various physiological irrigation strategies aimed at conserving significant amounts of water

Fig. 3 SLECI components before installation

and improving the use of this natural resource (water use efficiency (WUE) yield), as well as improving grape and wine quality. Strategies such as deficit irrigation (DI), regulated deficit irrigation (RDI), and partial root-zone drying (PRI) in vineyards are evaluated to enhance WUE yield, grape quality, and water management efficiency.

2.3 *Low-Flow Localized Irrigation*

Low-flow localized irrigation, also known as nano-irrigation, stands out as one of the best technologies for water retention in the soil. The drippers have a flow rate of around 0.7 L/h (Galego et al., 2021), an extremely low value, enabling the irrigation of large crop areas with minimal water and energy consumption. These technologies also have low installation and maintenance costs, thanks to their operation at very low pressure.

2.4 *Sub-surface Irrigation*

This type of technology is still under research, and one of the advantages already highlighted is that it allows direct irrigation at the plant's root, reducing water losses due to evaporation. On the negative side, this technology comes with higher installation costs, an increased risk of damage from wildlife in the area (such as rodents or moles), and a high likelihood of clogging due to the accumulation of salts and sediments in the drippers (Galego et al., 2021).

2.5 *Sprinkler or Centre-Pivot Irrigation*

This technology is very useful and versatile, particularly for irrigating large areas, especially if they have uneven topography. However, without proper sizing and protection, this system can become erosive to both the soil and the crops being irrigated. Given that most Mediterranean terrains have irregular topography, the limited water flow infiltrating the soil is a constant concern, leading to high rates of water dispersion, soil erosion, and crop damage.

In this technology, the importance of proper sizing is emphasized. While the installation cost is not particularly high, inadequate sizing and implementation by an irrigation engineer can result in major water losses (de Sousa et al., 1999). Research results presented by Ho et al. (2022) confirm that irrigation and its water efficiency are significantly low for coffee crops, suggesting that more advanced technologies could enhance this efficiency. Nevertheless, coffee crops utilizing this irrigation system demonstrate greater efficiency than those relying on micro-basin technologies.

2.6 Soil Oxygenation (CO_2 Nano-bubbles)

According to Wang et al. (2023), soil compaction affects about 68 million hectares globally due to the use of heavy machinery in agriculture. The technology discussed aims to balance soil properties and combat compaction rather than solely address water scarcity. Soil oxygenation, which is crucial for this approach, is achieved through irrigation water using techniques such as direct oxygenation or the introduction of macro-, micro-, or nano-bubbles (MNBs).

Khan et al. (2020) simplify the MNB irrigation system, where gas dissolves in pressurized water, and the key components include a gas tank for water oxygenation, a water pump, and an air compressor. High-pressure gas-water mixing allows supersaturation, and gas precipitates as MNBs through sudden decompression. There are three types of bubbles used in this technology: macro-bubbles, micro-bubbles (MBs), and common bubbles. Macro-bubbles quickly reach the water surface and collapse, while micro-bubbles, especially those <100 nm in size, remain suspended for an extended period, altering water properties. NBs are applied for soil oxygenation, a process that requires significant investment. Research focuses on generating and injecting O_2, CO_2, N_2, and O_3 NBs. They provide advantages, such as increased solubility and the promotion of plant growth, biomass, and soil quality. Zhu et al. (2021) show promising results in utilizing CO_2 for agriculture post-combustion, contributing to climate change mitigation. Achieving the efficient use of post-combustion CO_2 remains a challenge. The authors also provide final yield results using CO_2-based fertilizers in parameters, such as plant growth, CO_2 concentrations around leaf plant height, and biomass when compared to other fertilizers.

2.7 Responsive Drip®

The *Responsive Drip*® or Responsive Drip technology combines low-flow localized and subsurface irrigation. Owned by Responsive Drip Irrigation (RDI), it uses responsive piping *GrowStream*™ to meet plant irrigation and fertilization needs. Akbar et al. (2023) outline two essential stages: root signals and *GrowStream*™ response, which initiate the irrigation process. Below the surface, plants adapt to environmental signals for nutrient absorption. *GrowStream*™ technology creates symbiosis with plant roots, soil, and water resources through about 100,000 nano-pores per square centimetre. This low-pressure, cost-effective system ensures long-lasting, balanced irrigation.

Although it is successfully tested in various countries and climates, challenges remain, including high installation costs, limited availability of remote parts, and insufficient data on fauna durability, soil disturbances, pressure drops, and system clogging.

3 The SLECI Technology and Irrigation System

Fired clay has been around agricultural irrigation technologies for centuries. It all started in ancient Greece with the Oya Technology. SLECI stands for self-regulating low-energy clay-tube irrigation, and its clay components (shown in Fig. 3) utilize the same concept as the Oya's when it comes to irrigation terms, making use of its porous walls to properly interpret moisture humidity signals (and so providing self-regulation) and saving energy in the water flow process. In short, this technology combines low-flow localized irrigation, deficit irrigation, and underground irrigation all in one.

The essential components for this irrigation system are water deposit, 50 and 1 μm water filters to prevent system clogging, a thin-walled *DN16* irrigation hose, and one *SLECI* element per tree.

The *SLECI* elements must be buried and covered in a fine sand (Fig. 4a) borehole of 6–8mm diameter (Fig. 4b) drilled with a gasoline-powered auger (Fig. 4c). The main purpose of this sand is to allow water to penetrate deeper into the soil, reaching a root zone different from that of conventional irrigation systems—about 60 cm deep in the soil. In this way, only the fruit trees concerned are irrigated, and the probability of wasting water, by irrigating undesirable weeds in the fields, drops.

A schematic side view of the *SLECI* after installation is presented in Fig. 5.

Once these elements are all set according to Fig. 6 and a water pressure of around 0.2 bar (2.9 psi) at the end of the line is ensured, irrigation can properly start.

a) Fine sand use to fill the bore holes and embed the SLECI element.

b) Bore hole.

c) Gasoline-powered auger to drill the borehole.

Fig. 4 SLECI borehole drilling process and the components used. (**a**) Fine sand used to fill boreholes and embed the SLECI element. (**b**) Borehole. (**c**) Gasoline-powered auger to drill the borehole

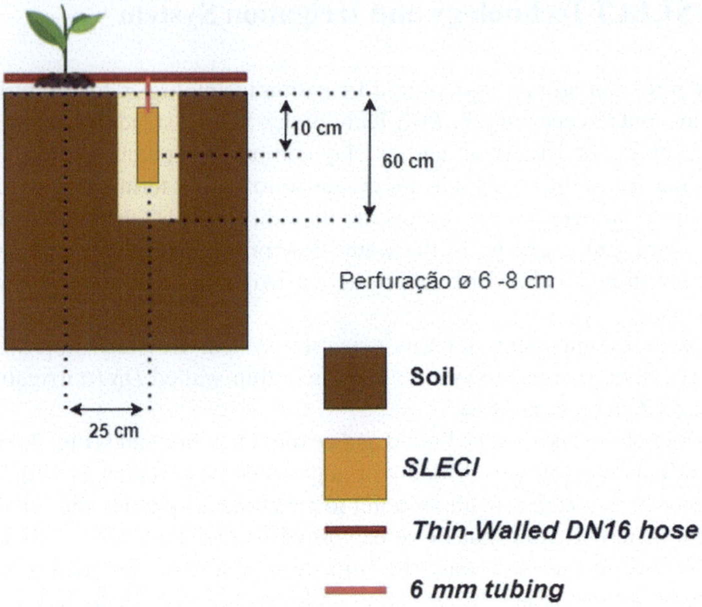

Fig. 5 Schematic side view of the SLECI after installation

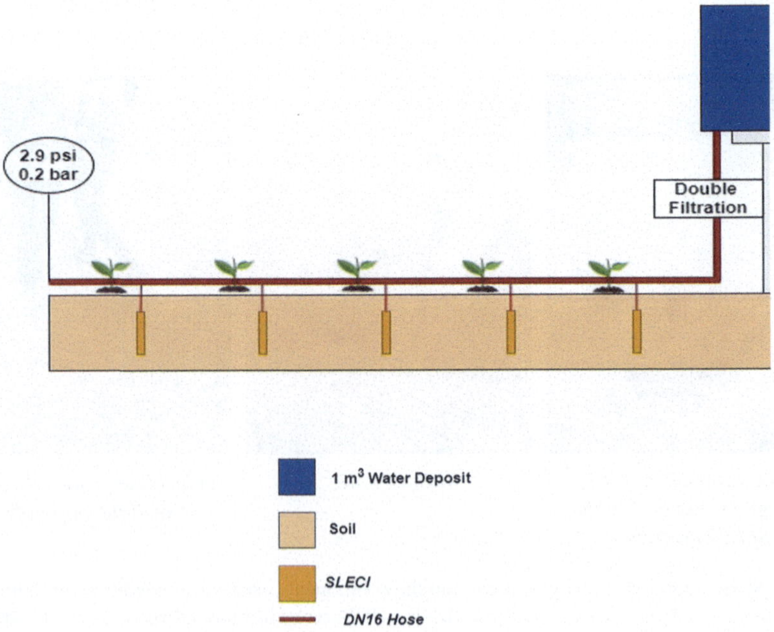

Fig. 6 Schematic representation of the SLECI after installation

Fig. 7 Extra accessories and components needed to get the system up and running on Portuguese test fields

It is important to note that depending on the situation, the characteristics of the terrain, and the presence of any pre-existing systems, as seen in Fig. 7, additional irrigation accessories and components, such as T connectors or taps, may be required in order to set up the system.

4 Field Tests and System Installation

The main purpose of both cherry and peach tree test fields is to compare the *SLECI* technology with the technology most found among agricultural producers—*DIS*. The test fields for cherry trees and peach trees were located at Seminário do Fundão (40.13347763178621, −7.483322202207046) (Fig. 8a) and Mata da Rainha in Catrão (40.1190023, −7.3749630) (Fig. 8c), respectively. Information for the test fields is summarized in Table 1, and the installation process occurred from 13 to 16 February 2023.

4.1 Cherry Trees

Installation at the cherry tree test field was carried out on 13 and 14 February 2023. The field consisted of 80 trees, around 4 years old, arranged in 4 rows with 20 trees per row, each with an irrigation line extension of roughly 20 m, with the trees spaced 2 m from each other, as shown in Fig. 9. The two irrigation systems (*SLECIS* and *DIS*) are distributed across the four irrigation lines, two for each, which means 40 trees per irrigation system.

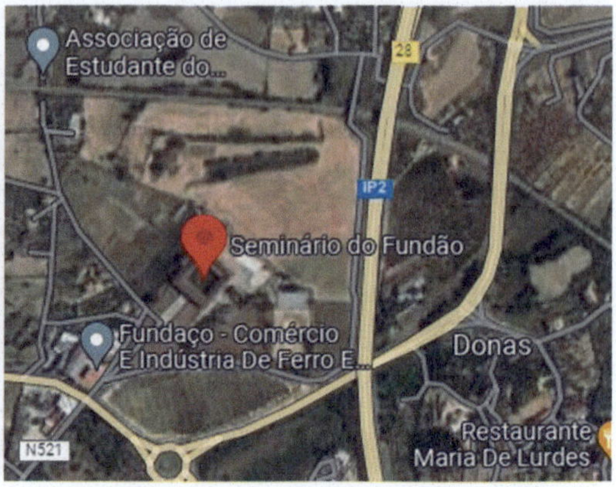

a) Cherry trees field test. *b) Cherry field test location.*

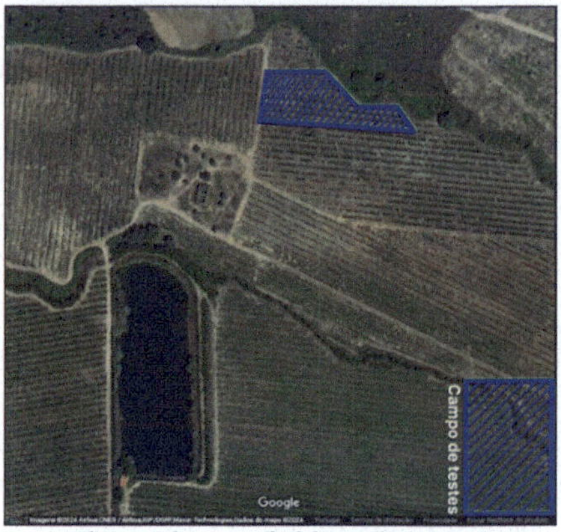

c) Peach trees field test. *d) peach trees field test location.*

Fig. 8 Portuguese test field site location. (**a**) Cherry tree test field at Seminário do Fundão. (**b**) Peach tree test field at Mata da Rainha. (**a**) Cherry tree field test. (**b**) Cherry field test location. (**c**) Peach tree field test. (**d**) Peach tree field test location

Table 1 Summary of the test fields

	Field test 1	Field test 2
Climate[a]	Hot-summer Mediterranean (Csa)	Hot-summer Mediterranean (Csa)
Soil type	Clay	Loamy clay
Water origin	Basin	Basin
Crop grown	Cherry	Peach

[a] Classification of Köppen-Geiger

The system's water source is a water basin, which is pumped for both systems after passing through a 130 μm water filter (no sand filter). Then, a T connection splits the flow between the *DIS*, with water being directly used for irrigation after

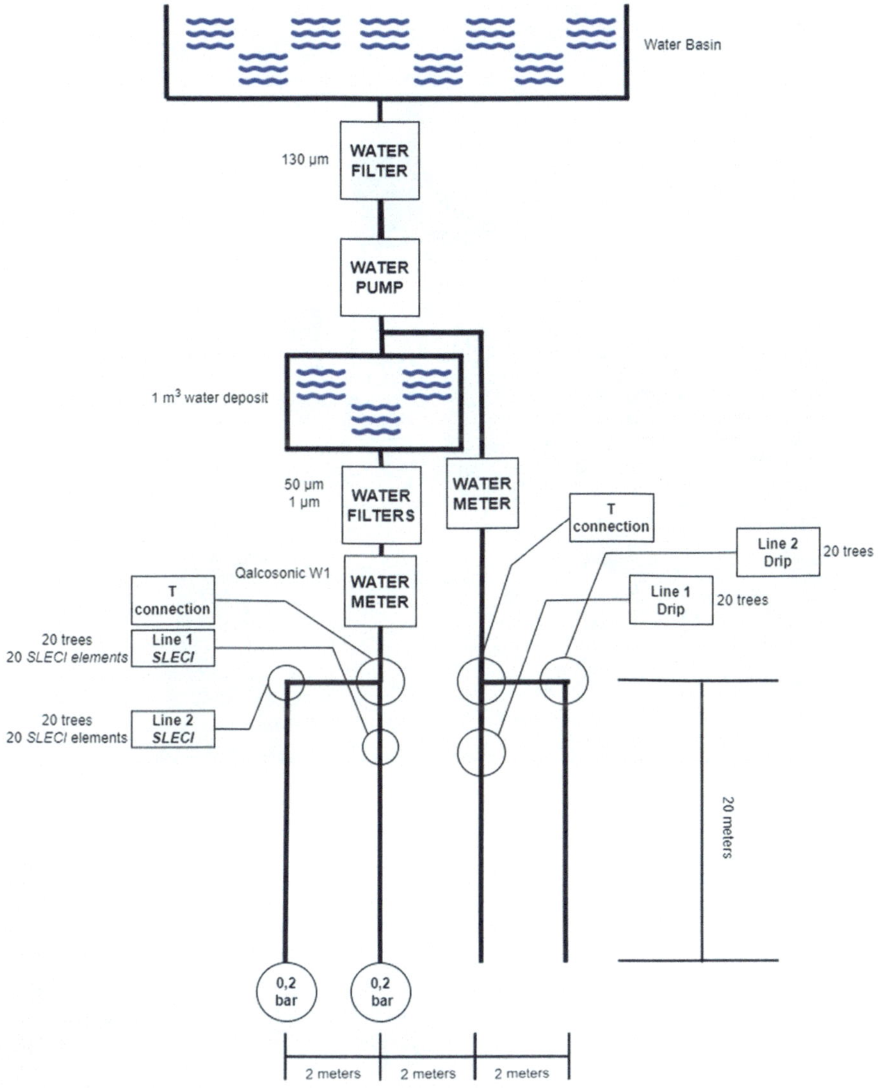

Fig. 9 Irrigation scheme for the cherry tree irrigation test field

filtration and passing through a water meter, and the *SLECIS*, which stores water in a 1 m^3 deposit (Fig. 10).

After the water deposit, a double filtration system is implemented, as shown in Fig. 11, so that water reaches the clay elements with the fewest particles possible, preventing system clogging during the irrigation process.

The outlet of the filtration system connects directly to the smart water meter, *Qalcosonic W1*, as shown in Fig. 12a, which monitors both systems for water

Fig. 10 1 m³ water deposit, assembled in the treated steel structure at a height of 2 m

Fig. 11 Double filtration cage, to be fixed at the outlet of the water deposit, with the respective 50 and 1 μm filters

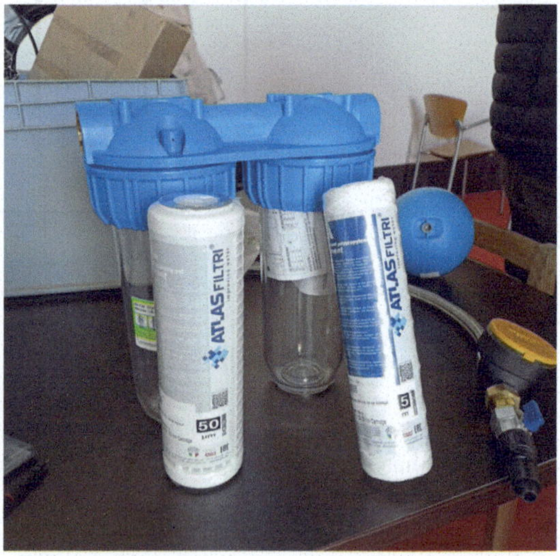

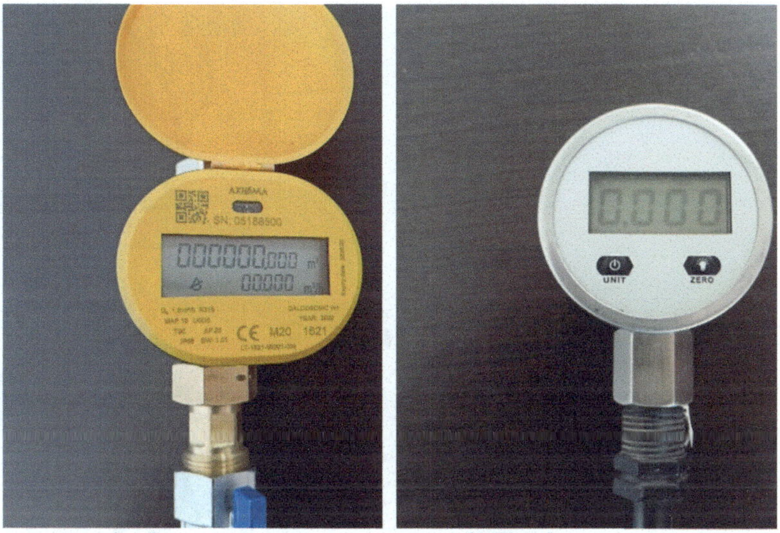

a) *Smart water meter.* b) *Digital pressure gauge.*

Fig. 12 Monitoring and installation of electronics. (**a**) Smart water meter. (**b**) Digital pressure gauge

consumption, water flows, and air temperature, among other parameters less relevant for this case study.

When we follow the *SLECI* technology water flow starting at the water deposit, it goes through 50 and 1 μm double filtration, then a smart water meter (Fig. 12a), followed by a T connection to split the flow between both irrigation main lines. At the time of installation, irrigation was ensured by the SLECI elements. In addition, the pressure was checked with a digital water pressure gauge (Fig. 12b), and everything was prepared for the irrigation season, which started in May 2023.

For this test field, the landscape had no slope, so the deposit was set up on a treated steel structure about 2 m high. In this way, the water deposit is able to perform its vital function for the system, reducing the hydraulic pressure to the desired 0.2 bar. The deposit is also filled via a float valve system that is directly connected to the water basin source.

Figure 13 provides a perspective on the *SLECI* elements before being buried, to perform the test previously described.

4.2 Peach Trees

Installation at the peach tree test field (see Fig. 14) was performed on 15 and 16 February 2023.

The test field segment for the monitoring tests consisted of three irrigation main lines, making a total of 97 already developed and pruned peach trees. It was decided

Fig. 13 SLECIS main line before installation underground, at the moment of the test run

Fig. 14 SLECIS installation on the peach tree test field

that two rows consisting of 49 trees should be irrigated with the *SLECI* technology, and then one row with a total of 48 trees should use the *DIS*. Figure 15 shows the detailed irrigation scheme of the test field, similar to the approach adopted in the cherry tree test field. Despite the similarities, there is a slight slope in the terrain.

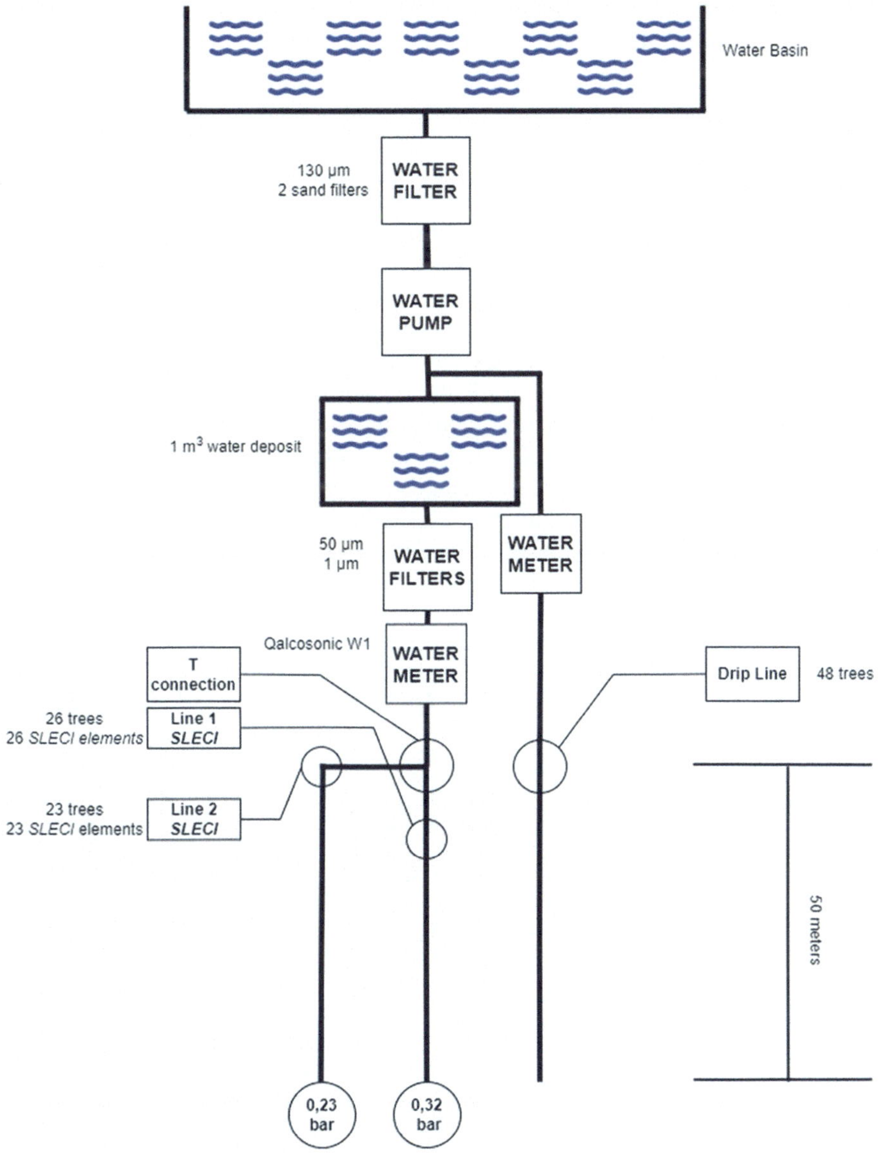

Fig. 15 Irrigation scheme for the peach tree test field

However, it was decided to fix the 1 m^3 water deposit at a height of 2 m, with an extension of 50 m on both *SLECI* lines.

For this test field, the water source is also a basin (Fig. 16) where water is pumped and flows through a 130 μm water filter and two sand filters (Fig. 17). For the rest of the system, the components and procedures were the same on both test fields. In the case of this test field, the water deposit was directly connected to the *DIS*, so every time the system went on, it started filling the 1 m^3 water deposit. It is worth mentioning that by the end of the installation process the two *SLECIS* irrigated lines presented different end-line pressures, as shown in Table 2.

5 Results and Discussion

In this section, the results regarding weather information and technical data provided by the smart water meters are presented, together with a status report on both crops for the 2023 irrigation period. This case study aims to evaluate the sustainability of a new SLECI micro-irrigation system under representative Mediterranean soil and climate conditions, using a test site located in Portugal. To achieve this, the approach adopted was to cross-check the weather information provided by the Castelo Branco weather station (available at the Instituto Português do Mar e da Atmosfera (IPMA))—such as soil water content (through *field capacity* (θ_{fc}) percentages), evapotranspiration (ET_0), average air temperatures, average air humidity, daily rainfall, and cumulative rainfall—against the technical data supplied by the water meters, such as water consumption for the drip irrigation system (*DIS*) and *SLECIS* and the water flow on the *SLECIS*.

Fig. 16 Peach tree basin, water source for irrigation of the test field

Fig. 17 Water filtration between the water source and the 1 m^3 water deposit

Table 2 Pressure differences in the SLECI lines in the peach tree test field

SLECIS line	Pressure
Line 1	0.32 bar
Line 2	0.23 bar

5.1 Weather and Technical Irrigation Data

Concerning weather information, the data considered have direct effects on agriculture and the hydric resources needed to satisfy each crop's needs, such as average air temperature, average air humidity, daily rainfall, and cumulative rainfall for the irrigation campaign period. Figures 18 and 19 show complete information on the cherry tree and peach tree test fields, respectively.

It must be noted that the 2023 irrigation campaign included out-of-season rainfall during the late spring/early summer period, which happened three times during the irrigation season in May, June, and September with a cumulative rainfall of 45.7, 47.3, and 86.5 mm, respectively. Air temperatures and air humidity values followed the naturally expected tendencies.

Throughout the irrigation season, there are some considerable heat peaks. At a heat peak, average air temperatures rise, and there is a natural drop in average air humidity percentage values, as shown in Fig. 20. Therefore, and because of the self-regulation and natural water flow regulation that the *SLECI* elements provide, above-average water consumption and water flow should be expected for these periods. On the other hand, in periods with more cumulative rainfall, less water and less water flow should be registered on the smart water meters.

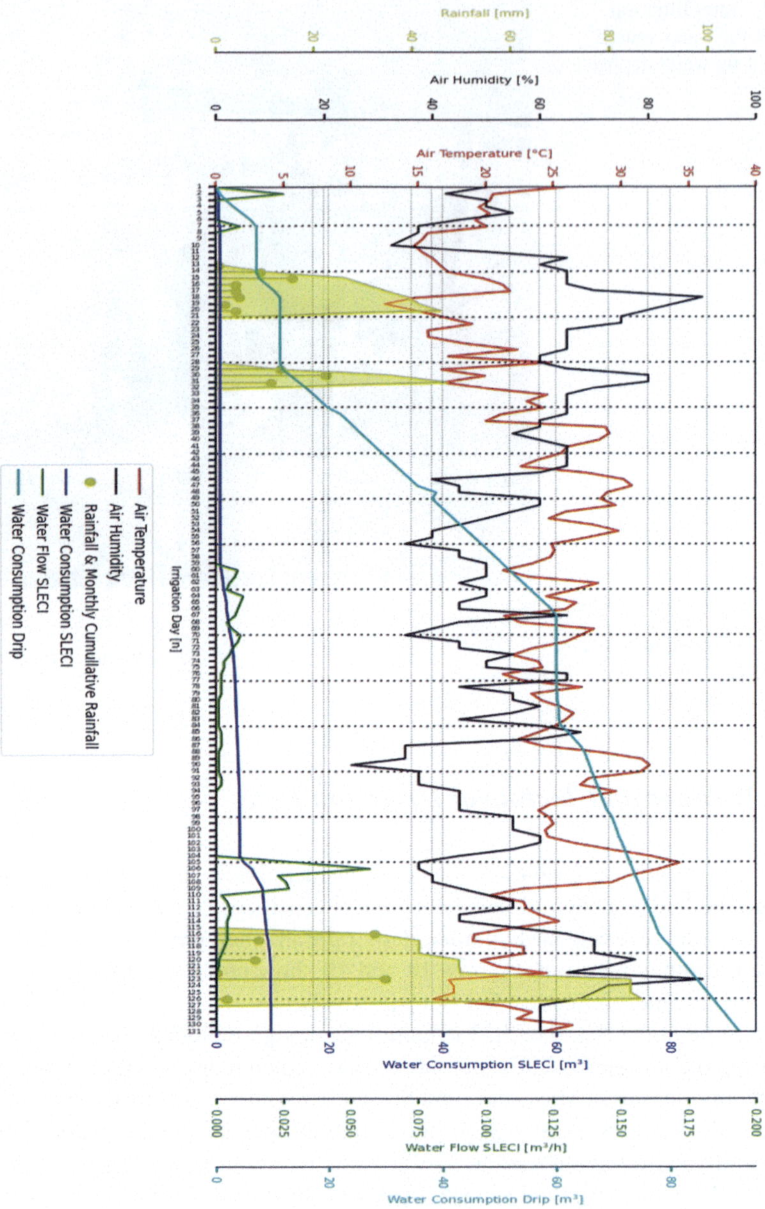

Fig. 18 Cherry tree test field and technical irrigation information

In light of this weather information, some key events must be analysed more carefully and compared with the irrigation technical data from the water meters, namely rainfall during irrigation weeks 3, 5, and 17/18 for the cherry crop but only week 12/13 for the peach crop, as well as the heatwaves in weeks 7, 13, and 16 for the cherries and weeks 2, 8, and 10 for the peach crop.

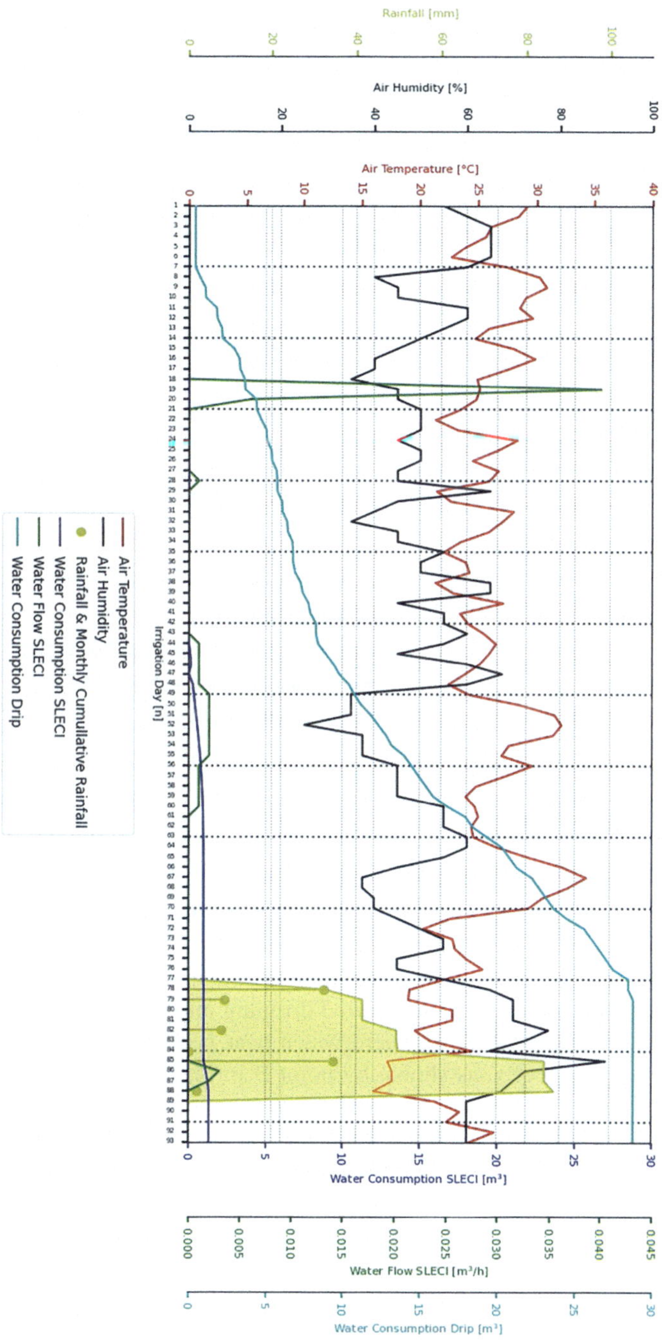

Fig. 19 Peach tree test field and technical irrigation information

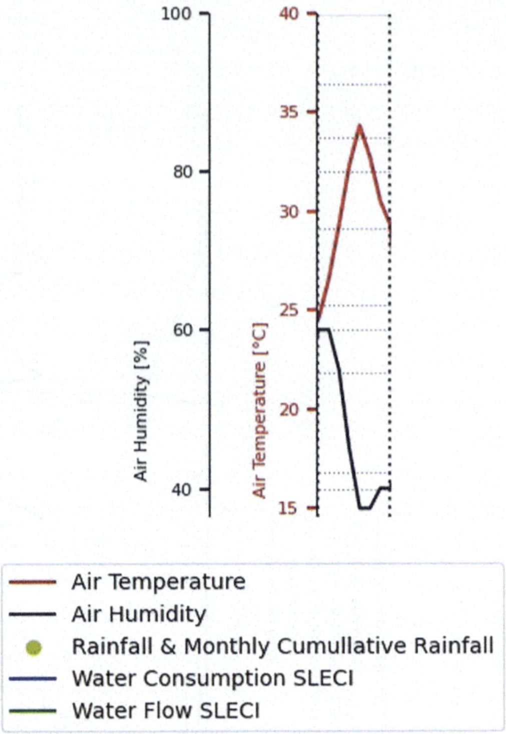

Fig. 20 Typical values for average air temperature and average air humidity during a heat peak

5.2 *Cherry Trees*

For the cherry tree test field, the irrigation period was extended from 10 May to 2 October 2023. After irrigation started at the cherry tree test field, a spike in the SLECIS water flow values, represented by point A in Fig. 21 is recorded. The system needed 0.005 m^3 of water to completely fill the main irrigation line, and once that happened, the *SLECI* elements started dripping into the fine sand underground. Also during this irrigation week, there was a heat peak, to which the *SLECI* elements responded properly, as shown by point B in Fig. 21, with an average water flow of 0.009 m^3/h.

An analysis of the weekly water consumption in the *SLECI* during the second irrigation week (19–26 May 2023) showed a system's misworking because that time was no water consumption for 10 days, and the weather reports indicated that need. Therefore, tests were carried out in the following weeks to solve the problem. Then, the third irrigation week (27 May to 2 June 2023) was marked by the first out-of-season rainfall, as mentioned before. Theoretically, Fig. 22 would show an unresponsive *SLECI* system as the soil was damp enough unrequiring any kind of irrigation, while the *DIS* kept consuming water and energy for expendable irrigation. However, in this case, the *SLECIS* was unresponsive due to the clogging of the

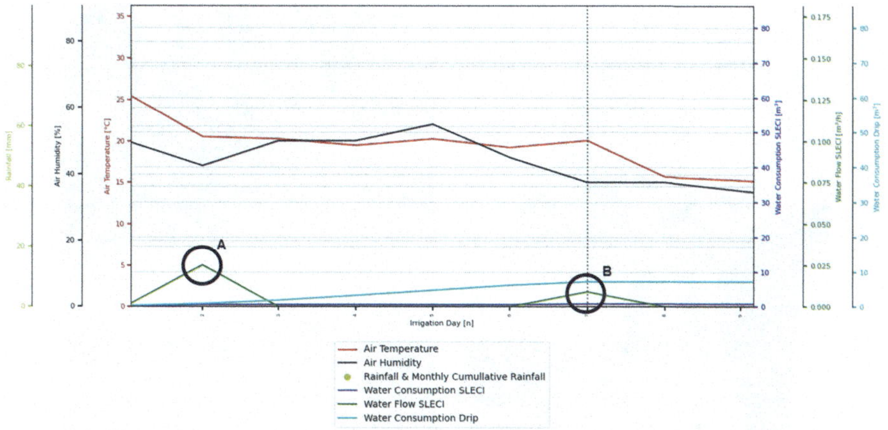

Fig. 21 SLECI behaviour under different conditions. Point A represents the initial water flow to fill the main irrigation line, and point B shows the SLECI response to a rise in air temperature and a drop in air humidity

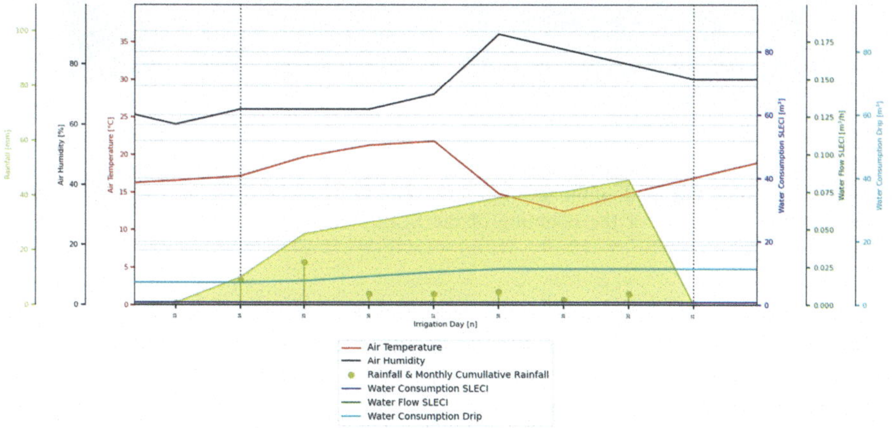

Fig. 22 SLECI system unresponsive during a rainfall period due to the clogging of the filtration system

filtering system, so further tests had to be done when evaluating the system's self-regulation reaction to this kind of meteorological event.

Following communication with the international project partners, during the fifth irrigation week (10–16 June 2023), a decision was made to double the *SLECI* elements on both irrigation lines with this system. The identified problems with the system seemed to raise from the inability of the water meter to detect a low water rate, the consistent creation of air bubbles inside the system, and the generation of faulty readings by the water meter ultrasonic sensor. The installation process and the result of the technical intervention are shown in Fig. 23a, b.

a) Drilling process of the bore hole for the doubling of the SLECI elements.

b) Result of the technical intervention.

Fig. 23 Installation and result of the technical intervention at the cherry tree test field. (**a**) Drilling process of the bore hole for the doubling of the SLECI elements. (**b**) Result of the technical intervention

The foliage from both systems was recorded, as shown in Fig. 24a, b, and even though the *SLECIS* showed no water consumption, the cherry trees irrigated by this system showed no sign of drying up or being in a state of hydric stress.

Thus, the *SLECI*-irrigated plants were without water for 6 weeks, while the *DIS* kept irrigating daily, which does not make sense when we analyse the foliage. Therefore, based on this information, it can be inferred that the *SLECI* system was irrigating the trees, but at a flow rate too low for the smart water meter to detect, as it requires a minimum of 1 L/h to record measurements accurately.

Still concerning the *SLECIS* system's smart water meter readings, in the eighth irrigation week (1–7 July 2023), it was decided to remove some faulty *SLECI* elements for further testing (see Fig. 25a). The holes left by the removed elements in the main line were then covered with electrical tape, as shown in Fig. 25b, so that there would be no water or pressure loss in the system.

Then on the 59th irrigation day (7 July 2023), a filter check was performed, which revealed that there was clogging, especially in the 50 μm filter, as shown in Fig. 26. At this time, there were no available spare filters of this type, so a decision was made to keep the system running with only 1 μm.

a) *SLECI irrigation system.* b) *Drip irrigation system.*

Fig. 24 Foliage comparison between irrigation systems, registered in the seventh irrigation week (24–30 June 2023). (**a**) SLECI irrigation system. (**b**) Drip irrigation system

This process of filter removal produced immediate changes in the system, as shown in Fig. 27a, where we can clearly see a damp stain at the top of the fine sand where the elements are embedded. Once the SLECI elements started irrigating properly, problems were observed because of three broken elements, as shown in Fig. 27b. This may have happened because of rodents in the area or due to high water temperatures at the outlet of the water deposit.

Figure 28 also illustrates the effects of filter replacement, and even though there was no irrigation or rainfall for 4 weeks in a row, *SLECIS* water flow did not go over 0.01 m^3/h in the next few days, as point A demonstrates. The figure also highlights the *SLECI* elements' response to average air temperatures ranging between 25 and 30 °C, during which they started dripping at a flow rate between 0.009 and 0.01 m^3/h. This response happened 2 days after the heat peak, with the exception of point C, where the response took place on the following day due to a streak of days with high average air temperatures and high evapotranspiration values (the last 8 days registered an average daily value of 6.34 mm/day).

On the third and fifth irrigation weeks (from 27 May to 16 June), the second and third rainfalls occurred. The *SLECIS* showed no water consumption, as Fig. 29 indicates, due to the clogging of the filtration system, so no conclusions could be drawn regarding the *SLECI* behaviour, and further tests had to be conducted to be able to

a) Removal of one SLECI element from each irrigation line. *b) Covering of the SLECI element hole, left by its removal.*

Fig. 25 SLECI element removal and hole covering process. (**a**) Removal of one SLECI element from each irrigation line. (**b**) Covering of the hole left by the removal of the SLECI element

Fig. 26 Clogged 50 μm water filter

a) SLECI element irrigating properly. *b) Broken SLECI element.*

Fig. 27 Demonstration of a working SLECI element and a broken one. (**a**) SLECI element irrigating properly. (**b**) Broken SLECI element

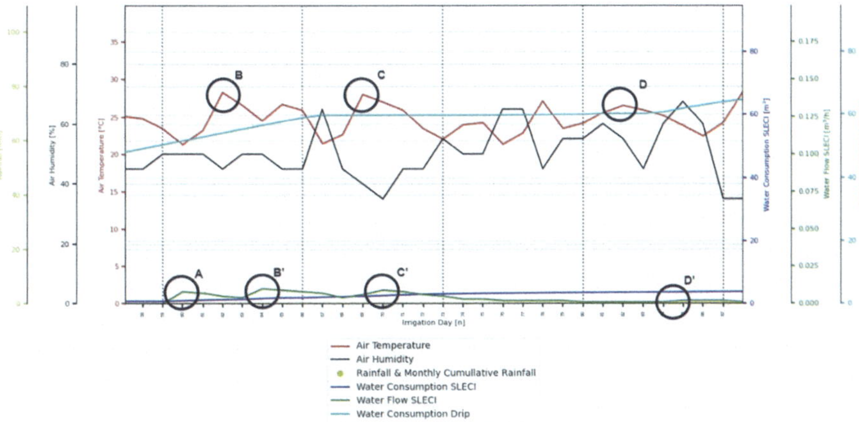

Fig. 28 SLECI response to the replacement of the filtering system and to the average air temperature of around 30 °C

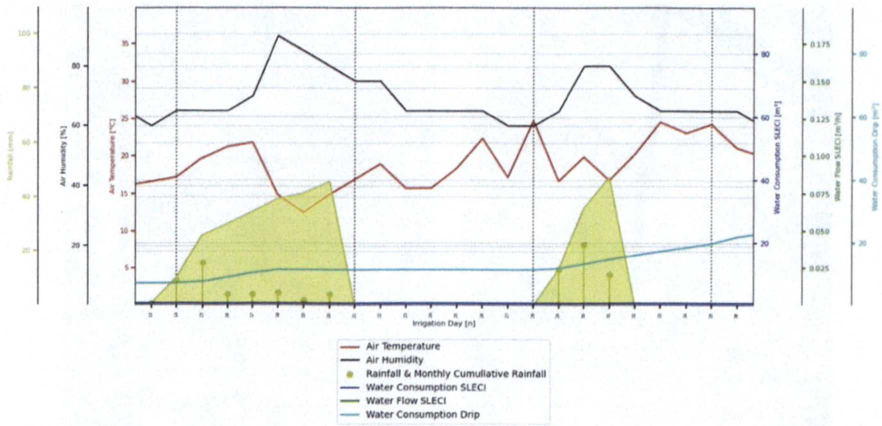

Fig. 29 SLECI irrigation system behaviour on the third and fifth irrigation weeks when there was rainfall

assess the clay elements' self-regulation properties and their efficiency under these weather conditions.

On the ninth irrigation week (8–14 July 2023), a 1 μm filter replacement was carried out due to clogging in the previous one, which indicates the high levels of residues and sustained particles in this water source. Once we replaced the filter, extra care was required to manage the formation of air sockets inside the main irrigation line. These air sockets could cause faulty readings on the water meters and lead to damage or malfunctioning of the *SLECI* elements. Figure 30a illustrates one of the effects of air sockets, where water is being prevented from flowing naturally to the second filter. Another broken *SLECI* element was recorded (Fig. 30b), making a total of 8 broken *SLECI* elements out of 53. Here, ants may have damaged the *SLECI* (Fig. 30c).

Then to tackle the possible problem of high water pressure inside the *SLECI* main irrigation lines, water pressure was monitored, as shown in Table 3

Additional information was gathered concerning the cherry trees irrigated with the *SLECIS* and the *DIS*, as shown in Figs. 31a, b and 32a, b.

Also, during this irrigation week, another heat peak took place on the 105th irrigation day (22 *August*), the hottest day of 2023, with an average air temperature of 34.3 °C and average air humidity of 37.5%. The *SLECI* responded promptly to this extreme event on the next day, as shown in Fig. 33, with an average flow rate of 0.057 m^3/h. The average air temperature remained between 25 and 30 °C until the 108th irrigation day (25 *August*), and given that weather condition, from the 106th (23 August) to the 109th irrigation day (26 August), the *SLECI* provided the crop with an average daily water flow rate of 0.033 $m^3/h/day$.

Bringing the irrigation season to a close for the cherry trees, there was a third and final rainfall period. As Fig. 34 shows, from the 116th irrigation day (2 September) to the 126th (22 September), the last rainfall of the irrigation season took place.

a) Formation of air socket in the SLECI water filtering system. *b) Broken SLECI element.* *c) Ant colony entrance hole near the clay element.*

Fig. 30 Report of different problems registered on irrigation week 9. (**a**) Formation of air socket in the SLECI water filtering system. (**b**) Broken SLECI element. (**c**) Ant colony entrance hole near the clay element

Table 3 Water pressure monitoring

Irrigation line	Pressure
Line 1—27 *SLECI* elements	0.354 bar
Line 2—26 *SLECI* elements	0.265 bar

Naturally, in a rainy period, the average air temperature decreases, whereas the average air humidity increases, so this would lead us to expect a drop in *SLECI* water consumption. During this time, a cumulative rainfall of 86.5 mm was recorded, and the average air temperature and air humidity were 17.5 °C and 71.8%, respectively. In response to this information, the *SLECIS* triggered a reaction on the 117th irrigation day (4 September), when cumulative rainfall was at 41.3 mm and the system's flow rate started decreasing at a daily rate of 8.61% until it completely stopped on the 122nd irrigation day (14 September). Meanwhile, the *DIS* kept consuming water during a period of heavy rainfall.

Besides water pumping to the 1 m^3 water deposit, another expense inherent to the *SLECIS* is the water filtration system. Data on the monitoring of filter replacements and the total costs for the cherry tree field tests are presented in Table 4.

Unfortunately, 2023 was a bad year for cherry trees, and on the test fields of Seminário do Fundão, a poor yield was observed among the 40 cherry trees. According to Sønsteby and Heide (2019), who conducted an analysis of the yield data and air temperature for cherry trees over a period of 40 years, these trees do not do so well in warmer weather. In the year following the collection of air temperature

a) *Close look at the SLECI irrigation foliage.* b) *General SLECI irrigation line 1.*

Fig. 31 Foliage monitoring for the SLECIS, registered on the 15th irrigation week (19–25 July 2023). (**a**) Close look at the SLECI irrigation foliage. (**b**) General SLECI irrigation line 1

information, no flowering was observed in trees grown at 23 °C, a few flowers were noted in trees grown at 20 °C, and an increasing abundance of flowering was seen in trees grown at even lower temperatures. Therefore, an analysis of air temperatures during the colder months must be conducted. Figure 35a–c shows air temperature monitoring from the chiller months of 2022, registering the most abnormal daily air temperatures compared to the monthly normal in 1971–2000.

As we can see, the months of January, October, and December registered above-average maximum air temperatures compared to the monthly norm in the period 1971–2000. These months also registered minimum air temperatures above the monthly norm, especially in October and December, when the minimum air temperature on most days is above the monthly norm. In the second and last third of December, the daily minimum temperatures were even above the normal maximum air temperature. Considering this information, we can see that 2023 would not give great cherry yields in the region of Castelo Branco, which includes Fundão, especially for small fields such as the one at Seminário do Fundão.

a) *Close look at the DIS irrigated foliage.* b) *General DIS irrigation line 1.*

Fig. 32 Foliage monitoring for the DIS, registered in the 15th irrigation week (19–25 July 2023). (**a**) Close look at the DIS irrigated foliage. (**b**) General DIS irrigation line 1

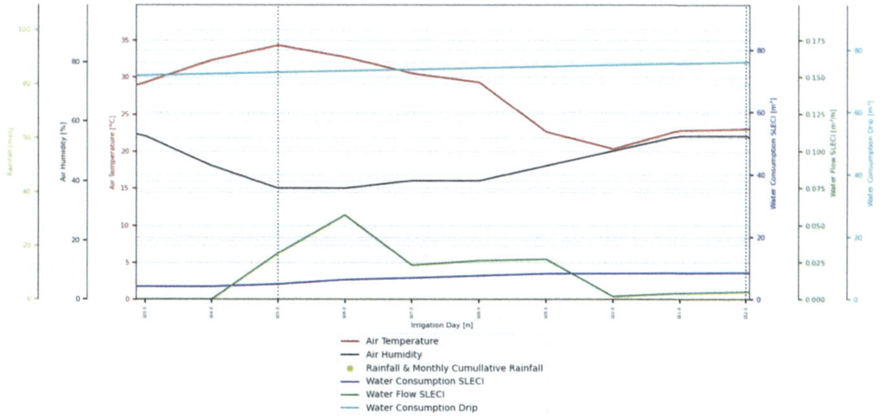

Fig. 33 SLECI behaviour for temperatures from 25 to 30 °C

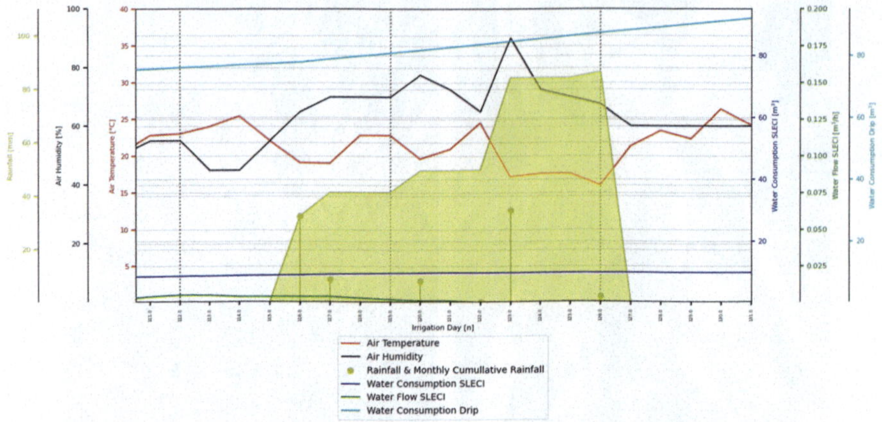

Fig. 34 SLECI behaviour during a period of heavy rainfall

Table 4 Filter replacement monitoring through the irrigation season on the cherry tree field test

Filter (*xx* µm)	Change number (*n* per type of filter)	Date of change (dd/month)	Days working (*n* days)
50	1	7/Jul	58
1	1	7/Jul	58
1	2	14/Jul	7
1	3	11/Ago	28

5.3 *Peach Trees*

For the peach tree test field, the irrigation period extended from 17 June to 2 October 2023. As Fig. 36 shows, once irrigation started on 17 June 2023, unlike the cherry crop, there was no water consumption at the *SLECIS*. This is due to the fact that the thin-walled *DN-16* main irrigation hose for the *SLECIS* was already filled with water and ready to initiate irrigation, but between the eighth and tenth irrigation days (7–9 June) rain fell, with a cumulative value of 47.3 mm, which is considerable for this region and this time of year. This, together with the loamy clay soil hydric dynamics, triggered no reaction on the *SLECI* elements until the eighth irrigation day (24 June), when the system provided an average daily water flow of 0.005 m^3/h/day for 2 days until the elements' self-regulation mechanism intervened.

Considering this information, it can be assumed that once the *DIS* was activated, the soil and trees were still draining the water from the previous rainfall. According to weather data provided by IPMA (Fig. 37a, b), by the time the *DIS* was triggered, soil water content values were still at [41,60]% θ_{fc} and remained so for a couple of days, until they started dropping on the fifth irrigation day (21 June) due to high air temperatures (average air temperature of 27.29 °C from the first irrigation day). The system provided irrigation water flow on the eighth irrigation day (24 June),

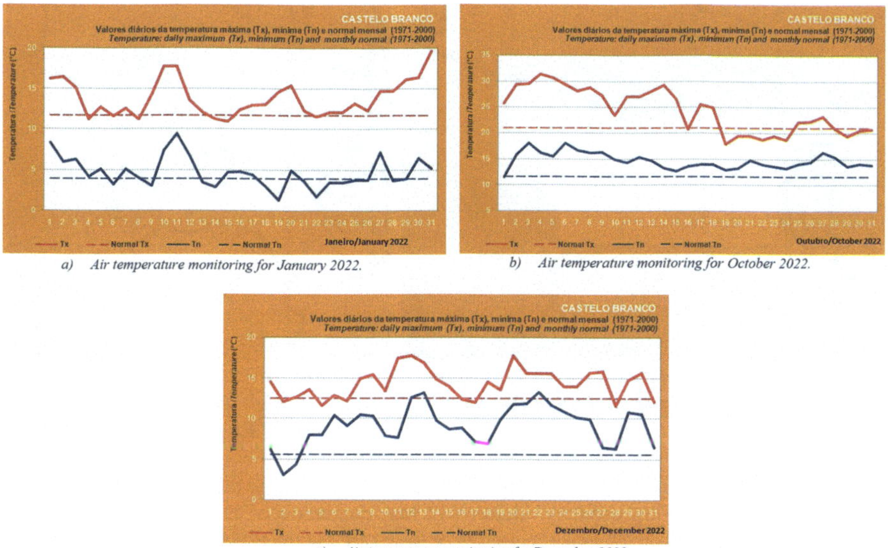

a) Air temperature monitoring for January 2022.

b) Air temperature monitoring for October 2022.

c) Air temperature monitoring for December 2022.

Fig. 35 Air temperature monitoring for the chillier months of 2022, with the daily maximum, daily minimum, and monthly normal (1971–2000). (**a**) Air temperature monitoring for January 2022. (**b**) Air temperature monitoring for October 2022. (**c**) Air temperature monitoring for December 2022

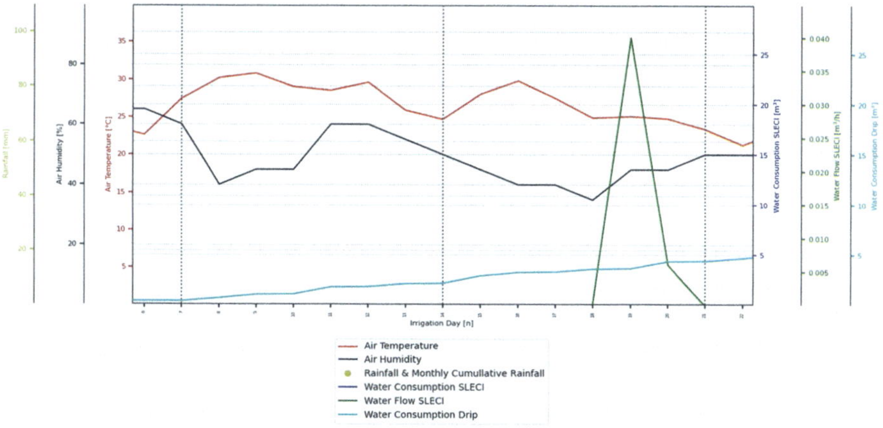

Fig. 36 SLECI system initial water flow on the third irrigation week

responding to a decrease in the soil water content, which was dropping to the interval [21,40]% θ_{fc} from the fifth irrigation day (21 June). Therefore, the *SLECIS* only activated after a spell of 3 days with this kind of evapotranspiration and water concentration in the soil. Girona et al. (2002), based on long-term experiments in fruit trees, suggest that these trees' (including peach trees') water threshold levels for

a) Field capacity values on 17th June (beginning of the DIS at the peach trees).

b) Field capacity values on 24th June (beginning of the SLECI at the peach trees).

Fig. 37 Difference between field capacity values at the beginning of the DIS and SLECI at the peach tree field test. (**a**) Field capacity values on 17 June (beginning of the DIS at the peach trees). (**b**) Field capacity values on 24 June (beginning of the SLECI at the peach trees)

irrigation scheduling should not be very different from those determined for herbaceous crops. The θ_{fc} value at which the *SLECIS* was triggered matches the conclusions of Theerawitaya et al. (2023) and Zotarelli et al. (2021), with regard to the water demands of herbaceous crops and peach trees. Cross-checking the conclusions of both studies suggests an ideal value of 25% θ_{fc} for this parameter in deficit irrigation, even though the studies were conducted in different kinds of soils: garden soil and sandy soil, respectively.

Foliage monitoring was performed at the peach tree test field, as Figs. 38a, b and 39a, b show for the *SLECI* and *DIS* technology systems, respectively.

Figure 40 shows that the *SLECIS* supplied no water flow for a full week, until the 28th irrigation day, when it provided the crops with an average water flow of 0.001 m^3/h in response to a week characterized by a daily average air temperature of 25 °C.

Then in the sixth irrigation week (22–28 July 2023), water pressure was monitored, with the results presented in Table 5.

Figure 41 provides information about the seventh irrigation week. After two and a half weeks with no irrigation, during a period with daily average air temperatures of 25.67 °C and the lowest daily average air humidity in the irrigation season (25% on the 52nd irrigation day, 7 August), the *SLECIS* system was again triggered on the 44th irrigation day (30 July). At that time, the system kept providing the plants with a daily average flow rate of 0.0014 $m^3/h/day$ for the next 16 irrigation days. In this situation, the *SLECIS* water flow curve followed the average air humidity, which dropped from the 49th irrigation day to the 55th irrigation day (4–10 August), and the *SLECI* irrigation water flow went from 0.001 to 0.002 m^3/h and then back to 0.001 m^3/h until the 60th irrigation day, when it once again interpreted that the soil and trees had no more need of water.

a) Close look at the SLECI irrigation foliage. *b) Overall view at the SLECI irrigation line 1.*

Fig. 38 SLECI peach tree foliage, recorded in the fourth irrigation week (8–14 July 2023). (**a**) Close look at the SLECI irrigation foliage. (**b**) Overall view at the SLECI irrigation line 1

After 8 weeks of irrigation, new monitoring metrics were introduced, "peach counting" in units and average "fruit calibre" in millimetres, at the 4th and 24th trees of both *SLECI* irrigation lines and then at the 4th, 24th, and 48th trees of the *DIS* line, with the results presented in Table 6.

In the following week, the foliage was monitored again, as shown in Figs. 42 and 43.

Figure 44 shows the period from the 78th until the 88th irrigation day (1 and 24 September), when the third and heaviest rainfall period occurred with a cumulative value of 86.5 mm. The *DIS* was suspended on the 79th irrigation day, so it was unnecessarily watering during the 78th irrigation day unnecessarily, because that day experienced a rainfall of 34.4 mm, and the *SLECIS* showed no consumption in this specific weather situation, until the 86th irrigation day (18 September), when it provided the crop with an average water flow of 0.003 m^3/h. This *SLECI* response happened during a heavy rainfall period, with a cumulative rainfall of 84.3 mm. Thus, more studies are needed to understand this behaviour as the natural weather conditions do not justify it.

Finally, on 31 August and 1 and 2 September, peach harvesting occurred for this test field, and so data on the fruit count and calibre were collected on 30 September 2023. The results are shown in Table 7.

As with the cherry tree field test, water pumping to the 1 m^3 water deposit is not the only expense related to the *SLECIS*. The water filtration system must be

a) Close look at the DIS irrigated foliage *b) Overall view of DIS irrigation line 1*

Fig. 39 DIS peach tree foliage, recorded on the fourth irrigation week (8–14 July 2023). (**a**) Close look at the DIS-irrigated foliage. (**b**) Overall view of DIS irrigation line 1

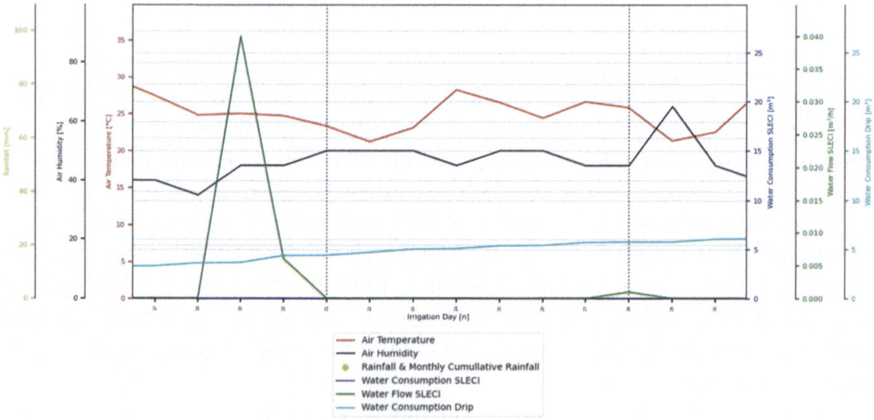

Fig. 40 SLECIS behaviour early in the fifth irrigation week

Table 5 Water pressure monitoring in the SLECI lines

Irrigation line	Pressure
Line 1—23 *SLECI* elements	0.53 bar
Line 2—26 *SLECI* elements	0.57 bar

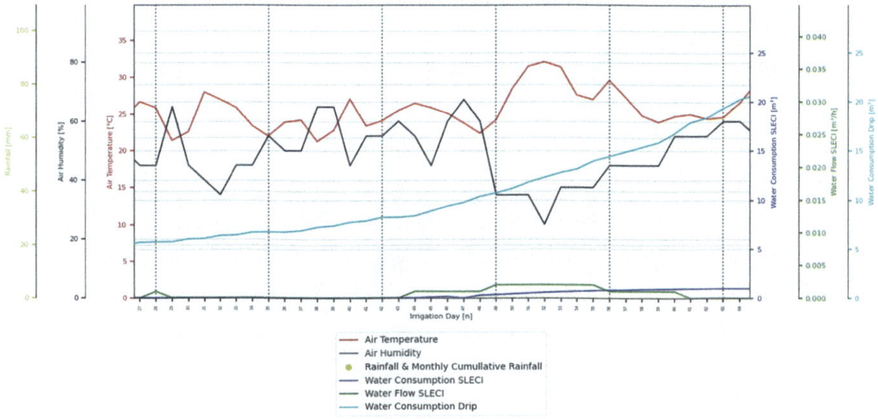

Fig. 41 SLECI response to a significant drop in average air humidity

Table 6 Week 8 data—fruit counting (units) and fruit caliber (mm)

Tree number (*units*)	Fruit count (*units*)	Calibre (*mm*)
4th tree	114	58.07
24th tree	232	49.84
4th tree	30	54.58
24th tree	89	45.02
4th tree	147	46.71
24th tree	86	47.36
48th tree	132	62.36

Line 1 *SLECIS* (yellow)
Line 2 *SLECIS* (green)
Line 1 *DIS* (orange)

considered, too. Thus, Table 8 presents data on filter replacement monitoring for the peach tree field test.

For both field tests, the maintenance, replacement and technical work on the water filtration system always depended on the water quality of the irrigation source being used—in this case, basin water source for each test field.

a) *Close look at the SLECI irrigation foliage.* b) *General SLECI irrigation line 1.*

Fig. 42 Foliage monitoring for the SLECI irrigation in the ninth irrigation week (12th to 18th August 2023). (**a**) Close look at the SLECI irrigation foliage. (**b**) General SLECI irrigation line 1

5.4 Peach Tree Yield and Water Use Efficiency (WUE)

Considering the final harvest, *water use efficiency (WUE)* can be calculated using Eq. 1, as presented by Wang et al. (2020). This was computed as the yield ratio (peach yield) divided by the total amount of irrigation and rain (IR) for the irrigation campaign.

$$\mathrm{WUE} = \frac{Y\left(\mathrm{kg}\right)}{\mathrm{IR}\left(\mathrm{m}^3\right)} \quad (1)$$

The summarized results are displayed in Table 9.

The peach trees covered by the *DIS WUE* results match the results of the work by Wang et al. (2020) during a 10-year period. Regarding the *SLECIS*, there are no values in this interval to be found in the literature, regarding this specific metric of *WUE*.

a) *Close look at the DIS irrigated foliage.* b) *General view of DIS irrigation line 1.*

Fig. 43 Foliage monitoring for the DIS in the ninth irrigation week (12th to 18th August 2023). (**a**) Close look at the DIS-irrigated foliage. (**b**) General view of DIS irrigation line 1

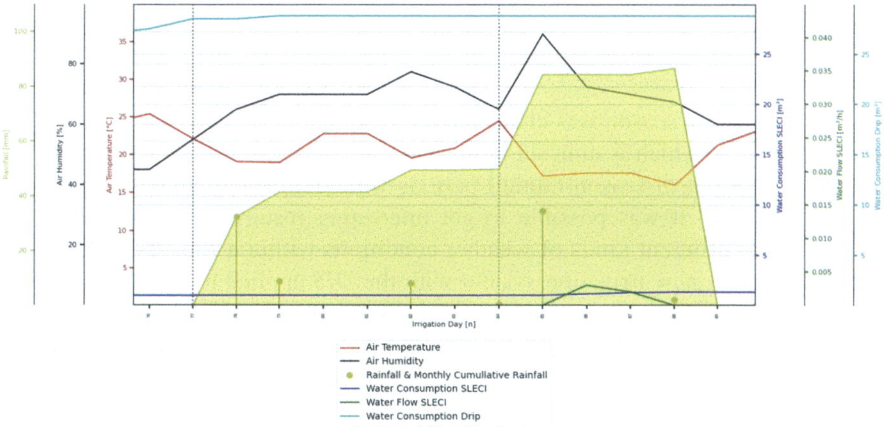

Fig. 44 SLECI behaviour for the third and heaviest rainfall period on the peach tree test field

Table 7 Week 11 data—fruit counting (units) and fruit calibre (mm)

Tree number (*units*)	Fruit count (*units*)	Calibre (*mm*)
4th tree	114	58.13
24th tree	232	50.01
4th tree	30	54.73
24th tree	89	45.56
4th tree	147	46.69
24th tree	86	47.57
48th tree	132	62.64
	Line 1 *SLECIS*	
	Line 2 *SLECIS*	
	Line 1 *DIS*	

Table 8 Filter change monitoring throughout the irrigation season on the cherry tree field test

Filter (*xx* μm)	Change number (*n* per type of filter)	Date of change (dd/month)	Days working (*n* days)
50	1	7/Jul	58
1	1	7/Jul	58

Table 9 Peach tree yield information

Irrigation tech	≃ Fruit count (units/tree)	≃ Fruit caliber (mm/tree)	≃ Fruit weight (g/peach)	≃ Fruit yield (kg)	Rainfall + irrigation (m^3)	WUE (kg/m^3)
SLECIS	116	52.1	66.4	377.398	13.195	28.601
DIS	121	52.3	66.65	387.143	41.143	9.4

6 Conclusion

The 2023 irrigation season was characterized by dynamic and constantly changing weather patterns, which is unusual in most periods of the late spring and summer, including three out-of-season rainfall periods with significant cumulative rainfall. Because of these, it was possible to get interesting results on the behaviour of *SLECIS* under different kinds of weather conditions within a short period of time and compare this new irrigation system with the *DIS* applied to cherry and peach trees. The *DIS* exhibited constant water consumption for the irrigation period, regardless of the average air temperature, average air humidity, soil water content (θ_{fc}), or evapotranspiration. On the other hand, the *SLECIS* demonstrated self-regulating properties on different occasions and under different conditions in both field tests by managing water flows according to weather conditions and soil water dynamics. The system was activated several times during periods when fruit trees typically felt hidric stress, characterized by average air temperatures between 25 and 30 °C, occasional heat peaks, average air humidity between 30% and 40%, and soil water content ranging from 21% to 40%. It also delivered increased water flow during extended periods of high evapotranspiration. The *SLECI* system also automatically halts irrigation when opposite conditions are detected, indicating that the soil is sufficiently moist and no additional water is needed.

Furthermore, unlike the *DIS* in terms of energy requirements and aside from the water savings comparison, the new micro-irrigation system does not require additional electronics, controllers, or electrical energy, as long as water is pumped into the reservoir and flow is maintained in the irrigation lines. According to Shahabuddin et al. (2023), e-waste generation is increasing globally, with over 53.6 million tonnes of e-waste produced in 2019 alone and only a small portion (17%) being collected and properly recycled. The rest is often burnt openly or dumped illegally, posing threats to both humans and the surrounding environment's biodiversity. In the end, the *SLECI* is undoubtedly the technology to consider for the future of irrigation and sustainability in the agricultural sector in a global context of water scarcity.

It should be noted that the evidence obtained only covers 1 year of implementation and irrigation. Although the *SLECIS* is promising when the micro-irrigation technology is applied to cherry and peach trees, for a more complete assessment, further tests must be conducted over 4–10 years, as suggested by Wang et al. (2020), Tong et al. (2022), and Pascual et al. (2016). Also, the cherry and peach trees monitored in these studies were 4 and >3 years old, respectively, so they were of a similar size. It would be interesting to compare tree growth rhythms across different irrigation methods over the years. Other metrics, like yield information and *WUE*, should also be monitored across multiple years and irrigation seasons. Concerning the monitoring of leaf and tree health and the existence of stress conditions, many studies use infrared canopy imaging and cross-check the data together with the weather station data. During the period that the tree should be irrigated, the smart water meters recorded a null (0 m^3/h) average daily flow rate on 61 days for the peach trees and on 74 for the cherry trees. Therefore, and because the *SLECIS* tree leaves presented no visual sign of hydric stress compared to the *DIS*, it can be assumed that the micro-irrigation clay elements were watering the trees at a lower rate than the smart water meter can read (1 m^3/h).

In future studies, it would be interesting to introduce these kinds of techniques and methods to get an even more detailed assessment of trees' reaction to this new micro-irrigation technology.

References

Akbar, G., Islam, Z., & Ahmad, B. (2023). Water productivity of garlic (Allium sativum L.) under the responsive drip irrigation (RDI) system compared to the drip and furrow irrigation systems in silty loam soil. *The Nucleus, 60*(2), 154–158.

Balogh, J. M., & Jámbor, A. (2020). The environmental impacts of agricultural trade: A systematic literature review. *Sustainability, 12*(3), 1152.

Bisht, N., & Chauhan, P. S. (2020). Excessive and disproportionate use of chemicals cause soil contamination and nutritional stress. In *Soil contamination-threats and sustainable solutions* (pp. 1–10). IntechOpen.

Blueberries Consulting Magazine, Science and technologies for the efficient use of water in agriculture, Blueberries Consulting Magazine, July, 2019. https://blueberriesconsulting.com/en/ciencia-y-tecnologias-para-eluso-eficiente-del-agua-en-la-agricultura/

Brychta, J., & Podhrázská, J. (2020). Evaluation of the erosion control effectiveness of cultivated crops and their optimal structure in agricultural landscape. In *Public recreation and landscape protection* (p. 153).

Darouich, H. M., Pedras, C. M., Gonçalves, J. M., & Pereira, L. S. (2014). Drip vs. surface irrigation: A comparison focussing on water saving and economic returns using multicriteria analysis applied to cotton. *Biosystems Engineering, 122*, 74–90.

De la Rosa, J. M., Conesa, M. R., Domingo, R., Aguayo, E., Falagán, N., & Pérez-Pastor, A. (2016). Combined effects of deficit irrigation and crop level on early nectarine trees. *Agricultural Water Management, 170*, 120–132.

de Sousa, P. L., Silva, L. L., & Serralheiro, R. P. (1999). Comparative analysis of main on-farm irrigation systems in Portugal. *Agricultural Water Management, 40*(2–3), 341–351.

European Commission. (2022a, April). *60 years of the CAP: From past to future*. https://agriculture.ec.europa.eu/document/download/f783ebe8-8405-4e86-8646-354c758e274a_en?filename=60-years-cap_en.pdf

European Commission. (2022b, June). *Overview of the CAP strategic plans—28 plans overview*. https://agriculture.ec.europa.eu/document/download/a376aab6-3a1d-4996-bb35-33c90b90c3bd_en?filename=csp-overview-28-plans-overview-june-2022_en.pdf

European Council. (n.d.). *Timeline: Common agricultural policy 2023–2027*. Retrieved February 8, 2024, from https://www.consilium.europa.eu/en/policies/cap-introduction/cap-future-2020-common-agricultural-policy-2023-2027/timeline-common-agricultural-policy-2023-2027/

Eurostat. (2023). *Regions in Europe: 2023 interactive edition*. European Commission. Retrieved February 8, 2024, from https://ec.europa.eu/eurostat/web/interactive-publications/regions-2023#agriculture

Food and Agriculture Organization of the United Nations. (2023). *Environmental sustainability in agriculture 2023*. https://www.fao.org/3/cc9065en/cc9065en.pdf

Food and Agriculture Organization of the United Nations (FAO) (2009). *The state of food and agriculture*. https://www.fao.org/publications/card/en/c/3aa4f41c-4316-5ddd-a656-22a00ef5d414/

Galego, D., Fernandes, I., Garcia, J., & Duarte, A. (2021). Contributos da tecnologia para a sustentabilidade da rega na fruticultura do Algarve. https://sapientia.ualg.pt/handle/10400.1/15407

García-Tejero, I. F., Durán-Zuazo, V. H., & Muriel-Fernández, J. L. (2014). Towards sustainable irrigated Mediterranean agriculture: Implications for water conservation in semi-arid environments. *Water International, 39*(5), 635–648, pp. 636–638.

Girona J., Behboudian M.H., Mata M., Del Campo J., Marsal J. (2010). Exploring six reduced irrigation options under water shortage for 'Golden Smoothee' apple: *Responses of yield components over three years, Agricultural Water Management, 98*(2), pp. 370–375 https://doi.org/10.1016/j.agwat.2010.09.011

Girona, J., Mata, M., Fereres, E., Goldhamer, D. A., & Cohen, M. (2002). Evapotranspiration and soil water dynamics of peach trees under water deficits. *Agricultural Water Management, 54*(2), 107–122.

González-Altozano, P., & Castel, J. R. (2003). Riego deficitario controlado en 'Clementina de Nules'. I. Efectos sobre la producción y la calidad de la fruta. *Spanish Journal of Agricultural Research, 1*(2), 81–92.

Hill, A. E., Ornelas, I., & Taylor, J. E. (2021). Agricultural labor supply. *Annual Review of Resource Economics, 13*, 39–64.

Ho, T. Q., Hoang, V. N., & Wilson, C. (2022). Sustainability certification and water efficiency in coffee farming: The role of irrigation technologies. *Resources, Conservation and Recycling, 180*(106), 175.

Instituto Português do Mar e da Atmosfera (IPMA). (n.d.). Retrieved January 28, 2024, from https://www.ipma.pt

Khan, P., Zhu, W., Huang, F., Gao, W., & Khan, N. A. (2020). Micro–nanobubble technology and water-related application. *Water Supply, 20*(6), 2021–2035.

Li, L., van der Werf, W., & Zhang, F. (2021). Crop diversity and sustainable agriculture: Mechanisms, designs and applications. *Frontiers of Agricultural Science and Engineering, 8*(3), 359–361.

Maheshwari, S. (2024). Bio-based business models: A review and new directions for future research. In *Frontiers in chemical, biological and pharmaceutical sciences Volume III* (p. 42). Bhumi Publishing. ISBN: 978-93-95847-85-8.

Maja, M. M., & Ayano, S. F. (2021). The impact of population growth on natural resources and farmers' capacity to adapt to climate change in low-income countries. *Earth Systems and Environment, 5*, 271–283.

Outhwaite, C. L., McCann, P., & Newbold, T. (2022). Agriculture and climate change are reshaping insect biodiversity worldwide. *Nature, 605*(7908), 97–102.

Pascual, M., Villar, J. M., & Rufat, J. (2016). Water use efficiency in peach trees over a four-years experiment on the effects of irrigation and nitrogen application. *Agricultural Water Management, 164*, 253–266.

Pereira, D., Leitao, J. C. C., Gaspar, P. D., Fael, C., Falorca, I., Khairy, W., Wahid, N., El Yousfi, H., Bouazzama, B., Siering, J., Hansmann, H., Zascerinska, J., Camilleri, S., Busuttil, F., Borg, M., Mizzi, J., Micallef, R., & Cutajar, J. (2023). Exploring irrigation and water supply technologies for smallholder farmers in the Mediterranean Region. *Sustainability, 15*, 6875. https://doi.org/10.3390/su15086875

Pérez-Pérez, J. G., Robles, J. M., & Botía, P. (2014). Effects of deficit irrigation in different fruit growth stages on 'Star Ruby' grapefruit trees in semi-arid conditions. *Agricultural Water Management, 133*, 44–54.

Purwanto, B. H., & Alam, S. (2020). Impact of intensive agricultural management on carbon and nitrogen dynamics in the humid tropics. *Soil Science and Plant Nutrition, 66*(1), 50–59.

Raven, P. H., & Wagner, D. L. (2021). Agricultural intensification and climate change are rapidly decreasing insect biodiversity. *Proceedings of the National Academy of Sciences, 118*(2), e2002548117.

Romero, P., Navarro, J. M., & Ordaz, P. B. (2022). Towards a sustainable viticulture: The combination of deficit irrigation strategies and agroecological practices in Mediterranean vineyards. A review and update. *Agricultural Water Management, 259*, 107216.

Rudel, T. K. (2020). The variable paths to sustainable intensification in agriculture. *Regional Environmental Change, 20*(4), 126.

Sadigov, R. (2022). Rapid growth of the world population and its socioeconomic results. *The Scientific World Journal, 2022*, 8110229.

Shahabuddin, M., Uddin, M. N., Chowdhury, J. I., Ahmed, S. F., Uddin, M. N., Mofijur, M., & Uddin, M. A. (2023). A review of the recent development, challenges, and opportunities of electronic waste (e-waste). *International Journal of Environmental Science and Technology, 20*(4), 4513–4520.

Shock, C. C. (2006). *Drip irrigation: An introduction*. Oregon State University.

Sønsteby, A., & Heide, O. M. (2019). Temperature effects on growth and floral initiation in sweet cherry (Prunus avium L.). *Scientia Horticulturae, 257*, 108762.

Stein-Bachinger, K., Preißel, S., Kühne, S., & Reckling, M. (2022). More diverse but less intensive farming enhances biodiversity. *Trends in Ecology & Evolution, 37*(5), 395–396.

Theerawitaya, C., Praseartkul, P., Taota, K., Tisarum, R., Samphumphuang, T., Singh, H. P., & Cha-Um, S. (2023). Investigating high throughput phenotyping based morpho-physiological and biochemical adaptations of Indian pennywort (Centella asiatica L. urban) in response to different irrigation regimes. *Plant Physiology and Biochemistry, 202*, 107,927.

Tong, X., Wu, P., Liu, X., Zhang, L., Zhou, W., & Wang, Z. (2022). A global meta-analysis of fruit tree yield and water use efficiency under deficit irrigation. *Agricultural Water Management, 260*(107), 321.

Wang, D., Zhang, H., & Gartung, J. (2020). Long-term productivity of early season peach trees under different irrigation methods and postharvest deficit irrigation. *Agricultural Water Management, 230*, 105940.

Wang, R., Shi, W., Kronzucker, H. J., & Li, Y. (2023). Oxygenation promotes vegetable growth by enhancing P nutrient availability and facilitating a stable soil bacterial community in compacted soil. *Soil and Tillage Research, 230*, 105686.

Zhu, Y., Li, P., Feng, X., Sun, D., Fang, T., Zhu, X. X., et al. (2021). Reversible CO2 absorption and release by fatty acid salt aqueous solutions: From industrial capture to agricultural applications. *Journal of CO_2 Utilization, 54*, 101746.

Zotarelli, L., Zambrano-Vaca, C., Barrett, C. E., Sharma, V., Popenoe, J., Sarkhosh, A., et al. (2021). A practical guide for peach irrigation scheduling in Florida: HS1413, 5/2021. *EDIS, 2021*(3).

The Impact of the Irrigation Project on the Health Status of Local Communities: An Analysis of the Teesta Barrage Irrigation Project in Bangladesh

Md. Abu Shahen and Md. Nurul Momen

Abstract The Bangladesh government launched the Teesta Barrage Irrigation Project (TBIP), channeling water through a canal from the Teesta barrage at Dalia point in Lalmonirhat district, with the goal of guaranteeing farmers' access to irrigation water. This irrigation project spans five districts in Bangladesh's north region. This study attempted to investigate the effects of the Teesta Barrage Irrigation Project on farmers' health conditions in the irrigated area as a result of increased crop production. As a result of the irrigation project, agricultural output has increased, financial resources have been enhanced, and the health of the farmers in the study region has improved compared to the past. Though both qualitative and quantitative methodologies are used in this study, the quantitative approach is used more often in practice. In order to do this, the study employed a social survey with face-to-face interviews and observation methods. The study discovered that there have been some notable improvements in health status, such as the majority of farmers receiving better primary health care, modern diagnostic, and Cesarean services; the use of private vehicles to transport patients; and notable improvements in living and hygienic conditions. The study also showed that middle-class and upper-class farmers benefited from the TBIP the most, receiving better health care and leading better lives than marginal farmers. Additionally, it was observed that marginal farmers in the research area continued to have poor health and did not receive enough benefits from the TBI Project.

Keywords Bangladesh · Crop production · Health · Teesta Barrage Irrigation Project

---

M. A. Shahen
Department of Social Work, University of Rajshahi, Rajshahi, Bangladesh

M. N. Momen (✉)
Department of Public Administration, University of Rajshahi, Rajshahi, Bangladesh
e-mail: nurulmomen@ru.ac.bd

J. C. Correia Leitão et al. (eds.), *Sustainable Water Management and Irrigation Systems*, Natural Resource Management and Policy 62,
https://doi.org/10.1007/978-3-031-93587-9_10

1 Introduction

There are 57 rivers that flow through Bangladesh and into the Bay of Bengal in the south, sharing boundaries with India. These rivers run through Nepal, Bhutan, China, and India in their upstream courses. The management of the regional river systems is significantly impacted by this global component. The main source of water resources in Bangladesh is transboundary flows, which originate in isolated catchments thousands of kilometers upstream. After the Ganges, Brahmaputra, and Meghna, Teesta is the fourth largest transboundary river in Bangladesh. The water of the Teesta River is essential for agricultural cultivation in Bangladesh's far northwest. In order to supply irrigation water to the TBIP area via a network of canals, Bangladesh built a barrage on the Teesta River in the Lalmonirhat district in 1990 (Islam et al., 2004). After the Teesta barrage was finished in 1990, the TBIP catchment area's canal networks could supply irrigation water from the river for three crop seasons (Momen & Rahman, 2023; Shahen & Momen 2024). It is a 615-meter-long concrete structure with 52 radial gates that can release material in 12,750 cusecs. The barrage uses a 110-meter-long canal head regulator with a 280-cusec discharge capacity to redirect water. There were two stages to the project's completion. The year 1998 saw the completion of phase I. Phase I comprises a net irrigable area of 111,608 ha and a command area of 154,250 ha. The remaining area is being developed under phase Il, which is now under construction (Sarker et al., 2011). However, Bangladesh is somewhat concerned about whether or not this phase can be completed in time. It was anticipated that the Teesta Barrage Irrigation Project (TBIP) would alter the ecology and socioeconomic situation of the farmers. It is also anticipated that this project will alter the irrigation system, enabling them to cultivate land conveniently and on schedule—even during the dry season. The region's farmers are in better health as a result of the increased production of rice. They can obtain better recreation, clothing, food, housing, and health facilities in order to assume a more advantageous place in society. It was anticipated that the TBIP would also have a good effect on the local environment and health. The Teesta barrage area is experiencing a daily increase in agricultural production, while the amount of irrigation water available through the TBIP is steadily decreasing. The cause of this water shortage is India's unilateral upstream extraction of Teesta water through the Gajoldoba barrage. Nonetheless, the influence of the Teesta Barrage Irrigation Project on the health of Nilphamari district farmers has been examined in this study.

According to research by Nagdeve (2006) and the International Water Management Institute (IWMI) (2017), farmers in irrigated areas invested their increased income and used family budgets to maintain health-care costs. By strengthening their financial stability, these farmers also helped slow the development of water-borne diseases. They also disclosed that, in order to receive better care, farmers in irrigated areas seek medical attention from reputable private hospitals. Additionally, irrigation projects increase farmers' awareness of good sanitation and other health-related practices, which in turn improves farmers' health. The arguments made by Stoeckel et al. (1983), Gbenga et al. (2015), Dhongade and Dangat (1985), Mukerji and Kulkarni (1989), Mohanty (1999), Roy (1983), Nagdeve (2006), Patil et al. (1978), and Prasaratkul et al. (1985) suggest that irrigation projects altered household fertility and contraceptive behavior, as well as the social and economic structure of households. These changes influenced family size norms, lowered infant and child mortality, facilitated the expensive bearing and raising of children, improved social mobility, and promoted the use of family planning and sterilization. They also looked at the fact that the irrigated area had better health status than the nonirrigated area and that irrigation and child health are significantly correlated. They also investigated the positive effects of the irrigation project on family health and nutrition, child health, health expenditures, living standards and demographic behavior, and access to diagnosis and treatment facilities for farmers in the irrigated area. It is evident that a well-functioning irrigation system is necessary to raise agricultural productivity and enhance the state of health in any part of the planet. However, finding out how the Teesta Barrage Irrigation Project is affecting the farmers' health in the Teesta barrage area is the primary goal of the research. The following particular goals help achieve the main goal: (a) to investigate farmers' health status and (b) to understand how the TBIP contributes to farmers in the study area receiving improved health care.

However, to fulfill the abovementioned objectives, the researchers followed a mixed-method approach. A social survey, key informant interviews, and focus group discussions were the primary tools used in this study to collect field data. The research findings were mostly based on primary data, although secondary data from books, journal papers, reports, and websites have also been used. Prior to the finalization of the questionnaire, a preevaluation was conducted. The study's questionnaire has a semi-structural design. Closed-ended and open-ended questions were included for the sake of the study. Data collection methods also included content analysis and observation. The Statistical Package for Social Science (SPSS) version 26 was utilized for the purpose of organizing, classifying, and analyzing the data for this investigation.

2 Nexus Between the Irrigation Project and Health Status

Numerous environmental, agricultural, and socioeconomic factors influence the complex interaction between irrigation projects and health conditions in Bangladesh. The widespread use of groundwater for irrigation, especially when it comes to rice farming, has a major effect on agricultural productivity and public health. About 79% of Bangladesh's agricultural area depends on groundwater for irrigation, making it the primary source of water (Chowdhury, 2021). Because of this dependence, the nation has been able to produce rice almost entirely on its own, especially in the northern areas (Krupnik et al., 2017). But excessive groundwater extraction has resulted in rising salinity and declining water tables, especially in coastal regions where groundwater salinity is a major issue (Ara et al., 2016; Mainuddin et al.,

2021). In addition, besides having an impact on the sustainability of agriculture, groundwater extraction puts human health at risk from arsenic poisoning. According to research, groundwater in Bangladesh is highly contaminated with arsenic, endangering public health, especially if contaminated rice is consumed (Jahan et al., 2021; Benner & Fendorf, 2010). Over 90% of groundwater extraction is said to be utilized for irrigation, and elevated levels of arsenic in groundwater have been connected to a number of health problems, including cancer (Benner & Fendorf, 2010).

The promotion of surface water irrigation by the Government of Bangladesh (GoB) seeks to lessen reliance on groundwater and reduce the health hazards that come with it (Krupnik et al., 2017; Paul & Hasan, 2021). But switching to surface water irrigation is not without its difficulties. Salinity problems and industrial pollution degrade surface water quality, necessitating treatment to meet irrigation requirements (Mahtab & Zahid, 2018). Smallholder farmers' use of surface water irrigation may be limited by the increased cost associated with this treatment (Mahtab & Zahid, 2018). In Bangladesh, irrigation practices and health outcomes are significantly influenced by climate change (Begum and Momen, 2021). It is anticipated that temperature and precipitation changes may modify irrigation water requirements, which could exacerbate already-existing water scarcity problems (Kirby & Mainuddin, 2022; Shahid, 2010). Flooding caused by the intensification of extreme weather events may further contaminate irrigation water sources and affect food safety (Jeuland et al., 2013). The interaction between irrigation practices, climate change, and health consequences highlights the need for integrated water management techniques that take into account both agricultural productivity and public health.

Irrigation projects have been linked to improvements in farmers' health conditions globally. For example, the Food and Agriculture Organization (FAO) discovered that in Burkina Faso, a portion of the country's increasing farm income was allocated to health care. Over a 3-year period, the number of visits to local clinics climbed by 50%, and the health-care spending of individual families increased from 5% to 12% of their family budget. Farmers are able to take precautions against a variety of illnesses, including filaria, malaria, and other waterborne illnesses. For example, FAO stated that increasing the financial stability of farmers through irrigation projects can help prevent the development of waterborne diseases like malaria and schistosomiasis, which are spread by mosquitoes and snails. According to Nagdeve's (2006) research, farmers in irrigated communities believe that their facilities offer superior medical care compared to nearby local medical facilities. The fact that they do not use local medical facilities may indicate that they are well-off enough to afford what they consider to be higher-quality care. It seems that the farmers go to local health clinics only to get basic prescription drugs and contraception. Additionally, enhanced water availability through irrigation canals may contribute to improved health and cleanliness. The construction of the irrigation canals have made it easier for people to obtain water in numerous locations (IWMI, 2017). It is found that farmers now have a greater understanding of health practices, which has raised their knowledge levels of health. Thus, it can be observed that irrigation has improved the farmers' health in the irrigated area (Nagdeve, 2006).

An investigation into how an irrigation project affects farmers' fertility and the use of contraception revealed that farmers in irrigated areas differ significantly in fertility rates and contraceptive use compared to farmers in non-irrigated areas (Nagdeve, 2006). The social and economic makeup of households is altered by the availability and use of irrigation projects, which has an impact on family size norms, newborn and child mortality, intermediary variables, and, eventually, fertility and contraception (Stoeckel et al., 1983). Cultivators and agricultural laborers are the primary beneficiaries of female sterilization programs.

It was also shown that childbearing costs were higher in irrigated villages. Also, children in the irrigated areas were found to receive better education. Nagdeve (2006) noted that childbirth and child-rearing involved various costs, such as transportation costs, hospital bills, medical expenses, store-bought clothing, children's play equipment, etc. These improvements in health status could be the result of socioeconomic developments brought about by irrigation projects.

Similarly, it was discovered that irrigation significantly impacted the health of the project region in the Indian province of Maharashtra. According to Nagdeve's (2006) research, irrigated areas had higher child survival rates and lower fertility rates. Farmers in irrigated communities seem to afford to purchase what they believe to be superior medical care because they shun local medical facilities. It was evident that both the number of children and the size of the family dropped (Stoeckel et al., 1983). Furthermore, farmers were not only becoming more knowledgeable about sterilization and family planning but were also practicing these methods. According to Nagdeve (2006), growing industrialization and development led the population in irrigated areas to modify their fertility patterns, enabling them to benefit economically from the irrigation project. Compared to nonirrigated areas, childbirth was more expensive in irrigated communities. With children's health improving, smaller family sizes became the norm and social mobility started to rise (Gbenga et al., 2015). The purchase of domestic goods like refrigerators, TVs, and radios became a routine occurrence and raised farmers' awareness of health issues (Nagdeve, 2006).

In their baseline study, Dhongade et al. (1985) examined the state of health in the irrigation project's command area. They discovered that the irrigated area had a better state of health than the nonirrigated area. After analyzing secondary data, Mukerji and Kulkarni (1989) discovered a strong correlation between irrigation and children's health. The irrigation variable has less of an impact than the amenities index. Mohanty (1999) came to the conclusion that farmers' health has improved as a result of irrigation investments. Productivity, total production, literacy, and living standards have all increased as a result, and it has also been shown to affect household and aggregate demographic behavior. A socioeconomic survey of the Girona irrigation project area in the Maharashtra district of Jalgaon was carried out by Patil et al. in 1978. They discovered that cash crops played a significant role and altered the health conditions of farmers in the project area. In his research, Roy (1983) discovered that irrigation directly influenced cropping patterns, improved agricultural practices, and increased returns per unit of farmed area. It also enhanced the health of the farmers in the study area.

According to Nagdeve's (2006) research, the cost of health-care expenses in irrigated areas is also very high because people now seek medical care and child-rearing services at district hospitals or private clinics. An anthropological study was carried out by Prasaratkul et al. (1985) in order to assess the effect of a major irrigation project on the general health of the surrounding community. They discovered that the population in the irrigated areas experienced changes in living standards as a result of the irrigation project. Thus, it can be said that the irrigation project has a significant and widely acknowledged impact on their health conditions (Nagdeve, 2006).

However, although irrigation projects in Bangladesh have greatly improved food security, they present serious health problems because of groundwater contamination and the difficulties associated with switching to surface water sources. A holistic strategy is needed to address these problems, one that involves enhancing water quality, responsibly managing groundwater resources, and adapting to the effects of climate change.

3 Irrigation Projects and Their Role in Bangladesh

In Bangladesh, irrigation projects are essential for raising agricultural output, guaranteeing food security, and bolstering rural livelihoods. The majority of the country's cultivated area depends on surface and groundwater supplies, and irrigation is the mainstay of the agricultural economy. Given Bangladesh's susceptibility to natural disasters and climate change, this reliance on irrigation is especially crucial. The fact that almost 80% of extracted groundwater is used for irrigation highlights the significance of irrigation in Bangladesh (Jahan et al., 2021). Maintaining agricultural output requires the intensive use of groundwater, particularly in a nation where rice is a staple crop and a major economic driver. In order to boost crop yield and guarantee food security, a number of irrigation technologies have been introduced, such as surface water irrigation systems and deep tube wells (DTWs) (Sumiya & Khatun, 2016). However, there is increasing concern about how sustainable these irrigation methods are, especially in view of groundwater depletion and contamination problems, like arsenic poisoning (Benner & Fendorf, 2010).

Along with that, the irrigation system in Bangladesh faces new difficulties as a result of climate change. Traditional irrigation methods may be disrupted, and agricultural output may be threatened by variations in the patterns of rainfall and the rise in the occurrence of severe weather conditions (Kirby & Mainuddin, 2022). Research shows that although Bangladesh can control how climate change affects the amount of water needed for irrigation, proactive steps are required to properly adjust to these changes (Kirby & Mainuddin, 2022). For example, there is potential to improve groundwater recharge and sustainability through the use of managed aquifer recharge (MAR) strategies (Rahman et al., 2018a).

Furthermore, there is a growing trend in Bangladesh to integrate renewable energy sources, including solar electricity, into irrigation systems. By installing

50,000 solar irrigation pumps by 2027, the Infrastructure Development Company Limited hopes to drastically cut the carbon emissions linked to irrigation systems that run on diesel (Salehin, 2023). In addition to advancing environmental objectives, this move to renewable energy sources strengthens irrigation systems' resilience in the face of climate change (Shoeb & Shafiullah, 2018). Moreover, irrigation projects can have major health impacts as well as agricultural benefits. Better agricultural output from improved irrigation can enhance rural people's nutritional outcomes and strengthen food security. In order to avoid negative health effects, irrigation-related health concerns, such as the use of arsenic-contaminated water, must be properly controlled (Benner & Fendorf, 2010).

In Bangladesh, promoting safe irrigation techniques and creating regulations to reduce health hazards are crucial elements of sustainable irrigation management. Irrigation projects are essential to Bangladesh's agricultural environment and support rural livelihoods and food security. However, groundwater depletion, climate change, and health issues related to water quality pose challenges to the sustainability of these programs. The future of irrigation in Bangladesh will depend on how well these issues are resolved through the use of cutting-edge technologies, the integration of renewable energy sources, and the implementation of sensible health regulations.

4 Assessing the Impact of Teesta Barrage Irrigation Project on the Health Status of Farmers

4.1 Sectors of Health Services Before and After the TBIP

The Bangladeshi government adopted a vision plan to raise the country's income from lower-middle to middle income by 2021 in order to meet the Sustainable Development Goal (SDG) targets. Although the Bangladeshi government has recently made some strides in the health sector, it is clear that the options for health care are still insufficient. For instance, it can be difficult to get admitted to a hospital, and there are never enough medications. There are also other issues that need to be addressed, such as improper use of medical equipment, poor health management, corruption, inadequate facilities, staff-to-patient ratio, and doctor-to-patient ratio. To improve health services, however, the government, nongovernmental organizations (NGOs), and private businesses in Bangladesh, operate a large number of clinics, health-care facilities, and various types of hospitals.

According to the above survey (Fig. 1), 20.0% of farmers received health-care services from union health centers prior to the inception of the TBIP, while currently, 9.2% receive care from union health centers or community clinics. Additionally, it was shown that prior to the TBIP, 10.0% of farmers received their medical care from Upazila Health Complex, whereas 15.7% currently receive their care from the same facility. According to the data, 28.0% of farmers received their

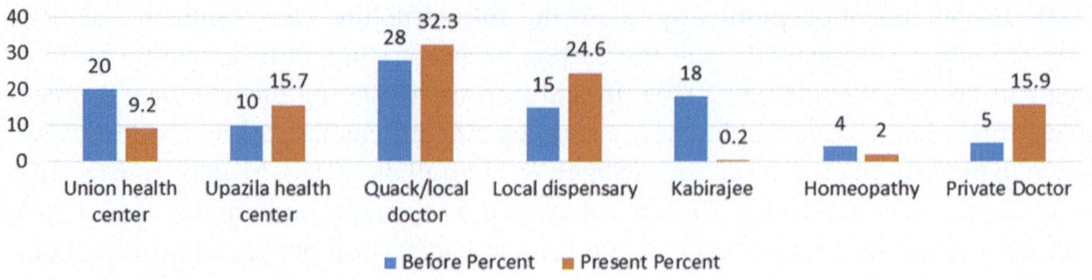

Fig. 1 Sectors of health services

medical care from a local quack prior to the TBIP, whereas 32.3% of farmers currently receive their medical care from a local quack before and after TBIP. Additionally, the research reveals that while 24.6% of the farmers in the study region receive their medical care from the same location as the local dispensary, 15.0% of them received it prior to the TBIP. It is surprising to learn that, among the farmers in the research region, only 0.2% receive primary health care services from Kabirajee therapy, whereas 18.0% received such treatment prior to the TBIP. Additionally, it was shown that 4.0% of farmers received medical care from a homeopathic doctor before the TBI Project, but 2.0% of farmers currently receive such care. The fact that only 15.9% of farmers currently receive medical care from private physicians or clinics, compared to 5.0% prior to the initiation of the TBIP in the research area, raises serious concerns. From the discussion above, it is clear that farmers are receiving their medical care from private clinics and the Upazila Health Center rather than from quacks, kabirajees, or homeopathic physicians. Furthermore, Nagdeve (2006) contended that the expense of child-rearing in irrigated areas is also relatively high because farmers now seek medical care and child-rearing services at district locations or private clinics. Furthermore, because of their poverty, farmers in developing nations like Bangladesh are increasingly concerned about the high expense of therapy. Therefore, it is evident from the data analysis above that farmers are growing more aware of their health and the services offered by both governmental and nongovernmental organizations (Rahman et al., 2005). Notably, this occurred as a result of improvements in their economic situation, which is currently better than it was prior to the TBIP. We can conclude that the TBIP contributes to higher agricultural yields. Consequently, this aids in providing improved health care inside or even outside of the farmers' communities for themselves and their families.

4.2 *Mode of Transportation*

Connectivity and patient transportation are essential components of a health system as they guarantee patient flow to the hospital. Ineffective patient mobility inside the facility can create treatment bottlenecks, undermine patient trust, and result in financial burdens. The first and most lasting impression that patients may have of a hospital is frequently shaped by their interactions with patient transporters. In

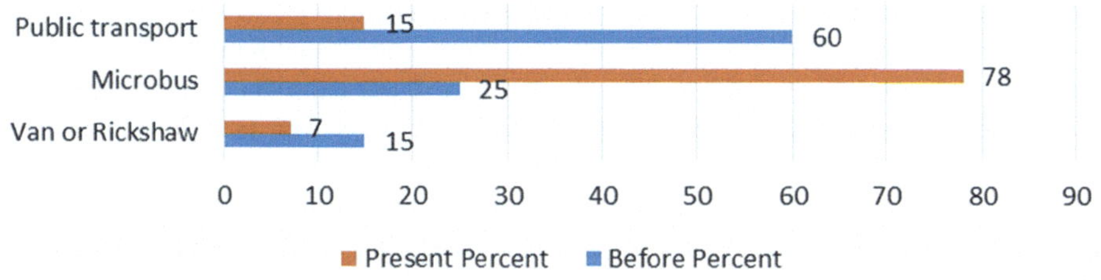

Fig. 2 Vehicles carrying sick family member

Bangladesh, various vehicle types are utilized for patient transportation in medical services, including rickshaws, buses, auto bikes, ambulances, vans, and human haulers. Also, ambulance services are supplied by both public and commercial health-care providers to tertiary public facilities at varying levels of primary health care. Unfortunately, due to some corrupt authorities, ambulance service is never provided within the public health-care system (Rahman et al., 2005).

According to the above data (Fig. 2), prior to the Teesta Barrage Irrigation Project (TBIP), 15.0% of farmers in the study area used a van or rickshaw to transport their patients to a distant health service center, whereas 7.0% of farmers still do so currently. Additionally, it was shown that in the research region, 25.0% of farmers utilized microbuses to transport their patients to health centers prior to the TBIP, whereas 78.0% of farmers currently use microbuses to transport their patients to health centers after TBIP. More intriguingly, the observation is made that 60.0% of farmers used public transportation to take their patients to health facilities prior to the TBIP, whereas currently, only 15.0% of farmers do so. Additionally, according to the above data, since the TBIP began or remains in effect in the study region, there has been a drop in the use of vans, rickshaws (15–7 = 8%), and public transportation (60–15 = 45%), while there has been an increase in the use of microbuses (78–25 = 53%). Because proper and adequate water delivery promotes crop output, which in turn improves the socioeconomic conditions of the farmers in the study region, it can be claimed that the TBIP water supply has improved the economic condition of the farmers in the study area.

4.3 *Diagnostic Facilities Before and After the TBIP*

In Bangladesh, the percentage of individuals living below the poverty line is 21.8% for rural residents and 11.3% for urban residents. Owing to the high poverty rate in the nation, a sizable fraction of the population lacks the financial means to pay for medical care, afford diagnostics at better health centers, receive proper nutrition, maintain cleanliness, and obtain health education (Rahman et al., 2005). It has been positively demonstrated in a study that poverty breeds helplessness and ill-health, and that ill-health breeds poverty (World Bank, 2000). Furthermore, a big reason why impoverished individuals perceive medical care as expensive is its cost. Patients in wealthy nations are now far more cautious about their medical expenses than their health insurance coverage. The Nordic nations, the United Kingdom, and other wealthy Western nations, such as New Zealand, have per capita health expenditures ranging from US$ 1000 to US$ 2000 (Jeremy, 2000). However, Bangladesh spends about US$ 26.60 per person (WHO, 2017), and as a result, people fail to access advanced diagnostic services at better locations (Rahman et al., 2005) since patients cannot receive free care at private hospitals and diagnostic centers (Andaleeb et al., 2007).

According to the findings of this study (Fig. 3), 3.5% of farmers received diagnostic services from local diagnostic centers prior to the TBIP, whereas 6.7% of farmers currently receive diagnostic services from local markets or local diagnostic centers. Additionally, it was discovered that in the research area, 25.0% of farmers received services from Upazila Health Complex prior to the TBIP, whereas 19.7% of farmers currently receive diagnostic services from Upazila Health Center. According to the study, 24.9% of farmers currently receive diagnostic services from Rangpur Medical College Hospital, whereas 45.0% of farmers received services from the hospital prior to the TBIP. Additionally, the study shows that in the research area, 0.7% of farmers sought services from Dhaka Medical College Hospital for their most serious and challenging health issues following the establishment of the TBIP, compared to 0.5% of farmers who sought these services prior to the TBIP. Surprisingly, however, only 26.0% of farmers received diagnostic services from private diagnostic centers prior to the TBIP, compared to 48.0% of farmers who received diagnostic services from centers like Popular, Apollo, Lab Aid, Update, Islami Bank Diagnosis Center, and Prime Medical Center. According to the above data, the number of farmers in the research area who received diagnostic services from government diagnostic centers decreased (70.5–45.3 = 25.2%), whereas the number of farmers who visited private diagnostic centers grew (48–26 = 22%) following the TBIP. According to research, the majority of farmers are now able to handle the costs of their diagnoses through their agricultural output because they are receiving their services from private diagnostic centers. They now lead better and healthier lives than they did before the arrival of the TBIP in the study region due to increased crop production made possible by the TBIP.

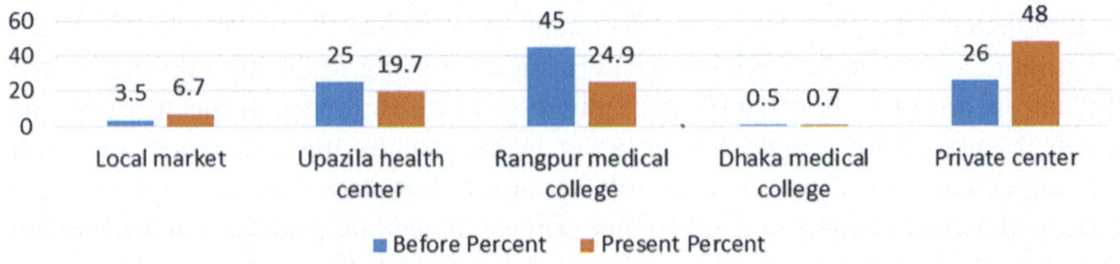

Fig. 3 Places of diagnosis facilities

4.4 *Caesar or Operation Places Before and After the TBIP*

Since the 1990s, the number of private-sector health-care facilities in Bangladesh has rapidly increased. Numerous contemporary private hospitals have been founded thus far, including the Apollo, Square, Lab Aid, Holy Family, Kumudini, and Ad-din hospitals, which are mostly located in Dhaka. As a result, due to travel difficulties and physical distance, persons who live outside of Dhaka occasionally have difficulty accessing these facilities. However, because private hospitals and clinics charge exorbitant fees for their services, the majority of the population in Bangladesh is unable to access these facilities. Eighteen percent of people live below the poverty line, compared to 24% in rural areas. A significant fraction of the population in the nation cannot afford to maintain their diet, cleanliness, or health care or undergo Cesarean sections because of the high prevalence of poverty. Therefore, it is evident that due to poverty and expensive costs, impoverished farmers do not have access to advanced operations or Cesarean treatments.

According to this study's findings (Fig. 4), 1.7% of farmers received surgery or Cesarean sections from district health centers or hospitals prior to the TBIP, whereas 2.5% of farmers currently receive these treatments from the same location. Additionally, it was discovered that while 18.9% of farmers currently receive their operations and Cesarean sections from Rangpur Medical College Hospital, 15.9% of farmers received these services from the hospital prior to the TBIP. According to the study, only 0.5% of farmers had a Cesarean section or operation performed at Dhaka Medical College Hospital prior to the TBIP, and 0% had one after the TBIP or currently have one. The study also looks into the fact that before the TBIP, 4.7% of farmers in the study area had operations or Cesarean sections performed by private clinics, whereas currently, 19.2% of farmers receive the same treatments from the same location. Furthermore, it was discovered that 77.1% of farmers did not receive any surgical or Cesarean care, nor did they have any family members undergoing surgery or Cesarean sections prior to the TBIP, whereas 59.5% of farmers did not receive any surgical or Cesarean care, nor did they have any family members undergoing surgery or Cesarean sections following the TBIP who are currently residing in the study area. Lastly, after the TBIP, there has been an increase in the

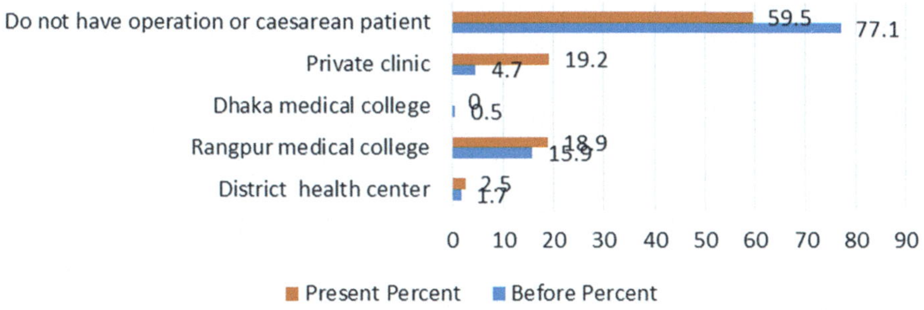

Fig. 4 Places of caesar or operation

number of patients attending private clinics (19.2–4.7 = 14.5%) and those undergoing operations or Cesarean sections (77.1–59.5 = 17.6%). Thus, it can be concluded that the TBIP has improved the farmers' economic circumstances in the project area, enabling them to obtain these services from better locations.

4.5 *Midwives at Childbirth Before and After the TBIP*

A skilled birth attendant (SBA) is "someone trained to proficiency in the skills needed to manage normal (uncomplicated) pregnancies, childbirth, and the immediate postnatal period, as well as in the identification, management, and referral of complications in women and newborns," according to the World Health Organization (WHO, 2004). Over 500,000 women globally lose their lives to obstetric problems each year (Thaddeus and Maine, 1994). Professional care during childbirth and in the initial postpartum phase can avert fatalities and severe birth complications (Liljestrand, 2000; Graham et al., 2001; Donnay, 2000). Additionally, it has been noted that the percentage of births aided by a trained birth attendant has become an indicator for assessing the decline in maternal mortality. The political leadership in Bangladesh is dedicated to lowering the death and morbidity rates among mothers and newborns, and the government has already begun training midwives.

According to this study's findings (Fig. 5), 20% of farmers responded that experienced midwives were present during childbirth prior to the start of the TBIP, but 85.3% of cases indicated that professional midwives were present during child delivery, either currently or after the TBIP. It means that the majority of farmers are now aware of their expectant mothers and their newborn children. The survey also reveals that currently, 14.7% of cases involve untrained midwives attending child delivery in the study area, whereas 80.0% of farmers contend that unskilled midwives were present during child delivery before the TBIP. The survey also indicates how infrequently midwives were called in to assist with deliveries before the TBIP was introduced in the study area. According to the above data, the study area's TBIP resulted in a decrease (80.0–14.7 = 65.3%) in the rate of unskilled midwives at childbirth and a rise (85.3 - 20 = 65.3%) in the rate of trained midwives at child delivery. Furthermore, because of the socioeconomic growth brought about by the

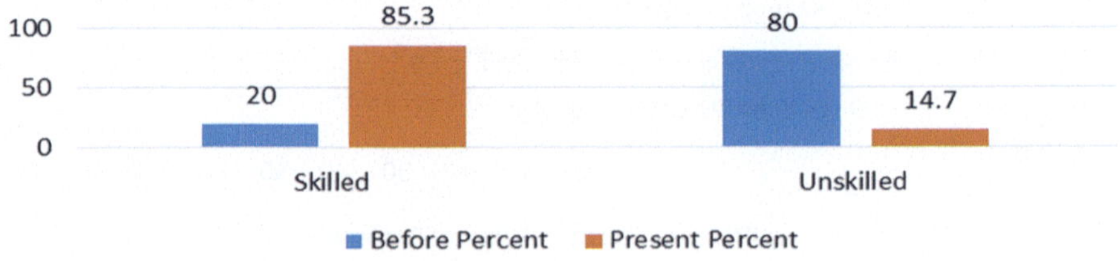

Fig. 5 Midwives attended during child born

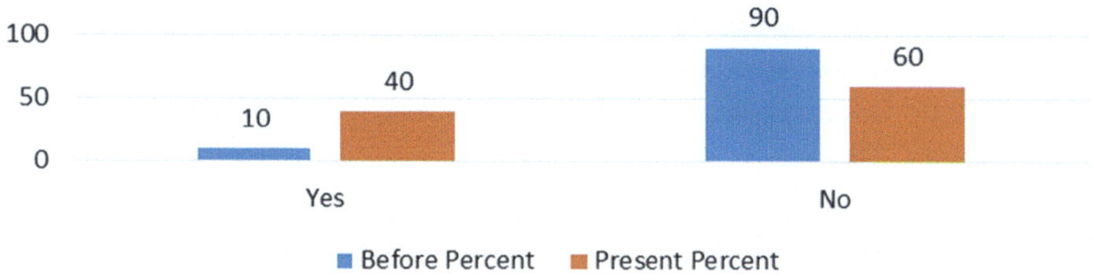

Fig. 6 Medicine for primary health

TBIP, it is also demonstrated that farmers are cognizant and aware of the care provided to expectant mothers and newborn infants in the research area.

4.6 *Medicine for Primary Health Care*

Primary health care (PHC) offers preventive, curative, and rehabilitative services to address the community's primary health concerns. One of the eight fundamental elements of PHC, as stated in the Alma-Ata Declaration, is the supply of critical medications. The World Health Organization (WHO) defined essential medicines as those that meet the population's priority health care needs. They are chosen with consideration for comparative cost-effectiveness, evidence of safety and efficacy, and public health importance. Essential medications should always be accessible within the framework of operational health systems, in sufficient quantities, in the right dose forms, with guaranteed quality and sufficient information, and at a cost that both the community and the individual can pay.

According to the findings of this study (Fig. 6), 10% of farmers kept primary medications in-house prior to the TBIP, whereas currently, 40% of farmers do so for the treatment of their families. Additionally, it was discovered that 60% of farmers in the research area do not currently keep primary medicine for their family members, whereas 90% did not do so before the establishment of the TBIP. The aforementioned data indicates that following the TBIP, there was a 30% rise in the number of primary medicine stockers among farmers in the research area (40.0–10.0). In conclusion, it can be stated that farmers' awareness of primary health care is still lacking, but things are gradually getting better, and the TBIP is helping them manage their families' primary health-care needs.

4.7 *Categories of Houses*

According to Schoenauer (2000), a house is a structure with plumbing, ventilation, and electrical infrastructure that serves as a place of residence. In Bangladesh, around 85% of households reside in houses with tin roofs. The use of cement or

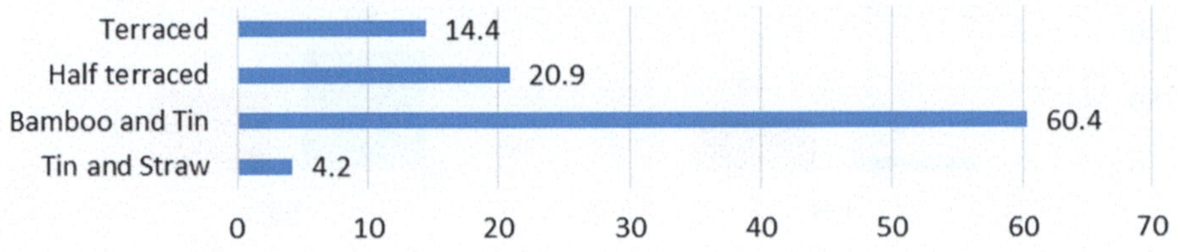

Fig. 7 Categories of house

ceramic tiles for roofing varies significantly between urban and rural areas. Cement tile use is more than five times as common in urban households as it is in rural ones. In 2014, tin made up 43% of outer walls, whereas in the 2007 Bangladesh Demographic and Health Survey (BDHS), 40% of homes had walls built from natural materials, including cane, palm, trunks, soil, or bamboo mixed with mud (NIPORT et al., 2009). Walls composed of cement are seen in 26% of homes. Tin wall construction is more common in rural households (48%) than in urban ones (30%). On the other hand, compared to 16% of rural families that have cement walls, the percentage of urban households with cement walls is three times higher.

According to the above data (Fig. 7), farmers' houses are composed of tin and straw in 4.2% of cases, bamboo and tin in 60.4%, half-patinated sections in 20.9% of cases, and terraced areas in 14.4% of cases. Based on the study above, one could infer that bamboo and tin make up the majority of farmers' homes. In the Dimla upazila, however, the Population and Housing Census (2011) revealed that 2.5% of general households reside in jhupri, 90.5% in kutcha houses, 6.1% in half-terraced houses, and 0.9% in pucca houses. Additionally, it is discovered that in Jaldhaka upazila, 91.9% of general households reside in kutcha houses, 6.7% in half-terraced houses, 0.9% in pucca houses, and the remaining 0.5% in jhupri houses. Thus, it was found that the bulk of farmers' houses (60.4%) are constructed of tin and bamboo. Additionally, a relationship between the farmers' house category and occupation is observed. For instance, their home is constructed entirely of bamboo, tin, and straw reflecting the lack of alternative employment opportunities for those without access to farming.

4.8 Categories of Sanitation

The provision of facilities and services for the secure disposal of human waste and urine is commonly referred to as sanitation. Improving sanitation is recognized to have a big positive influence on health in both homes and communities as inadequate hygiene is a major global cause of disease. Additionally, sanitation includes the upkeep of hygienic conditions through services like wastewater disposal and garbage collection (WHO, 2017). Around the world, 4.9 billion people used upgraded sanitation facilities in 2015, while 2.4 billion did not. Nine hundred forty-six million people who had no access to any facilities at all were among those who

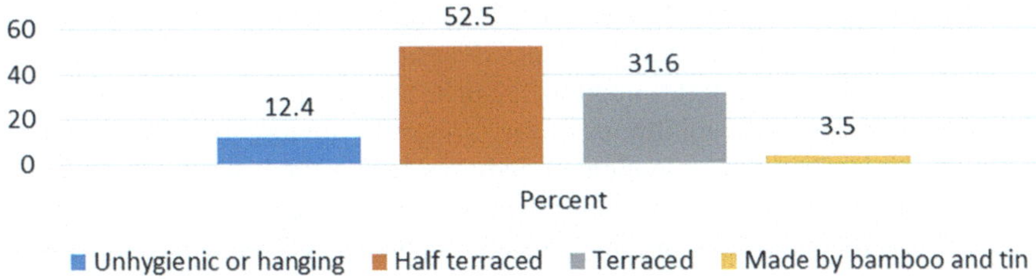

Fig. 8 Categories of sanitation

lacked proper sanitation and continued to defecate outside. The improper handling of wastewater and feces still poses a serious threat to public health and the environment (UN Economic and Social Council, 2016). According to this study, 12.4% of farmers use unsanitary restrooms refer to unhygienic (Fig. 8). Unhygienic refers to these toilets posing health risks due to poor sanitation and that the majority are located in inconvenient locations for small children and the elderly.

In addition, it has been noted that 52.5% of farmers use half-terraced restrooms, which feature tin or straw for the roof and cement and brick for the walls. Additionally, data indicates that 31.6% of farmers make use of terraced toilets. A terraced toilet is one made up of brick and cement walls and roof. An additional data point shows that 3.5% of farmers utilize tin-and-bamboo toilets. However, the 2011 Population and Housing Census revealed that in the Dimla upazila, general families use sanitary latrines in 27.7% of cases, nonsanitary latrines in 60.5% of cases, and no toilets in 11.8% of cases. Additionally, it was discovered that in the Jaldhaka upazila, 15.8% of general families lack toilets, 50.3% use nonsanitary latrines, and 33.9% use sanitary latrines. Thus, it can be said that up until 2016, the number of people using restrooms has increased and the facilities have improved. According to the analysis, the majority of farmers (52.5%) use half-terraces, which consist of tin or straw for the roof and cement and brick for the walls. Because the TBI Project increased crop output, which improved the farmers' economic situation, it can be claimed that the TBIP has a role in improving the sanitation conditions of farmers in the research area.

5 Research Synopsis

A few criteria have been chosen, and information has been gathered from the study region in order to satisfy the research purpose. Following the examination of field data, it was discovered that the bulk of farmers' houses (60.4%) were constructed with tin and bamboo. Additionally, a relationship between the farmers' house category and profession was observed. Their home is made entirely of bamboo, tin, and straw since it was found that those who did not work in agriculture had no other occupation except farming. It is also evident that the majority of farmers (52.5%)

used half-terraced toilets, which have a tin or straw roof and a wall composed of cement and brick. Since the TBIP enhanced agricultural output, which in turn improved the farmers' economic situation, it can be claimed that the TBIP had a role in improving the sanitation conditions of the farmers in the research area. Farmers in the study region had better hygienic conditions when the economy was doing well. Additionally, the study looked into the growing awareness of farmers' health concerns. It noted that health consciousness among the farmers increased as a result of improvements in their economic ability, which was better than after the TBIP. Additionally, the research highlighted that the TBIP contributed to higher crop yields, which benefited farmers and their families by enabling them to access better health care inside or outside of their communities. In an attempt to ascertain the level of mother-and-child health consciousness within the family, the study found that while the majority of farmer families (64.2%) were aware of the advantages of breastfeeding, the union parishad was not operating at optimal efficiency.

Prior to the TBIP, 60.0% of farmers used public transportation, whereas after the establishment of the TBIP, the majority of farmers (78.0%) used microbuses to transfer their patients to health centers. A sufficient and appropriate supply of TBIP water increased crop yield, thereby enhancing the socioeconomic status of farmers. According to the survey, before the TBIP began, the majority of farmers (45.0%) received their diagnoses from Rangpur Medical College Hospital, while after the TBIP, the majority (44.5%) obtained their diagnoses from private diagnostic centers, indicating the growing financial ability to pay private health bills. Additionally, it noted that the 80% of delivery instances involving unskilled midwives were present prior to TBIP, whereas the majority (85.3%) involving professional midwives were present during child delivery. It also showed that the majority of youngsters (71.1%) received all necessary immunizations, making them immune to all infections. Due to the farmers' socioeconomic development through the TBIP, they became cognizant and aware of the care provided for expectant mothers and newborn children in the research area. The study also investigated how farmers' awareness of primary health care was growing daily and how the TBIP helped them handle primary care for their families. When comparing the circumstances surrounding Ceasarian sections among the farmer's family members' before and after the TBIP, it was discovered that 22.9% of the farmers received these services from various health centers, with the majority (15.9%) receiving them from private clinics or health centers prior to the TBIP. Additionally, it was shown that 41.1% of the farmers received surgery or Cesarean sections from various types of health centers, with the majority (19.2%) receiving these treatments from private clinics or health centers. One could argue that the TBIP improved the farmers' financial circumstances in the project region, enabling them to obtain these services from more advantageous locations. Thus, it is possible to draw the conclusion that the TBIP improves the health of the farmers in the research regions, demonstrating the potential for irrigation projects to enhance the health of the farmers' families.

6 Conclusion

Five districts in the northern region of Bangladesh—Nilphamari, Rangpur, Dinajpur, Jaipurhat, and Gaibandha—were included in the Teesta Barrage Irrigation Project. It was believed that the majority of Bangladesh's northern region was susceptible to drought (Begum & Momen, 2020). Because of the region's dryness, irrigation from rivers, subterranean water, and rain was essential to agricultural operations. It was noted that the production of paddy and other crops in that area needed proper irrigation. Thus, the government launched the TBI Project to provide irrigation water during the dry season in order to meet agricultural needs. The project significantly benefited the farmers in the study area. Proper and sufficient irrigation during the Aus and Rabi seasons enhanced agricultural output and farmers' revenue. The primary beneficiaries of this project were the farmers, who were able to meet the health and educational demands of their families with ease.

In addition to being able to produce more crops than they previously could, the farmers in that area were able to save more money for their ongoing medical needs. Even though health care was very expensive, people made every effort to cope by saving as much as possible and earning money from farming. The TBIP supported their ability to spend money on bettering their health and increasing their income. The study discovered notable changes in the farmers' health status, which could be attributed to the TBIP in that area. For example, farmers were able to use private diagnostic centers for diagnostic services, and they also transported patients to more reputable health-care providers, such as private clinics or individual doctors using private vehicles. The Teesta Barrage Irrigation Project in the study region has enabled farmers to access improved health facilities from more reputable treatment centers.

References

Andaleeb, Syed Saad, Siddiqui Nazlee and Khandakar Shahjahan (2007). Patient satisfaction with health services in Bangladesh. Health Policy and Planning 2007; 22:263–273. *Advance Access publication*, 15 June 2007. Available at- https://academic.oup.com/heapol/article/22/4/263/620111

Ara, I., Lewis, M., & Ostendorf, B. (2016). Spatio-temporal analysis of the impact of climate, cropping intensity and means of irrigation: An assessment on rice yield determinants in Bangladesh. *Agriculture & Food Security, 5*(1). https://doi.org/10.1186/s40066-016-0061-9

Begum, M. M., & Momen, M. N. (2020). Drought. In S. Romaniuk, M. Thapa, & P. Marton (Eds.), *The Palgrave encyclopedia of global security studies*. Palgrave Macmillan. https://doi.org/10.1007/978-3-319-74336-3_323-1

Begum, M. M., & Momen, M. N. (2021). Global climate change and its response. In F. Maggino (Ed.), *Encyclopedia of quality of life and Well-being research*. Springer. https://doi.org/10.1007/978-3-319-69909-7_104663-1

Benner, S., & Fendorf, S. (2010). Arsenic in south asia groundwater. *Geography Compass, 4*(10), 1532–1552. https://doi.org/10.1111/j.1749-8198.2010.00387.x

Chowdhury, A. (2021). Irrigation decisions and use of groundwater in Bangladesh: Perspectives on some evolving crisis. *American International Journal of Multidisciplinary Scientific Research*, 1–6. https://doi.org/10.46281/aijmsr.v1li1.1400

Dhongade, M. P., & Dangat, S. B. (1985). *Socio-economic bench-mark survey of the Sina medium irrigation project command area in Maharashtra.* Department of Agricultural Economics, Mahatma Phule Agricultural University. Available at- https://find.uoc.ac.in/Record/65720

Donnay, F. (2000). Maternal survival in developing countries: What has been done, what can be achieved in the next decade? *Int J GynaecolObstet* 2000; 70: 89–97. Available at- https://pubmed.ncbi.nlm.nih.gov/10884537/#:~:text=Priority%20interventions%20include%3A%20(1),district%2Dlevel%20planning%20with%20community

Gbenga, O., Babatunde, R. O., Adenuga, A. H., & Olagunju, F. I. (2015). Irrigation and income-poverty alleviation: An assessment study of Kampe Irrigation Dam in Kogi State, Nigeria. *The Journal of Agricultural Sciences, 10*(2), 76–86. Available at- http://www.sab.ac.lk/media/agri-journal/volume-10/no-2/Vol-10-No-02-A01.pdf

Graham, W. J., Bell, J. S., & Bullough, C. H. (2001). Can skilled attendance at delivery reduce maternal mortality in developing countries? In V. De Brouwere & W. Van Lerberghe (Eds.), *Safe motherhood strategies: A review of the evidence; studies in health services organization and policy, 17* (pp. 97–130). ITG Press. Available at- https://www.researchgate.net/publication/239924607_Can_Skilled_Attendance_at_Delivery_Reduce_Maternal_Mortality_in_Developping_Countries

International Water Management Institute. (2017). *Managing irrigation water for safe domestic use*. Available at-http://www.iwmi.cgiar.org/issues/water-and-health/irrigation-and-health/

Islam, N. M., Azam, M. A., & Islam, Q. R. (2004). Teesta River water sharing: A case study in Teesta Barrage project. 2nd Asian regional conference. *International Commission on Irrigation and Drainage.* Moama NSW, Australia. Available at- https://eurekamag.com/research/004/341/004341322.php

Jahan, I., Abedin, A., Islam, M., Hossain, M., Hoque, T., Quadir, Q., & Rahman, M. (2021). Translocation of soil arsenic towards accumulation in rice: Magnitude of water management to minimize health risk. *Water, 13*(20), 2816. https://doi.org/10.3390/w13202816

Jeremy, H. (2000). *Challenges for health systems in member countries of the organisation for economic co-operation and development.* Bulletin of the World Health Organization. Available at- https://www.researchgate.net/publication/12402827_Challenges_for_health_systems_in_Member_Countries_of_the_Organisation_for_Economic_Co-operation_and_Development

Jeuland, M., Harshadeep, N., Escurra, J., Blackmore, D., & Sadoff, C. (2013). Implications of climate change for water resources development in the ganges basin. *Water Policy, 15*(S1), 26–50. https://doi.org/10.2166/wp.2013.107

Kirby, M., & Mainuddin, M. (2022). The impact of climate change, population growth and development on sustainable water security in Bangladesh to 2100. *Scientific Reports, 12*(1), 22344. https://doi.org/10.1038/s41598-022-26807-6

Krupnik, T., Schulthess, U., Ahmed, Z., & McDonald, A. (2017). Sustainable crop intensification through surface water irrigation in Bangladesh? A geospatial assessment of landscape-scale production potential. *Land Use Policy, 60*, 206–222. https://doi.org/10.1016/j.landusepol.2016.10.001

Liljestrand, J. (2000). Strategies to reduce maternal mortality worldwide. *Current Opinion in Obstetrics and Gynecology, 12*(6), 513–517. Available at- https://journals.lww.com/co-obgyn/abstract/2000/12000/strategies_to_reduce_maternal_mortality_worldwide.10.aspx

Mahtab, M., & Zahid, A. (2018). Coastal surface water suitability analysis for irrigation in Bangladesh. *Applied Water Science, 8*(1). https://doi.org/10.1007/s13201-018-0650-9

Mainuddin, M., Karim, F., Gaydon, D., & Kirby, J. (2021). Impact of climate change and management strategies on water and salt balance of the polders and islands in the ganges delta. *Scientific Reports, 11*(1), 7041. https://doi.org/10.1038/s41598-021-86206-1

Mohanty, Sanjay Kumar (1999). Agricultural Development and Demographic Response: A Study of Salandi Irrigation Project (Unpublished Ph.D. Thesis), *IIPS*, Mumbai. Available at-http://epc2006.princeton.edu/papers/60007

Momen, M. N., & Rahman, M. M. (2023). The impoverishment of political discourse and legal rights in transboundary Teesta water management. *World Water Policy, 9*(1), 11–28. https://doi.org/10.1002/wwp2.12093

Mukerji, S., & Kulkarni, S. (1989). *Demographic consequences of a development project- A study of Girna Irrigation Project, Maharashtra State, India*. IIPS Project Report. Available at-http://epc2006.princeton.edu/papers/60007

Nagdeve, D. A. (2006). Impact of an irrigation project on fertility and contraceptive behaviour in rural Maharashtra. Paper presentation at European Association of Population Studies. Available at http://epc2006.princeton.edu/papers/60007

National Institute of Population Research and Training. (2009). *Bangladesh Demographic and Health Survey 2007*. NIPORT, Mitra and Associates, and Macro International. Available at-https://dhsprogram.com/pubs/pdf/FR207/FR207[April-10-2009].pdf

Patil, R.G; Suryawanshi, S.D. and Kapse, P.M. (1978). The socio-economic survey of Girna irrigation project area in Jalgaon District (Maharashtra), *Department of Agricultural Economics, Mahatma PhuleKrishiVidyapeeth, Rahuri, Maharashtra*. Available at-http://epc2006.princeton.edu/papers/60007.

Paul, S., & Hasan, K. (2021). The impact of the proposed rubber dam facilitated surface water irrigation on adjacent groundwater at chapai Nawabganj district. *Bangladesh. Applied Water Science, 11*(2). https://doi.org/10.1007/s13201-021-01369-6

Population and Housing Census. (2011). Community report: Nilphamari. Bangladesh Bureau of Statistics (BBS), Ministry of Planning, Government of the People's Republic of Bangladesh. Available at- http://203.112.218.65:8008/WebTestApplication/userfiles/Image/PopCen2011/COMMUNITY_Nilphamari.pdf

Prasaratkul, P., Chayovan, N., & Thongthai, V. (1985). Demographic impacts of rural development related to an irrigation project. In R. E. Bilsborrow & P. F. DeLargy (Eds.), Impact of rural development projects on demographic behaviour (Policy Development Studies No. 9, pp. 205–220). United Nations Fund for Population Activities. The online version isn't available, you can access it through the New York Public Library: libraryopac.searo.who.int

Rahman, M. S., Ashaduzzaman, M. A. S., & Rahman, M. M. (2005). Poor people's access to health Services in Bangladesh: Focusing on the issues of inequality. In *NAPSIPAG annual conference 2005*. Available at- https://www.semanticscholar.org/paper/Poor-People%27s-Access-to-Health-Services-in-Focusing-Rahman-Ashaduzzaman/decf07cd5a1e616a3091cf1f1153573f9f04346f

Rahman, A., Hosono, T., Mazumder, Q., & Jahan, C. (2018a). *Sustainable groundwater management in context of climate change in northwest Bangladesh*. https://doi.org/10.5772/intechopen.73305

Roy, T. K. (1983). *Impact of Rajasthan Canal Project on social, economic and environmental conditions*. Allied Printers. Available at-http://epc2006.princeton.edu/papers/60007

Salehin, M. (2023). Utilization of solar energy in irrigation systems in Bangladesh. *Energy and Power Engineering, 15*(12), 468–481. https://doi.org/10.4236/epe.2023.1512026

Sarker, D. C., Pramanik, B. K., & Zerin, A. I. (2011). Climate impact assessment: A case study of Teesta barrage irrigation project in Bangladesh. *International Journal of Civil and Environmental Engineering IJCEE-IJENS, 11*(01) Available at- http://ijens.org/Vol%2011%20I%2001/1110501-7474%20IJCEE-IJENS.pdf. Retrieved 25 Mar 2015

Schoenauer, N. (2000). *6,000 years of housing* (rev. Ed.). W.W. Norton & Company.. Available at- https://www.amazon.com/000-Years-Housing-Revised-Expanded/dp/0393731200.

Shahen, M. A., & Momen, M. N. (2024). Assessing the nexus between irrigation project and crop production. In *The Palgrave handbook of environmental policy and law. Palgrave studies in sustainable futures*. Palgrave Macmillan. https://doi.org/10.1007/978-3-031-30231-2_35-1

Shahid, S. (2010). Impact of climate change on irrigation water demand of dry season boro rice in Northwest Bangladesh. *Climatic Change, 105*(3–4), 433–453. https://doi.org/10.1007/s10584-010-9895-5

Shoeb, A., & Shafiullah, G. (2018). Renewable energy integrated islanded microgrid for sustainable irrigation—A Bangladesh perspective. *Energies, 11*(5), 1283. https://doi.org/10.3390/en11051283

Stoeckel, J., Prasith-rathsint S., Piampiti S., & Yongkittikul, T. (1983). Fertility impacts of irrigation and electrification in Northeast Thailand. *USAID*. Available at-http://www.popline.org/node/398964.

Sumiya, N., & Khatun, H. (2016). Groundwater variability in Bangladesh: Assessment based on rainfall variation and use of water in irrigation. *Journal of the Asiatic Society of Bangladesh Science, 42*(2), 177–189. https://doi.org/10.3329/jasbs.v42i2.46221

UN Economic and Social Council. (2016). Progress towards the Sustainable Development Goals. *Report of the Secretary-General.* Available at-https://sustainabledevelopment.un.org/sdg6. Source: Report of the Secretary-General, "Progress towards the Sustainable Development Goals". E/2016/75*.

World Health organization. (2004). *Making pregnancy safer: The critical role of the skilled attendant.* A joint statement by WHO, ICM and FIGO. Available at- https://www.who.int/publications/i/item/9241591692

World Health Organization (2017). Geneva, Switzerland. Available at- http://www.who.int/topics/sanitation/en/

World Bank. (2000). World development report 2000/2001: Attacking poverty. Washington, DC: World Bank. Retrieved from https://documents1.worldbank.org/curated/en/230351468332946759/pdf/World-development-report-2000-2001-attacking-poverty.pdf

World Health Organization. (2017). Bangladesh National Health Accounts: An overview on the public and private expenditures in health sector. https://www.who.int/bangladesh/news/detail/05-10-2017-bangladesh-national-health-accounts-an-overview-on-the-public-and-private-expenditures-in-health-sector

Thaddeus, S., & Maine, D. (1994). Too far to walk: Maternal mortality in context. Social Science & Medicine, 38(8), 1091–1110. https://doi.org/10.1016/0277-9536(94)90226-7

Inventories of Existing Financial Support Mechanisms and Identifying Financing Tools Promoting Irrigation Efficiency for Smallholder Farmers in the Mediterranean Region

Nadya Wahid, Joseph Piscopo, Reno Micallef, Wael Khairy, Mariana Salvado, and Regina Krause

Abstract To encourage irrigators to adopt water-saving irrigation technology, it is necessary to provide clear information on financing mechanisms and investment tools that support these innovations. The aim of this work is to analyze existing financial mechanisms to adopt new, efficient irrigation technologies studied in some Mediterranean countries. The collection of data on financing options was conducted through qualitative interviews using a questionnaire. In this work, different financing mechanisms are revealed; offer a range of options to meet the needs of each farmer for use a case study technology. The strategy of implementation roadmap financial planning must be checking the specific funding programs and the matrix of demand funding by country studied. This could help ensure the economic and social development of farmers and their families.

Keywords Financing mechanisms · Funding programs · Efficient irrigation systems · Water-saving irrigation technology

N. Wahid (✉)
University Sultane Moulay Slimane, Beni Mellal, Morocco
e-mail: n.wahid@usms.ma

J. Piscopo · R. Micallef
EcoGozo Directorate, Victoria, Malta
e-mail: joseph.d.piscopo@gov.mt; reno.micallef@gov.mt

W. Khairy
Faculty of Engineering, Heliopolis University, Cairo, Egypt
e-mail: wael.khairy@hu.edu.eg

J. C. Correia Leitão et al. (eds.), *Sustainable Water Management and Irrigation Systems*, Natural Resource Management and Policy 62,
https://doi.org/10.1007/978-3-031-93587-9_11

1 Introduction

Controlling water demand for urban, agricultural, and industrial uses appears to be a priority today for the sustainable management of a rare and fragile resource in the Mediterranean region. The scarcity of water resources and the growing demand for water, particularly in the agricultural sector, calls for improved efficiency and productivity in water use (FAO, 2011). Stakeholders in the irrigation sector, particularly decision-makers and irrigators, need new technologies for enhancing irrigation efficiency and water productivity in order to implement appropriate strategies for sustainable resource management (Tsur et al., 2004; Ward et al., 2008). The transition to a green economy that respects ecological balances and is likely to create opportunities for wealth and sustainable jobs is now a major objective of new strategic approaches to sustainable development, which are currently being adopted by some Mediterranean countries (Mélanie, 2007; Molle & Sanchis-Ibor, 2019). Financial reforms and incentive policies are already being implemented to improve the integration of environmental dimensions and promote the development of strategic areas, such as enhancing irrigation water efficiency (Ward, 2010; Plan Bleu, 2008).

Mediterranean sustainable development strategies focus on managing water demand by improving both agricultural efficiency and economic returns (Plan Bleu, 2008; Fernandez & Mouliérac, 2010). The long-term economic viability of irrigation infrastructure investments depends on the additional income farmers generate or the broader economic benefits these investments create for the nation (Wahid, 2023). Farmers, driven by economic incentives, will only invest in advanced irrigation technologies when they expect a positive return on their investment—meaning the projected income gains exceed the costs (Merriam & Freeman, 2007; Chopart et al., 2007a, 2007b). This economic framework motivates farmers to adopt water-saving irrigation technologies and adjust their farming practices accordingly. Decision-support tools that help farmers evaluate both short-term and long-term investment options are therefore essential for improving both the economic performance and environmental sustainability of their operations.

In this context, technological advancement in irrigated agriculture faces two major challenges at both environmental and economic levels, which is why it receives substantial public support (Malchev et al., 2022; Wahid et al., 2022). These technologies must simultaneously achieve two critical objectives: conserving water resources while increasing farm productivity. To address these dual requirements, public incentive programs aim to improve the cost-benefit ratio of adopting new

M. Salvado
Município do Fundão, Fundão, Portugal
e-mail: mariana.salvado@cm-fundao.pt

R. Krause
Hochschule Wismar, University of Applied Sciences: Technology, Business and Design, Wismar, Germany
e-mail: regina.krause@hs-wismar.de

irrigation techniques. This support takes multiple forms, including subsidizing farmers' equipment purchases and providing training and advisory services to help irrigators implement these technologies effectively.

Building a green economy will not be possible without establishing innovative partnerships with the private sector, local authorities, and civil society. These partnerships should make it possible to mobilize the necessary investments and technological solutions, promote local skills, and strengthen everyone's commitment and solidarity (Nation Unies, 2014; PAGE, 2022). Financing eco-innovative and efficient irrigation water technologies will be the real drivers of the green economy. The capacity to adopt new, efficient irrigation water technologies must benefit from new green financing tools and appropriate support measures. Particular attention should be paid to the integration of the informal sector, given its importance. The role of the banking sector in financing the green economy must be strengthened; banks are called upon to develop financial products geared toward the green economy (PAGE, 2022). Furthermore, highlighting the mechanisms and financing options for efficient irrigation water technologies in Mediterranean countries is necessary for small farmers, decision-makers, and field operators. However, a considerable challenge arises from the lack of information on financing mechanisms and investment tools to support key technologies for efficient irrigation. Therefore, the aim of the present work is to provide baseline inventories of existing financial mechanisms and to analyze financing options—focusing especially on public financing instruments, the involvement of banking sector funds, and public-private partnerships—to adopt new, efficient irrigation technologies studied in some Mediterranean countries (self-regulating, low-energy, clay-tube irrigation (SLECI) technology for smallholder irrigation; the reuse of treated conventional wastewater utilizing instream wetland technology; and solar desalination of salt water in the greenhouse).

2 Methodology

The analysis of financing mechanisms is based on three study models that examine technologies designed to promote irrigation efficiency for smallholder farmers. These irrigation technologies are examined within the framework of the MedWet project. The case studies describe exploratory technologies for efficient irrigation as follows:

Model Case Study 1 Self-regulating low-energy clay-based irrigation (SLECI) technology serves as a micro-irrigation system designed specifically for smallholder farmers. The main objectives of SLECI are to minimize water consumption by crops, reduce energy usage, and consequently lower pumping costs per unit of yield for farmers. Additionally, this technology optimizes farmers' time management by improving yield efficiency per hour spent in the field. The technology is currently being tested at eight pilot sites across three countries: Morocco, Malta, and Portugal.

Model Case Study 2 Reuse of treated conventional wastewater utilizing instream wetland technology—this method is implemented in Egypt. Public authorities and/or private entities can use a collective treatment network to generate additional water at a lower cost. The treated effluent water can be used to irrigate non-fruit trees or nonedible crops, such as cactus.

Model Case Study 3 Solar desalination of salt water in a greenhouse setting—this technology is implemented in Malta. The water produced is used for crop irrigation.

The collection of data on financing options promoting agricultural water efficiency and best water management practices in Morocco, Malta, Egypt, and Portugal is conducted through qualitative interviews using questionnaires (Fig. 1). The questionnaires contain necessary information on the nature and diversity of irrigation efficiency financing programs for small farmers, financial options, and financial models promoting irrigation efficiency for smallholder farmers.

We carried out a semi-descriptive interview with the partners of the Med-Wet project, which established pilot sites in the countries of Malta, Portugal, and Egypt. The questionnaires were sent to them via electronic mail. For the SLECI technology, partners from Portugal and Malta participated. With regard to the reuse of treated wastewater, we communicated with the partner from Egypt. For the desalination technology, we exchanged information with the partner from Malta. At the Morocco level, direct interactions were made with the following establishments: Crédit Agricol bank, the National Agricultural Council Office (ONCA), the Chamber of Agriculture, and the National Research and Valorization Agricultural Office of Tadla (ORNVAT).

3 Results

3.1 Financing Options for the SLECI Technology Study Model

3.1.1 Case of Morocco

3.1.1.1 Public Investment Funds

The promotion of economic growth and the valorization of water mobilized in irrigation are focused on the massive conversion to localized irrigation within the framework of the National Program for Saving Water in Irrigation (PNEEI) in Morocco (PMV, 2018). As part of this program, Morocco has made a colossal effort to develop public financing for major hydraulic developments and infrastructure, followed by a transition to localized irrigation. Also, financial aid from the state for the promotion of agricultural investments is provided, thanks to the irrigation extension program (PEI). There are also some ministerial projects that can finance water-saving initiatives in irrigation, such as the case of the National Solar Pumping

Questionnaire on financing programs promoting irrigation efficiency for smallholder farmers

1. Information on the nature and diversity of irrigation efficiency financing programs for small farmers

1.1 Is the irrigation efficiency financing program for small farmers national at the country level?

☐ Yes

☐ No

1.2 If the answer is no, please provide the irrigation efficiency financing program for small farmers by region where the pilot sites are installed (3 in Malta, 2 in Portugal and 1 in Egypt).

1.3 The nature and source of funds to finance irrigation efficiency for small farmers are:

☐ Public,

☐ Private,

☐ International (public-private).

For each type of irrigation investment fund present, please respond by document on the financing options 2 and 3 below.

2. Financial options

2.1 Financial Mechanism (e.g. Rates and ceilings according to the individual or collective irrigation project, etc.)

2.2 Financial instrument (includes financial securities and financial contracts),

2.3 Eligibility criteria (For each type of irrigation investment fund),

2.4 Access modalities (Creation of the financial aid application file),

2.5 Application procedures (Stages and deadlines for making investments and submission of funding application files).

3. Matrix with suitable financial models promoting irrigation efficiency for smallholder farmers

Thank you

Fig. 1 Questionnaire on financing programs promoting irrigation efficiency for smallholder farmers

Program in Water Saving Projects in Irrigation. This program is carried out through a partnership between the Ministry of Energy, Mines, Water and Environment; the Ministry of Agriculture; the National Agency for Development of Renewable Energies and Energy Efficiency (ADEREE); and the bank "Crédit agricole du Maroc." With an overall budget of 400 million dirhams (DH), the project aims to install a network of 3000 photovoltaic pumping systems with a total power of 15 MW.

The Minister of Agriculture, Maritime Fisheries, Rural Development and Water and Forests; the Minister of the Interior; and the Minister Delegate to the Minister of the Economy and Finance have established the modalities for obtaining and granting appropriate public financial assistance for localized and supplementary irrigation (GG, 2023).

3.1.1.1.1 Public Financing Ceiling for Localized Irrigation Projects

For areas with a size of 5 ha, farmers can have an investment fund capped at €3800 per net hectare equipped with agricultural operations, including an irrigation network (well, pumping device, head station, pipe, dripper, etc.). If it is necessary to resort to the construction of basins for the storage of irrigation water, this ceiling may be increased by an amount not exceeding €1200 per net equipped hectare.

For areas of more than 5 ha and less than or equal to 20 ha, localized irrigation projects carried out by operating farmers may have an investment fund capped at €2850 per net equipped hectare. If it is necessary to resort to the construction of basins for the storage of irrigation water, this ceiling may be increased by an amount corresponding to €900 per net equipped hectare.

3.1.1.1.2 Grant Application Documentation Requirements

Before the investment is made, the applicant submits the application file for examination of the localized irrigation project or supplement in electronic and paper version. This file must include the following documents:

- A request for prior approval. Documents and pieces allowing the applicant to be identified:
  - Copy of the applicant's identification document;
  - Copy of the mandate and copy of the agent's identification document if applicable;
  - A copy of the statutes or the extract from the commercial register or the cooperative register,
  - A copy of the document designating the legal representative as well as a copy of the document identifying him;
  - Commitment of the farmers concerned to maintain the subsidized investment.
- Authorization for digging wells or drilling, if necessary.
- Copies of the necessary documents and authorizations justifying the availability of water resources to meet the project's water needs.
- The technical file of the project includes a technical calculation note making it possible to justify:
  - The adequacy between the available water resources and the water needs of the crops to be irrigated;

  - The sizing of the irrigation project;
  - The stability of the basin as well as the process of its construction and the devices for its security for water storage basins with a total storage capacity exceeding 10,000 m^2.

- The drawn parts of the project:

  - A detailed plan of the installation at an appropriate scale, signed and sealed with the mention approved by the engineer or technician who established the study;
  - If the relief is significant, a drawing of the detailed plan of the project must be drawn up, composed of contour curves showing the existing height differences and slopes (dimensional plan). A plan of the shelters specifying their dimensions and the construction materials provided.
  - Estimated quotes for equipment, materials and arrangements grouped by component of the project.
  - The supplier's commitment to equip the project with equipment complying with Moroccan Standard No. 12.1.100-2007 (Localized irrigation installation—General requirements).
  - The equipment test bulletins provided for the project equipment, issued less than 2 years ago.
  - Catalogs mentioning the operating characteristics of the main equipment of the installation: pumps, motors, control and adjustment valves, electrical and automation equipment, fertigation equipment and water distributors.
  - A copy of the documents justifying the technical qualifications of the companies or the approval of the design office which carried out the studies and which will supervise the completion of the project.

- Subsidy application: After approval of the investment, the applicant submits the subsidy application file for the localized irrigation project or supplement in electronic and paper version. This file includes the following documents:

  - Subsidy application;
  - The original final invoices detailed in quantity, unit price and total price and by component. The small farmers can present, instead of final invoices, the statements justifying the earthworks of the trenches for the pipes, when they are carried out by the farmers using their own means;
  - Copies of proof of payment of final invoices;
  - The applicant's RIB certificate.

3.1.1.2 International Investment Funds

International cooperation development through Public-Private Partnerships (PPP) represents a critical priority in the water, irrigation, and environmental sectors. This collaborative approach is exemplified by two flagship projects in Morocco: the localized irrigation project in the Guerdane perimeter of the Souss basin and the hydraulic energy production project in the Oum Er Rabiaa basin.

3.1.1.3 Private Investment Funds

Individually supplied capital. Internal funding and more conventional financial tools, like agricol bank loans (Banque de Crédit Agricol), are primarily used to meet the financial needs of the agricultural industry. At the same time, the majority of the funds required to pay for operating expenses come from sales of goods and services. The terms used by a traditional financial intermediary alone are insufficient to calculate risk.

3.1.2 Case in Malta

Malta's long dry season, 550 mm of annual rainfall that is concentrated in a few months, small land surface and catchment area, uneven and insufficient topographic relief, and aquifer characteristics all contribute to the country's lack of natural water resources. Furthermore, national water resources have been under increasing strain in recent decades due to population growth and economic expansion. Infrastructure and water resources. Of the 35.5 million cubic meters of potable water produced in 2022, 64% came from seawater desalination, with the remaining portion coming from groundwater. The second-largest water user on the islands, agriculture, also depends heavily on groundwater. As of 2017, highly polished reclaimed water (New Water) plants were put into service, increasing the capacity for water production. In 2022, 1.6 million cubic meters of new water were produced, mostly for use in agriculture.

Similar trends are anticipated for the commercial and agricultural sectors, which rely to differing degrees on the use of stressed groundwater resources to meet their water demands, in addition to the growing demand for drinkable water in the residential sector. The Water Services Corporation (WSC), the country's water and wastewater utility, makes sure that the demand for high-quality water services is consistently and sustainably satisfied. Malta's National Investment Plan (NIP) 2023–2033 aims to give a high-level overview of the country's current and expected water and wastewater service delivery challenges, identify any gaps in EU compliance, and prioritize investments required to upgrade and renew the country's infrastructure.

3.1.2.1 Public Investment Funds

The Maltese government will invest in the agricultural sector in the upcoming years (2022–2027) in accordance with the EU funding programming process. At the national level, small farmers can access financing mechanisms for irrigation efficiency.

There are different mechanisms for financing efficiency in agriculture in Malta. Among these financing mechanisms linked to localized irrigation for small farmers are the following:

1. Eco-Scheme 6: this financial instrument aims to promote the use of certified disease- and pathogen-resistant plant materials, along with the cultivation of crop and plant varieties with less demand for water. Farmers will receive compensation for the costs incurred on certified propagating materials, as well as be assisted in developing a crop plan to minimize the demand for water. Prespecified minimum and maximum direct payments per annum are set for the term ending 2027. The eligibility criteria for farmers are outlined in Malta's CAP Strategic Plan. During the commitment period of 1 year, farmers shall use certified disease- and pathogen-resistant plant materials, as well as propagating materials, to strengthen crop resilience, with at least one crop commitment included, as indicated in the crop plan at the application stage. The committed parcel is deemed eligible if it (i) consists of arable land, (ii) A minimum amount of agricultural activity must be conducted on the plot each year, and (iii) is kept in good agricultural condition all year round. Each year, the Agriculture and Rural Payments Agency (ARPA) will open short time windows, during which eligible parties may express their interest in applying for specific interventions.
2. AECC 1 (Integration and Maintenance of Autochthonous Maltese Species): the aim of this financial mechanism, which is subdivided into three measures, is mainly to conserve Malta's biodiversity. Its main objective is to support indigenous species, namely, carob and mulberry trees. To meet the eligible criteria, farmers need to maintain all relevant records, which will be presented to ARPA inspectors; also, the species needs to be identified and accredited. Each year, ARPA will open short time windows during which eligible parties may express their interest in applying for specific interventions. In relation to the carob and mulberry trees, these have to be found on eligible agricultural land, of which owners must own a minimum of 0.1124 ha.
3. AECC 2 (Maintenance of Trees): this scheme aims to support the maintenance of newly planted recommended tree species. The support rate is €44.49/tree/year, with a maximum of 18 trees per hectare. To meet the eligibility criteria, farmers must (1) have a minimum of 1124 m^2 of agricultural land, (2) keep records of their holdings for inspection, (3) be registered in the Integrated Administration and Control System (IACS) Farmer Registry, and (4) declare the type and location of the trees to be maintained. Each year, ARPA will open short time windows during which eligible parties may express their interest in applying for specific interventions.
4. Measure 4.1 (Support for Investments in Agricultural Holdings): a total of €4.3 million in public expenditure was allocated toward supporting investments in environmentally efficient systems and equipment, in buildings, and in irrigation through improved water management and storage. This list is not exhaustive. Notable examples may include the financing of new rainwater harvesting cisterns/reservoirs, investments in the use of recycled water, and improvements to existing irrigation installations (only if they offer potential water savings of 5–25%). For the eligibility criteria, farmers may be natural or legal persons, with legal persons being limited to small and medium-sized enterprises (SMEs).

5. Measure 4.4 (Support for Investments Linked to Agri-Envi-Climate Objectives): the scope of this measure was to support nonproductive investments (investments that do not seek to improve farmers' productivity) that have a positive impact on the environment. Particular focus was to incentivize the restoration of habitats and landscapes, soil conservation, and water management. For instance, this scheme covered the building and restoration of rubble walls, the planting of trees, and investments in Public Amenities of Natura 2000 sites. Farmers (part time or full time), groups of farmers, businesses, nongovernmental organizations (NGOs), public entities, land managers, and partnerships can apply for this financial instrument.
6. Measure 6.1 (Business Start-Up Aid for Young Farmers): this measure was intended to finance projects presented through a business plan by young farmers under the age of 41 years, facilitating their entry into the agricultural business. With this scheme, applicants had the opportunity to receive a percentage of their capital costs, up to €70,000, to start a business. A scheme with similar intentions is set to reopen in February 2024 under the Common Agricultural Policy (CAP) Strategic Plan 2023–2027. The budget allocated to this measure was €1 million in total public expenditure. Beneficiaries had to possess a holding or farm, be recognized as a farmer by the Employment and Training Corporation, and hold all necessary licenses/permits.

3.1.2.2 Private Investment Funds

Young farmers claim that the most significant push factors driving them to leave farming are the irregularity of income and the lack of capital to invest in modern machinery and equipment (Government of Malta, 2021). Other challenges include difficulties in accessing land due to the high costs of purchasing or renting, unless they inherit it; land fragmentation, making it difficult to acquire a sufficient area of land for a viable business; difficulties in obtaining loans for business investment due to the lack of collateral; resistance from older farmers—often those from whom they will eventually inherit land but must first work with—who are not open to new ideas; and limited opportunities for continuous professional development (Government of Malta, 2021).

3.1.3 Case in Portugal

Irrigation efficiency financing programs are carried out at the country level. The public financing programs are as follows.

3.1.3.1 PDR 2020 (Rural Development Program 2014–2020)

PDR 2020 is the financial instrument that, through FEADER/EAFRD (European Agricultural Fund for Rural Development), supports the agroforestry sector and rural development in the mainland. Its guiding principles aim for the sustainable growth of agroforestry activities across the national territory through integrated measures, actions, and operations in four main areas of intervention:

1. Innovation and knowledge.
2. Competitiveness and the organization of production.
3. Climate, environment, and irrigation efficiency.
4. Local development.

3.1.3.2 National Irrigation Program (PN Regadios)

The creation of the National Irrigation Program (PN Regadios) aims to develop sustainable irrigation in Portugal as part of a medium to long-term strategy, doubling the funds allocated to irrigation financing for the mainland from 2014 to 2020 under PDR 2020.

Thus, through loans obtained by the Portuguese government from the European Investment Bank (EIB) and the Council of Europe Development Bank (CEB), PN Regadios includes irrigation area interventions (new, rehabilitation, modernization, and pumping reinforcements) that make use of recent community initiatives to encourage productive investment and the corresponding financing conditions.

A resolution has been approved to increase the National Irrigation Program's ambition by expanding the intervened area and extending its duration until 2028.

3.1.3.3 Strategic Plan for the Common Agricultural Policy (PEPAC) 2023–2027

The Strategic Plan for the Common Agricultural Policy (CAP) for Portugal for the period 2023–2027 integrates support measures to achieve the specific objectives of the EU's Common Agricultural Policy (CAP). Through direct payments, sectorial measures for fruits and vegetables, vineyards, beekeeping, and rural development instruments, it implements CAP instruments funded by the EU through the European Agricultural Guarantee Fund (EAGF) and the European Agricultural Fund for Rural Development (EAFRD).

The Portuguese CAP Strategic Plan is structured into six intervention axes, with differentiated application at the regional level. The Rural Development axe includes the subtopic "Environmental and Climate Management," which encompasses the measure "Efficient Water Use."

3.2 *Financing Options for Model Case Study 2 in Egypt: The Reuse of Treated Wastewater Utilizing Instream Wetland Technology*

3.2.1 Public Investment Funds in Egypt

The Egyptian Ministry of Housing, Utilities, and Urban Communities (MHUUC) created the wastewater treatment reuse code (ECP 501/2015), which was enacted in 2015 (MHUUC, 2017). According to UNDP Egypt Accelerator Lab (2023), the first findings in Egypt discuss the application of key engineering constructed wetlands technology (CWT) for irrigation and water supply for smallholder farmers in the desert environment (El-Wahat El-Baharyea). In Egypt, all irrigation and drainage systems are governed by the Ministry of Water Resources and Irrigation (MWRI). Nonetheless, substantial expenditures are made in water supply, transfer, distribution, delivery, drainage, and irrigation infrastructure rehabilitation. These projects are considered public goods and are funded by the national budget as well as World Bank donor support (Abbas Saleh, 2018; Helmy & Roushdy, 2022). Therefore, irrigation and drainage infrastructure investments, operations, maintenance, and rehabilitation costs have historically been covered by the Ministry of Water Resources and Irrigation (MWRI), which does not charge farmers for irrigation water delivery other than the cost of pumping water from the source to the field. Additionally, the majority of farms in Egypt are small, and the conventional irrigation technique entails splitting a farmer's land into tiny basins so that he can sufficiently irrigate his plots.

3.2.2 International Investment Funds

International funds, such as those from the FAO, PENUD, the Netherlands, and others, are given to the SEKEM Development Foundation (SDF) in order to support the rural mode. The following are part of the trustworthy network of friends and partners that SEKEM has been expanding over the years: SEKEM friends in Germany, Austria, the Netherlands, and Scandinavia (SEKEM, 2024). By generating funds for SEKEM schools, Heliopolis University, and other cultural endeavors, they have been advancing SEKEM's mission for decades.

For example, in 2019, SEKEM Development Foundation (SDF) reinitiated SEKEM's old project for the development of villages in Belbeis. With the support of the staff and students of Heliopolis University for Sustainable Development (HUSD), the data collection phase was concluded successfully. Based on the information gathered, SEKEM was able to determine key areas for development and to plan for projects accordingly. To address the unemployment challenge in this area, SDF developed the "SEKEM Bedaya – Marilou Fund" project.

The project will support the residents of the Blebeis villages in starting their own business in order to enhance their economic and social conditions while establishing productive, lucrative, and environmentally friendly work. The support offered will be in the form of microcredit loans. The loan amount will be determined based on the type of project, starting from €37 going up to €3700.

As a trusted and reliable shareholder, the Triodos Bank in the Netherlands supports SEKEM's activities and vision. Marilou van Golstein was, among other roles, the chair of the management board of Triodos Investment Management, who contributed to the establishment of several microcredit programs with a focus on supporting women. She shares SEKEM's vision for rural prosperity. Thus, to honor her contributions to the SEKEM vision, especially in the context of microcredits, SEKEM named its new microcredit project "SEKEM Bedaya – Marilou Fund."

In the years 2007–2009, SDF-SEKEM acquired three new plots of desert land (with a total area of 6000 feddans), with the aim of establishing model farms, proving that biodynamic agricultural methods are the optimal choice for desert land reclamation, food security, and sustainable development in Egypt and to meet SEKEM's growing pure, biodynamic raw material needs. From SEKEM's first year until 2013, about 1600 feddans have been reclaimed and developed, including all necessary infrastructure and community centers for sustainable development on all three farms.

3.2.3 Private Investment Funds

Individually supplied capital: Achieving sustainable development, empowering women and economically marginalized groups, and making the best use of natural resources are all goals that the government has set for economic development in light of Egypt's Vision 2030 and its strategic goals to raise the standard of living for its citizens (Central Bank of Egypt, 2024). Through the development of people in the agricultural sector, the Central Bank of Egypt (CBE) has been pushing the banking industry to concentrate on sustainable financing while taking environmental, social, and governance factors into account.

Through a comprehensive approach consisting of four parallel tracks, the CBE supports smallholders: lining agricultural irrigation canals to improve water and land use efficiency; diversifying and increasing sources of income; building property and institutional capacity to ensure sustainability; and consolidating smallholdings to increase the productivity of their owners.

3.3 Financing Options for Model Case Study 3: Solar Desalination of Salt Water in the Greenhouse

With regard to water productivity in agriculture, in areas such as Malta, where water is scarce but has a high economic value, the cultivation of crops that have low water requirements but high economic return should be encouraged (Attard et al., 2007). It should be noted that in Malta, approximately 56% of the total water production for potable use comes from very expensive desalination plants (Attard et al., 2007).

It has been estimated that 80% of this water ends up in local sewers. This implies that artificial water available for irrigation has an even higher economic value. Moreover, as mentioned previously, the excessive extraction of groundwater for irrigation, which has led to increased salinity levels, has exacerbated matters and raised the economic value of water even further. Thus, water productivity in agriculture is of utmost importance (Attard et al., 2007).

Water desalination is a potential solution to present and future water shortages; it has gained a larger foothold in the water recycling market and is an effective irrigation technique that can benefit from the same financing instruments cited in detail in the case study of the self-regulating low-energy clay based irrigation (SLECI) technology established in Malta (see Sect. 3.1.2: Case of Malta). Solar desalination of salt water in the greenhouse technology has a helpful double effect for the production of treated water as well as for the production of salinity-resistant crops.

4 Discussion

For the models of effective irrigation technologies and systems under study, funding is essential (Bierlen & Barry, 2000; Ward, 2010). These would never come to pass without funding because they rely heavily on it. As a result, inadequate financial structures and conditions may make these water supply projects less competitive. It is important to distinguish between case study models that encourage irrigation efficiency. Depending on the nation (Egypt, Morocco, Morocco, Malta, and Portugal), different types of investments are used to finance technologies that promote efficient irrigation (Briscoe, 1999; Bierlen & Barry, 2000; Ward, 2010). The risks taken on by public financing (barrier financing) and the scope of international investment funds' coverage (barrier and risk financing) determine which financing instruments should be used in each nation. By creating rules and specifications for irrigation infrastructure, public financing, also known as barrier financing, can generally contribute to increased water and economic efficiency (Briscoe, 1999; Cuesta et al., 2005; Ward, 2010). Private investment projects (risk projects) are also hampered by a number of issues that need to be resolved with capital interventions.

In this work, various financing mechanisms offer a range of options to meet the unique needs of each farmer for use a case study technology (SLECI technology for smallholder irrigation, reuse of treated conventional wastewater using instream wetland technology and solar desalination of salt water in the greenhouse). However, financial planning strategy of implementation roadmap calls for check funding programs and matrix of demand funding by country (Fig. 2). For instruments funding programs, it's necessary to:

- Define the priority funding programs for application form per country:
  - Public and international funds for Morocco and Egypt, and
  - public funds for Malta and Portugal.
- Develop of financial institutions engagement model. The public financing programs are:

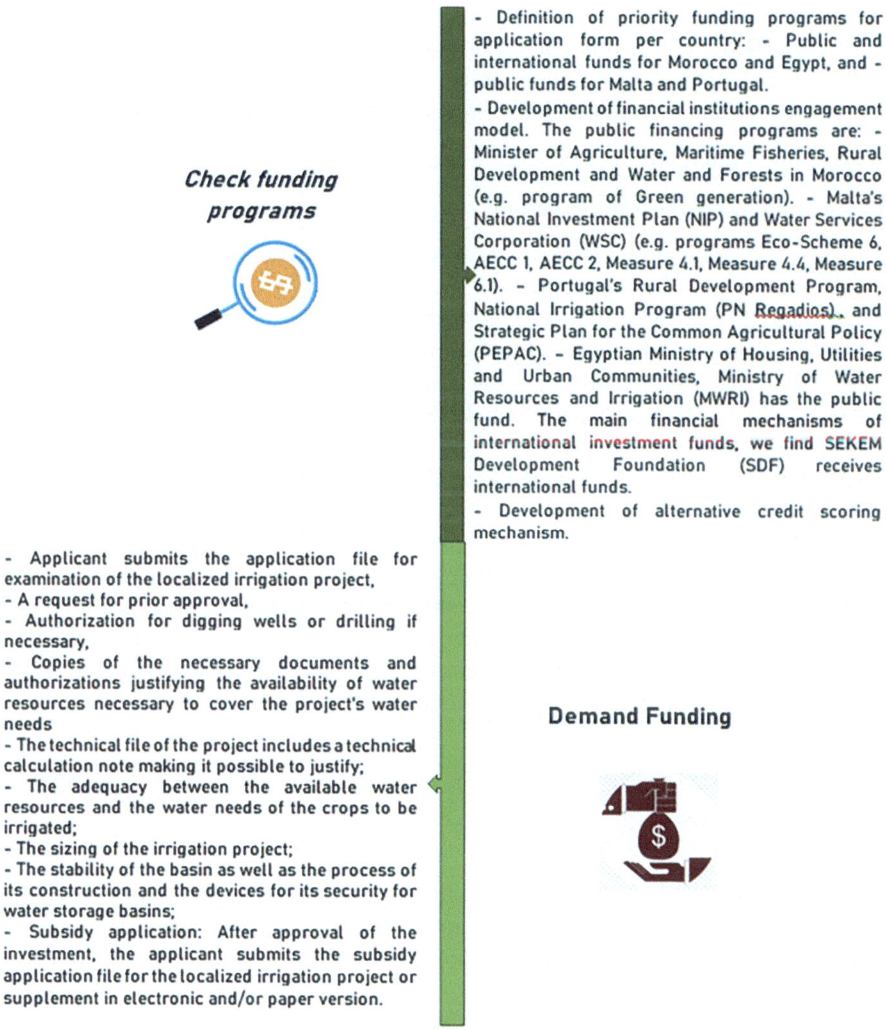

Fig. 2 Strategy implementation roadmap financing instruments

- Minister of Agriculture, Maritime Fisheries, Rural Development and Water and Forests in Morocco (e.g. program of Green generation).
- Malta's National Investment Plan (NIP) and Water Services Corporation (WSC) (e.g. programs Eco-Scheme 6, AECC 1, AECC 2, Measure 4.1, Measure 4.4, Measure 6.1).
- Portugal's Rural Development Program, National Irrigation Program (PN Regadios), and Strategic Plan for the Common Agricultural Policy (PEPAC).

- Egyptian Ministry of Housing, Utilities and Urban Communities, Ministry of Water Resources and Irrigation (MWRI) has the public fund. The main financial mechanisms of international investment funds, we find SEKEM Development Foundation (SDF) receives international funds.

– Develop of alternative credit scoring mechanism.

To demand funding, it is necessary to:

– Submit the application file for the examination of the localized irrigation project.
– Provide a request for prior approval.
– Obtain authorization for digging wells or drilling, if necessary.
– Submit copies of the necessary documents and authorizations justifying the availability of water resources to cover the project's water requirements.
– Check whether the technical file includes a technical calculation note making it possible to justify.
– Check the adequacy between the available water resources and the water needs of the crops to be irrigated.
– Provide the sizing of the irrigation project.
– Verify the stability of the basin and describe the process of its construction and the devices necessary to maintain the security of the water storage basins.
– Submit a subsidy application: after the approval of the investment, the applicant must submit a subsidy application file for the localized irrigation project or its supplement, in electronic and/or paper version.

5 Conclusions

The sustainable development of the Mediterranean: By increasing agronomic and economic efficiency, sustainable development strategies help manage water demand. Furthermore, incentive programs and financial reforms are already in place to enhance the environmental dimension's integration and encourage the growth of key industries like irrigation water efficiency. Efficiency of water use: By implementing water-saving irrigation technology, these policies encourage irrigators to modify their farming methods. The sustainable development of the Mediterranean: By subsidizing farmers' equipment and providing training and advisory support for irrigators, the sustainable development strategy seeks to increase the cost-benefit ratio of farmers' adoption of new irrigation techniques. Thus, this work's objective is to examine the financing mechanism. Tools for investments and financing to implement new, effective irrigation: Some Mediterranean countries are researching efficient irrigation systems, such as solar desalination of salt water in greenhouses, reusing treated conventional wastewater through instream wetland technology, and self-regulating low-energy clay-based irrigation (SLECI) technology for smallholder irrigation. The MedWet project addresses these irrigation technologies. Data gathering on financing options that support water efficiency in agriculture: The best water management practices and water efficiency Water management practices in Morocco, Malta, Egypt, and Portugal are based on questionnaires and qualitative interviews.

Several funding sources are used in this project. For example, self-regulating low-energy clay-based irrigation (SLECI) technology for smallholder irrigation, instream wetland technology for the reuse of treated conventional wastewater, and solar desalination of salt water in a greenhouse are examples of case study technologies that provide a variety of options to meet the specific needs of each farmer. Additionally, the funding sources It is necessary to tailor financing mechanisms to the unique characteristics of the countries under study, which include Egypt, Malta, Morocco, and Portugal. The financing tools to be employed at the national level are determined by the degree of international investment fund coverage and the risk taken on by public financing. Water efficiency can be positively impacted by public funding. By creating options and resources for irrigation infrastructure investment, water efficiency and economic efficiency are impacted. In general, public funding can have a positive impact on water efficiency. By creating rules and guidelines for irrigation infrastructure, water efficiency and economic efficiency are impacted. Without a doubt, one of the cornerstones of the industry's long-term resilience and economic expansion will be a carefully considered and successfully executed financial plan. Checking the funding programs and the demand funding matrix by nation is an essential part of the implementation roadmap financial planning strategy. Through improved water-saving irrigation technology, including SLECI technology, which uses self-regulating low-energy clay-based irrigation (SLECI) for smallholder irrigation, instream wetland technology to reuse treated conventional wastewater, and solar desalination of salt water in a greenhouse, as well as more stringent management of its scarcity for present and future generations, this could help ensure the economic and social development of farmers nationwide.

Acknowledgments This research was carried out within the Med-Wet project "Improving MEDiterranean irrigation and Water supply for smallholder farmers by providing Efficient, low-cost and nature-based Technologies and practices" (Project ID 1646), funded by the Partnership for Research and Innovation in the Mediterranean Area (PRIMA) program.

References

Abbas Saleh, E. M. (2018). The economic value and cost recovery of water in the Egyptian irrigated agriculture. *Egyptian Journal of Agricultural Research, 96*(2), 787–813.

Attard, G., Mangion, J., & Micallef, P. (2007). Water use efficiency and water productivity in Malta. In N. Lamaddalena, M. Shatanawi, M. Todorovic, C. Bogliotti, & R. Albrizio (Eds.), *Water use efficiency and water productivity: WASAMED project* (pp. 153–162). CIHEAM.

Bierlen, R., & Barry, P. J. (2000). Financial structure, technological change, and farm size. *Journal of Agricultural and Applied Economics, 32*(3), 549–560.

Briscoe, J. (1999). The financing of hydropower, irrigation, and water supply infrastructure in developing countries. *International Journal of Water Resources Development, 15*, 459–491.

Central Bank of Egypt. Transforming the livelihoods of smallholder farmers. 2024. https://www.cbe.org.eg/en/financial-inclusion/projects/support-small-farmer

Chopart, J. L., Mézino, M., Aure, F., Le Mézo, L., Mété, M., & Vauclin, M. (2007a). OSIRI: A simple decision-making tool for monitoring irrigation of small farms in heterogeneous environments. *Agricultural Water Management, 87*, 128–138.

Chopart, J.L., Mézino, M., Le Mézo, L., Fusillier, J.L. (2007b). FIVE-CoRe: A simple model for farm irrigation volume estimates according to constraints and requirements. Application to sugar cane in Reunion (France). In: "*Proceedings ISSCT vol. 26, XXXVI ISSCT congress*", Durban, South Africa, abstracts book p. 98-99 and poster paper, pp. 490, 493.

Cuesta, T. S., Cancela, J. J., Dafonte, J., Valcarcel, M., & Neira, X. X. (2005). Social aspects influencing water management in the Lemos Valley Irrigation District, Spain. *Irrigation and Drainage, 54*, 125–133.

FAO. (2011). *L'état des ressources en terres et en eau pour l'alimentation et l'agriculture dans le monde. Gérer les systèmes en danger*. Rapport de synthèse du FAO.

Fernandez, S., & Mouliérac, A. (2010). *Évaluation économique de la gestion de la demande en eau en Méditerranée* (p. 44). Plan Bleu: Centre d'activités régionales, PNUE, Environnement et Développement en Méditerranée.

GG (Green generation). (2023). https://www.agriculture.gov.ma/fr/ministere/generation-green-2020-2030

Government of Malta. (2021). *CAP strategic plan 2023–2027* (68p). Public consultation document.

Helmy, I., & Roushdy, R. (2022). Household vulnerability and resilience to shocks. In C. Krafft & R. Assaad (Eds.), *The Egyptian labor market: A focus on gender and economic vulnerability* (pp. 295–325). Oxford University Press. https://journals.ukim.mk/index.php/jafes/article/view/1892

Malchev, S., Kornov, G., & Hansmann, H. (2022). Innovative clay-based micro-irrigation system "SLECI" (Self-regulating, Low Energy, Clay-based Irrigation): Preliminary results from cherry orchard trials. *Journal of Agricultural, Food and Environmental Sciences, 76*(5), 45–55.

Mélanie, T. (2007). Politiques environnementales et agriculture. *L'agriculture nouvelle défis (édition)*, 207–212.

Merriam, J. L., & Freeman, B. J. (2007). Operational cost benefits study of flexible on-farm irrigation supply systems. *Journal of Irrigation and Drainage Engineering, 133*, 12–16.

MHUUC (Ministry of Housing, Utilities and Urban Communities). (2015). Egyptian Code of practice for the reuse of treated wastewater for agricultural purposes (ECP 501), 2017. Ministry of Housing, Utilities and Urban Communities of Egypt, revised in 2017.

Molle, F., & Sanchis-Ibor, C. (2019). Irrigation policies in the Mediterranean: Trends and challenges. Chapitre 10, research gate. https://doi.org/10.1007/978-3-030-03698-0_10

Nation Unies. (2014). L'économie verte du Maroc: Un objectif stratégique qui nécessite une dynamique partenariale et dynamique et une coordination des efforts. *Commission économique pour l'Afrique, Bureau pour l'Afrique nord*, 19.

PAGE. (2022). *La Transition du Maroc vers une Economie Verte: Etat des Lieux et Inventaire* (446 p). Organisation des Nations Unies pour le Développement Industriel.

Plan Bleu. (2008). *Les Perspectives du Plan Bleu sur le développement durable en Méditerranée*. Sophia Antipolis.

PMV (Plan Maroc Vert). (2018). Le Plan Maroc Vert bilan et impacts. Ministère d'agriculture, de pêche maritime et de développement rural, d'eau et de forêt, Maroc, p 43.

SEKEM. SEKEM Bedaya—Marilou Fund endorses small business owners in the 13 villages. 2024. https://sekem.com/en/sekem-bedaya-marilou-fund-endorses-small-business-owners-in-the-13-villages/.

Tsur, Y., Dinar, A., Doukkali, R. M., & Roe, T. (2004). Irrigation water pricing: Policy implications based on international comparison. *Environment and Development Economics, 9*, 735–755.

UNDP Egypt Accelerator Lab. (2023). Rural resilience in Egypt: Exploring finance as a tool. *United Nations Development Programme, 51p.*

Wahid, N. (2023). Cost-benefit analysis blueprint for water efficient irrigation in selected Mediterranean countries. MED-WET project, D 3.2 report.

Wahid, N., Bouazzama, B., Yousfi, H. E., & Zascerinska, J. (2022). A histrocial overview of evolution of the irrigated agriculture sector in Morocco. *Journal of Regional Economic and Social Development, 14*, 109–118. https://doi.org/10.17770/jresd2022vol14.6896

Ward, F. A. (2010). Financing irrigation water management and infrastructure: A review. *Water Resources Development, 26*(3), 321–349.

Ward, F. A., Kealy, M. J., Guy, G., & Manning, L. M. (2008). Efficient conservation measures in irrigated agriculture to sustain urban and environmental water demands. *Review of Agricultural Economics, 30*, 425–434.

Analysis of the Performance and Productivity of "Regadio" Irrigation Systems and Water Management in the Cova da Beira Region, Portugal

Dina Pereira and João Carlos Correia Leitão

Abstract The effectiveness and productivity of Regadio irrigation systems and water management techniques in Portugal's Cova da Beira region are assessed in this study. Maintaining agricultural activity and resolving the region's water scarcity issues depend heavily on the Regadio system. The research finds areas for improvement and suggests methods to increase the irrigation system's productivity, efficiency, and sustainable water use by examining the current infrastructure and management practices.

This study used 7 years of observations (2016–2022) to evaluate system performance, drawing on a comprehensive dataset spanning the operating points of the "Regadio" system—specifically, Belmonte, Covilhã, Fundão, Penamacor, and Sabugal. To evaluate the impacts of farm size—differentiating between smaller holdings (less than 20 ha) and larger ones (more than 20 ha)—on irrigation system performance, as indicated by the ratio of irrigated area to benefitted area, the analy-

D. Pereira (✉)
CEG-IST, University of Lisbon, Lisboa, Portugal
e-mail: dina@ubi.pt

J. C. Correia Leitão
Department of Management and Economics, University of Beira Interior, Covilhã, Portugal
e-mail: jleitao@ubi.pt

J. C. Correia Leitão et al. (eds.), *Sustainable Water Management and Irrigation Systems*, Natural Resource Management and Policy 62,
https://doi.org/10.1007/978-3-031-93587-9_12

sis uses panel data and quantile regression techniques. The findings show that the size of agricultural holdings has a major impact on how well the irrigation system works, with larger units having a beneficial effect on efficiency. Regional economic returns have a significant and favorable impact on system performance, according to the panel data study. However, in the quantile regression, where environmental threats have a more noticeable detrimental influence, this effect is not verified. These results highlight the significance of contextual and structural elements in determining the sustainability and efficacy of irrigation systems in a certain spatial area.

Keywords Environmental risks · Irrigation systems · Sustainable development · Water efficiency

1 Introduction

1.1 Overview of the Cova da Beira Region, Highlighting Its Agricultural Importance and Reliance on Irrigation Systems

Portugal's Cova da Beira region is an important agricultural hub for the nation, and it is situated in the country's center. Several municipalities are located in this region, including Belmonte, Covilhã, Fundão, Penamacor, and Sabugal. The region is distinguished by its varied topography, which includes fertile valleys, undulating hills, and mountainous regions.

Cova da Beira's economy heavily depends on agricultural operations, and the region is renowned for its high agricultural output and extensive farming history. Numerous commodities, including cereals, vegetables, fruits, vineyards, and olive groves, can be grown in the region due its rich soils and suitable temperature.

Irrigation systems are crucial for maintaining and increasing agricultural productivity in the region. The availability of water becomes an essential component for effective crop production, given the semi-arid Mediterranean environment that predominates in the area, which is marked by hot, dry summers and very low precipitation. Agricultural irrigation systems offer a way to supplement rainfall, reduce water scarcity, and guarantee a steady supply of water in the region.

Besides increasing crop yields, irrigation is also essential for enhancing crop quality and fostering a diverse agricultural industry. Localized irrigation, sprinkler systems, drip irrigation, and other irrigation methods are used by farmers in the area to supply water to certain crops and various soil types. By efficiently delivering water to the root zones of plants, these irrigation systems reduce water loss due to evaporation or runoff.

Additionally, the development of irrigation systems has aided in the expansion of crop farming in the area by enabling farmers to plant crops that require large amounts of water, which was previously impractical due to the scarcity of available natural water sources. This has, consequently, contributed to the region's agricultural advancement, economic expansion, and food security.

The Cova da Beira region's irrigation management techniques seek to maximize water usage and advance sustainable agricultural methods. Farmers work to embrace effective irrigation scheduling, precision agriculture practices, and water-saving measures with the backing of local agricultural institutions and government programs. These initiatives are essential to ensure the prudent use of water resources, reduce negative environmental effects, and support long-term agricultural sustainability.

To summarize, Portugal's Cova da Beira region is a large agricultural area distinguished by its reliance on irrigation infrastructure. The effective use of irrigation systems, which supplement limited water supplies and enable sustainable crop cultivation, is crucial for the region's agricultural output, economic growth, and food security. In order to ensure the region's agricultural success and resilience in the face of water scarcity concerns, continued attention must be paid to optimizing irrigation infrastructure and encouraging prudent water management practices.

1.2 Explanation of the "Regadio" Irrigation System and Its Significance to the Region's Agricultural Landscape

The region's agricultural terrain relies heavily on the "Regadio" irrigation system, which is commonly used throughout Portugal. It is a system designed to manage and distribute water to agricultural areas collectively. It is a system designed to manage and distribute water to agricultural areas collectively. It implies a combination of public investment and producers' cooperative management to ensure their continued efficiency and productivity in areas where a large number of small farmers and a smaller number of large producers coexist, and both face water shortage issues.

Water from rivers, lakes, or dams is collected, stored, and distributed through a network of canals, channels, and reservoirs, making up the "Regadio" irrigation system. The method works on the idea of diverting and managing water flows to effectively irrigate crops. Water is provided to agricultural fields for crop production by a system of canals and distribution networks, offering a dependable and controlled water supply.

The importance of the "Regadio" irrigation system lies in its capacity to support agricultural activities while overcoming the problems associated with water scarcity. The system aids farmers in optimizing crop development, increasing yields, and maintaining agricultural production throughout the year by guaranteeing a steady and adequate water supply. It lessens reliance on rainfall alone, which can be erratic and insufficient, especially in areas with arid or semi-arid climates.

Due to problems with water scarcity, farmers in the Cova da Beira region can cultivate a range of crops, such as cereals, fruits, vegetables, and vineyards, even in areas that would not be suitable for maintaining agricultural output without the usage of the Regadio system. It offers the required water resources to fulfill the unique irrigation needs of various crops, promoting their development and overall quality.

The "Regadio" irrigation system also encourages water efficiency and conservation. Water losses due to evaporation, seepage, or ineffective irrigation techniques are minimized by using regulated distribution devices, such as water gates and flow management. This ensures the wise use of water resources, cutting down on waste and promoting the sustainable use of finite water resources.

The "Regadio" system also helps the socioeconomic development of the area. It helps rural livelihoods, generates employment opportunities, and promotes economic growth by supplying dependable water for agriculture. A robust agricultural sector supported by the "Regadio" system also improves food security, lessens reliance on outside food supplies, and encourages homegrown food production.

In summary, the "Regadio" irrigation system is essential for the Cova da Beira region's agricultural environment. Its importance stems from its capacity to address issues with water scarcity, offer a dependable water supply, support various agricultural cultivation, encourage effective water use, and support socioeconomic growth. The "Regadio" system is an essential part of the agricultural landscape in the Cova da Beira region since it promotes the area's agricultural sustainability and resilience in the face of water resource constraints.

2 Water Availability and Scarcity

2.1 Assessment of the Water Availability Situation in the Cova da Beira Region, Including Annual Precipitation Patterns and Water Sources

In order to evaluate the Cova da Beira region's water resources and handle water scarcity issues, it is crucial to comprehend the region's water availability condition. This evaluation takes into account the numerous water sources used in the area as well as annual precipitation patterns.

2.1.1 Annual Precipitation Patterns

A Mediterranean climate with separate rainy and dry seasons prevails in the Cova da Beira region. Although annual precipitation patterns vary across the region, a broad tendency may be seen. The dry season lasts from April to September, and the wettest months are normally from October to March.

Its geography, height, and proximity to the coast are some of the elements affecting the amount of precipitation in the Cova da Beira region. Due to orographic influences, higher altitudes, like the Serra da Estrela mountain range, frequently experience higher precipitation rates. In comparison to inland locations, coastal areas may see somewhat more precipitation and higher humidity levels.

2.1.2 Water Sources

To supply its agricultural, household, and industrial water needs, the Cova da Beira region depends on a number of water sources. Among these sources are the following.

2.1.2.1 Surface Water

The region's water supply is significantly influenced by surface water. The Zêzere River, a tributary of the Tagus River, is the primary surface water supply in the area. It is a key source of water for agriculture, drinking water, and industrial usage, coming from the Serra da Estrela and flowing through the Cova da Beira region.

2.1.2.2 Groundwater

Another significant water source in the Cova da Beira region is groundwater. Groundwater may be stored in the area, thanks to its granitic and schist formation-dominated geology. In order to replenish water supplies, especially during dry spells or when surface water supplies are scarce, wells and boreholes access groundwater reserves.

2.1.2.3 Reservoirs and Dams

The Cova da Beira region has a number of carefully placed dams and reservoirs to store water for diverse uses. These reservoirs, like those at Lagoa Comprida and Vale do Rossim, aid in regulating the availability of water, especially during the dry season. They support the region's irrigation, water supply, and hydropower production.

2.1.2.4 Rainwater Harvesting

On a lesser scale, rainwater harvesting techniques, like collecting and storing rainwater from roofs or paved surfaces, are used to supplement water requirements for nonpotable uses, like irrigation or livestock watering.

It is significant to remember that the Cova da Beira region, like many other places, has problems related to water scarcity. The region's water needs may not always be met by the region's yearly precipitation, particularly during extended dry spells. To ensure sustainable water usage and lessen the effects of water shortage, water management strategies, such as effective irrigation systems and conservation measures (which are still a huge need in the region), are essential.

To summarize, the Cova da Beira region has a Mediterranean climate with distinct rainy and dry seasons. The wettest months are from October through March. Annual precipitation patterns vary. To supply its water needs, the area depends on surface water, groundwater, reservoirs, and dams, as well as rainwater collection. However, to solve issues of water shortage and guarantee sustainable water use in the area, appropriate water management methods are required.

2.2 Identification and Analysis of Water Scarcity Issues Affecting Agricultural Productivity

In several areas in Portugal, especially the Cova da Beira region, lack of water causes serious problems for agricultural productivity. Understanding how these water scarcity issues affect agricultural activities and developing practical mitigation methods depend on the identification and analysis of these issues. The main problems caused by water scarcity in agricultural production can be summarized as follows.

2.2.1 Limited Water Availability

The Cova da Beira region has a Mediterranean climate with low yearly precipitation. The amount of rainfall may not always be enough to satisfy the water needs of agricultural crops all year round. Water stress in plants caused by insufficient water can have an impact on their development, yield, and quality. This restriction becomes more important during dry seasons or protracted droughts.

2.2.2 Seasonal Variability in Precipitation

Seasonal variations in the precipitation patterns can be seen in the region, with wetter intervals in the winter and drier circumstances in the summer. Because of this fluctuation, farmers must rely on water storage, irrigation systems, and effective water management techniques to fill up the gaps between rainfall events, which presents issues for crop growth. The effects of seasonal unpredictability in precipitation might be made worse by inadequate water storage capacity and restricted access to alternate water sources.

2.2.3 Unequal Distribution of Water Resources

The distribution of water resources might vary, even within the Cova da Beira region. Areas near rivers or reservoirs may have easier access to water for irrigation than those farther away. As a result of infrastructural and water transport constraints, regions farther distant or at higher elevations may experience larger water scarcity issues. Disparities in agricultural productivity and regional economic development may result from this unequal allocation. The fact that the "Regadio" is still sourcing only a part of the region is also a key issue to bear in mind.

2.2.4 Competition for Water Resources

Competing demands for water from agriculture, industry, and domestic consumption might make the water shortage in the Cova da Beira region worse. Urbanization and industrialization may place more strain on water supplies, limiting the amount of water available for agriculture. To maintain agricultural output in the face of rising competition, it becomes crucial to balance the needs of various sectors and ensure fair water allocation.

2.2.5 Climate Change Impacts

Climate change exacerbates problems with water scarcity by changing precipitation patterns, raising temperatures, and worsening drought conditions. The amount of water available for agricultural use in the Cova da Beira region may be altered due to variations in the distribution of rainfall and increased rates of evapotranspiration. Additionally, as a result of higher temperatures, crops may require more water, placing additional strain on water supplies. To lessen the effects of climate change on agricultural productivity, resilient water management measures must be put in place.

There are several solutions to the problem of water scarcity, including the following.

2.2.5.1 Adoption of Efficient Irrigation Techniques

By encouraging the utilization of water-efficient irrigation methods, like drip irrigation and precision irrigation, water use can be maximized and water losses caused by evaporation or runoff can be reduced. This is a point to be addressed within the scope of our project "Med-Wet."

2.2.5.2 Water Conservation and Management Practices

Reduced water demand and increased total water usage efficiency in agricultural activities can be achieved by putting into practice water conservation techniques, including crop rotation, mulching, and soil moisture monitoring.

2.2.5.3 Integrated Water Resource Management

A fair and sustainable distribution of water resources can be achieved by implementing integrated water resource management strategies, as well as efficient monitoring practices, that take into account the demands of all water users, including those in the agricultural, industrial, and home sectors.

2.2.5.4 Investment in Water Infrastructure

Enhancing water availability and reducing the effects of water scarcity on agricultural output can be achieved by updating/expanding irrigation infrastructure, increasing water storage capacity, and investing in water harvesting and recycling systems.

2.2.5.5 Climate Change Adaptation

Farmers can adjust to changing climatic circumstances by developing climate change adaptation techniques that focus on resilient agricultural practices, water-saving technologies, and drought-tolerant crop varieties.

In summary, the water scarcity problems in the Cova da Beira region have an impact on agricultural productivity by limiting water availability, introducing seasonal variability in precipitation that hinders local production diversification, causing unequal distribution of water resources, intensifying competition for water, and being made worse by the effects of climate change, which lower total yield and make it more difficult to plan the introduction of new crops. To address these issues and ensure sustainable agricultural productivity in the area, it is imperative to implement efficient irrigation techniques, adopt water conservation practices, promote integrated water resource management, invest in water infrastructure, and adapt to climate change.

2.3 Evaluation of the Current "Regadio" Irrigation Systems in the Region, Including Their Design, Infrastructure, and Coverage

The "Regadio" irrigation systems in the Cova da Beira region are essential for sustaining agricultural activity and guaranteeing water accessibility for crop cultivation. The design, infrastructure, and coverage of the "Regadio" irrigation systems are all evaluated below.

2.3.1 Design and Performance

Performance and water conservation should be given top priority when designing "Regadio" irrigation systems in the area. A design evaluation should take into account elements like irrigation methods, water distribution strategies, and system structure. To reduce water losses from evaporation and runoff, efficient irrigation methods, like drip irrigation and localized irrigation, should be promoted. Analyzing the efficacy of irrigation scheduling technologies and water distribution strategies can help maximize water usage and boost system performance as a whole.

2.3.2 Infrastructure

The performance and efficiency of "Regadio" irrigation systems are greatly influenced by their infrastructure. The condition and capacity of reservoirs, canals, channels, and other water storage and conveyance systems should be the main focus of the evaluation. To guarantee a steady supply of water and avoid water losses, it is crucial to evaluate the infrastructure's needs for maintenance and repair. If necessary, infrastructure upgrades can improve water delivery effectiveness and reduce losses throughout the distribution network.

2.3.3 Coverage and Accessibility

The extent to which agricultural lands can take advantage of water supply from "Regadio" irrigation systems depends on their coverage. The systems' coverage area and the availability of water resources for farmers should both be evaluated. It is crucial to assess whether all agricultural regions in Cova da Beira have access to the "Regadio" systems or if some are underserved. In fact, it must be noted that there are important areas in the Cova da Beira region that could be more productive if the "Regadio" were territorially extended to them. It can be helpful to prioritize infrastructure building and expansion in areas with limited access to maintain fair water distribution throughout the region.

2.3.4 Modernization and Technological Integration

The degree of modernization and technological integration of "Regadio" irrigation systems should also be considered when evaluating them. Assessing the application of cutting-edge irrigation technology, including remote monitoring, soil moisture sensors, and automated irrigation systems, among others, can reveal the efficacy and efficiency of the systems. Technology integration can improve irrigation scheduling, water management procedures, and system performance as a whole.

2.3.5 Sustainability and Environmental Impact

The sustainability and environmental impact of "Regadio" irrigation systems must be assessed. The degree to which the systems conform to sustainable agricultural practices can be assessed by looking at how well they regulate pesticide inputs, conserve water, and apply water-saving techniques. Assessing the systems' environmental impact might reveal areas for improvement to reduce unfavorable ecological effects, such as water contamination from agricultural runoff or habitat damage.

2.3.6 Stakeholder Engagement and Support

Stakeholder support and engagement are essential for the success of "Regadio" irrigation systems. It is critical to assess the degree of cooperation and communication among farmers; agricultural organizations; water associations, such as the Associação de Regantes of Cova da Beira; and water management authorities. Insights into the general efficacy of the systems and the aid given to the agricultural community can be gained by evaluating the availability of training programs, technical support, and financial support for farmers, as it is one of the targets of the 'Med-Wet' project.

Finally, assessing the design, infrastructure, coverage, modernization, sustainability, and stakeholder involvement of the "Regadio" irrigation systems in the Cova da Beira region is necessary to determine their current state. Finding areas for development can direct efforts to boost system performance, broaden coverage, incorporate technology, guarantee sustainability, and give farmers the support they require. To encourage effective water management, boost agricultural output, and address the difficulties of water shortage in the area, ongoing evaluation and monitoring of the "Regadio" systems is crucial.

3 Analysis of the Performance and Productivity of "Regadio" Irrigation Methods, Considering Factors Such as Small Explorations Versus Bigger Exploration Irrigation Performance, Environmental Impact, and Economic Returns

Based on variables such as small versus bigger exploration irrigation performance (measured by the ratio of irrigated area to benefited area), environmental impact (assessed through the collected residues per area), and economic returns (using the region VAB), "Regadio" irrigation techniques in the Cova da Beira region can be assessed for their overall performance and productivity. Analyzing these factors can provide insights into the performance and sustainability of the irrigation methods.

Using a dataset created with the annual reports provided by Associação de Regantes da Cova da Beira, covering the territorial points where the "Regadio" irrigation system operates (Belmonte, Covilhã, Fundo, Penamacor, and Sabugal), we have gathered 7 years of observations (from 2016 to 2022) to evaluate its performance. We have estimated a panel data model with fixed and random effects, along with a quantile regression, to assess the effects of irrigated agricultural explorations of smaller dimensions (less than 20 ha) and bigger dimensions (more than 20 ha), as well as the region's economic returns and environmental impact, on the performance of the irrigation system.

3.1 Overall Performance

It is crucial to take a comprehensive perspective when evaluating the overall performance of "Regadio" irrigation techniques. This involves assessing technology use, farmer knowledge and practices, irrigation schedules, and water management methods.

An analysis of efficiency and effectiveness should take into account crop output, crop utilisation efficiency, and economic returns from the irrigation systems. Comparing the effectiveness of various irrigation methods, including drip irrigation, sprinkler irrigation, and furrow irrigation, can shed light on the best strategies to be applied for particular crops and soil types.

Additionally, stakeholder opinions, surveys, and monitoring data can offer important information on how farmers feel about using "Regadio" irrigation techniques. This input can aid in locating problem areas, correcting difficulties, and fostering the use of sustainable irrigation techniques.

To summarize, assessing the performance of "Regadio" irrigation techniques entails looking at variables, including water loss, uniformity of application, and energy usage. The "Regadio" irrigation methods in the Cova da Beira region can be made more efficient and effective by identifying improvement areas and putting the right strategies in place. This will improve water management, boost agricultural productivity, and ensure the sustainable use of water resources.

In the current analysis and due to the lack of existent information, we have retrieved data from a set of annual reports, and we created a dependent variable to assess the performance of the "Regadio," that is, the ratio of the irrigated area to the benefited area. For the explanatory variables, we have tested small irrigated explorations (less than 20 ha) and big irrigated explorations (more than 20 ha), economic returns of the region (VAB), and environmental impact (collected residues per unit of area).

In a posterior analysis to assess the performance of the solutions tested within the scope of the "Med-Wet" project, we will make use of additional information that the partners are currently monitoring with the use of sensors.

3.2 Analysis of the Performance and Productivity of "Regadio" Irrigation Methods: Results

When using panel data, the same cross-sectional unit is polled repeatedly over time (the performance variable is the ratio of the irrigated to the benefitted area), with the assumption that panel data has both a spatial and temporal dimension. We may notice heterogeneity in these units over time because we are working with panel data from exploratory units. The panel also provides more relevant data, increased variability, reduced collinearity across variables, and greater efficiency by combining time series with cross-sectional observations. The analysis is based on both random and fixed effect panel estimations, as well as pooled ordinary least squares (OLS).

The next table shows the results obtained for both fixed and random effect panel data estimations. The F-test in panel data regression models provides a statistical test of whether all the coefficients on the independent variables are jointly different from zero. In other words, it tests the null hypothesis that all explanatory variables have no effect on the dependent variable, that is, their coefficients are equal to zero (see Table 1). The econometric specification can be illustrated as follows:

Performance (irrigated area/benefited area)$_{it}$ = β1 (small irrigated explorations)$_{it}$ + β2 (big irrigated explorations)$_{it}$ + β3 (economic returns of the region)$_{it}$ + β4 (environmental impact)$_{it}$ + αi + εit

Table 1 Estimation of panel data regression models

Dependent variable Performance (irrigated area/benefited area)	Random effects	Fixed effects	Between effects	ML random effects
Independent variables				
Small irrigated explorations	−0.0642356 (0.0000)	−2.652218 (0.085)	−0.0468402 (0.0001)	−0.0505386 (0.0000)
Big irrigated explorations	2.537021 (0.0000)	0 (omitted)	1.716036 (0.0000)	1.847332 (0.0000)
Economic returns of the region	0.0444327 (0.0000)	0.0478037 (0.167)	−0.0126479 (0.473)	−0.053483 (0.203)
Environmental impact	−0.0271229 (0.149)	−0.0342222 (0.311)	−0.066057 (0.166)	−0.0076345 (0.595)
Observations	49	49	49	49
Wald	766.26			42.82
F test		367.49	7.90	
R^2	0.0000	0.0005	0.0001	0.0000

Robust standard errors are presented in brackets. Wald tests are used to evaluate the null hypothesis of the noncommon significance of the parameters of the explanatory variables against the alternative hypothesis of the common significance of the parameters of the explanatory variables. *F* tests are used to assess the null hypothesis of the noncommon significance of the estimated parameters against the alternative hypothesis of the common significance of the estimated parameters.
*Significant at 10%, **significant at 5%, ***significant at 1%

Analyzing the results obtained for the methods above, we can state that in the random effect method, small irrigated explorations negatively affect, in a significant way, the performance of the "Regadio." On the other hand, big irrigated explorations both positively and significantly affect the performance of the system. For the control variables, the economic returns of the region favorably impact, in a significant manner, the performance of the "Regadio," and the environmental impacts denote no significant effect.

For the fixed effect model, only the variable small irrigated explorations negatively influence, in a significant way, the performance of the "Regadio," although to a lesser extent.

When assessing the between effect and ML random effect models, the only variables that influence the performance of the system are small irrigated explorations and big irrigated explorations.

We can conclude that the size of the explorations is important for assessing the performance of the "Regadio," in a way that the bigger the size of the units, the greater the impact on the system's performance. In addition, the region being more economically developed also affects the performance of the irrigation system, as greater productivity enables the system to perform better and more efficiently.

Next, we illustrate the usefulness of the quantile regression approach and implement the estimation using Stata. Quantile regression may be preferred among the several alternative statistical twists provided by the approach since it does not presuppose a parametric distribution, and it estimates the whole conditional distribution of the independent variable. This approach is typically thought to be more flexible and informative. Table 2 shows the main results of the quantile regression.

We have performed simultaneous estimations for different quantiles, specifically for $q = 0.25$, 0.50, 0.75, and 0.90. The coefficients vary across quantiles. The highly statistically significant variables for quantile 25 (small irrigated explorations in the negative mode and big irrigated explorations in a positive way) have a much greater impact on the irrigation system's performance at lower conditional quantiles of performance. Looking at quantile 50, the same effects are observed, although with higher coefficients. Conversely, the highly statistically significant variables for quantile 75 (small irrigated explorations negatively and big irrigated explorations positively) have a higher effect on the irrigation system's performance as the conditional quantiles of performance are increasing. However, in quantile 75, the environmental impact has a detrimental effect on the 'Regadio''s performance in terms of irrigated area. On the highest quantile, Q90, the effects of small irrigated explorations and big irrigated explorations are also significant, with smaller-sized units having a negative impact and bigger-sized units having a positive impact. Additionally, environmental impact appears to have a significant influence on the performance of the irrigation system in a negative manner, but its significance gets more intense at higher conditional quantiles of performance.

The analysis of quantiles can derive important inferences for the present study as it indicates that as the conditional quantiles of performance increase, the statistical significance of the effects of the variables increases as well. Thus, the size of the

Table 2 Quantile regression model estimation

Dependent variable Performance (irrigated area/benefited area)	Coef.	Std. err.	*t*	*P* > [*t*]	95% [conf. interval]	
Q25						
Small irrigated explorations	−0.0554369	0.0243738	−2.27	0.028	−0.1045591	−0.0063147
Big irrigated explorations	2.282566	0.7012605	3.25	0.002	0.8692684	3.695864
Economic returns of the region	0.0070261	0.0285943	0.25	0.807	−0.050602	0.0646541
Environmental impact	−0.0442566	0.0835721	−0.53	0.599	−0.212685	0.1241719
Q50						
Small irrigated explorations	−0.0558562	0.0161981	−3.45	0.001	−0.0885014	−0.023211
Big irrigated explorations	1.801501	0.466038	3.87	0.000	0.8622629	2.740738
Economic returns of the region	−0.0030158	0.019003	−0.16	0.875	−0.0413138	0.0352822
Environmental impact	−0.0560376	0.0555396	−1.01	0.319	−0.1679704	0.0558952
Q75						
Small irrigated explorations	−0.0494952	0.0150312	−3.29	0.002	−0.0797886	−0.0192019
Big irrigated explorations	1.632411	0.4324635	3.77	0.000	0.7608377	2.503984
Economic returns of the region	−0.0169677	0.017634	−0.96	0.341	−0.0525066	0.0185712
Environmental impact	−0.1126786	0.0515384	−2.19	0.034	−0.2165475	−0.0088097
Q90						
Small irrigated explorations	−0.0469474	0.0118662	−3.96	0.000	−0.0708622	−0.0230327
Big irrigated explorations	1.220266	0.3414029	3.57	0.001	0.5322132	1.908318
Economic returns of the region	−0.0133013	0.0139209	−0.96	0.345	−0.041357	0.0147544
Environmental impact	−0.1367183	0.0406864	−3.36	0.002	−0.2187163	−0.0547203

explorations is important for the performance of the irrigation system, while environmental impact significantly affects performance but in a negative way.

When comparing the two estimation methods, either for quantile regression or panel data estimations, the size of the units of exploration has a significant impact on irrigation system performance, becoming positive as the size grows. However, the panel estimation indicates that the region's economic returns have a positive and significant impact on the system's performance, whereas the quantile regression does not show this effect, making the environmental impact significant in this context in a negative way.

In future analyses regarding the efficiency and productivity of irrigation systems, we intend to add more variables into the equation, for instance, water use efficiency (WUE), which is a key indicator of how effective irrigation systems are. It calculates the amount of water that crops need to use to produce one unit of yield. In “Regadio” systems, determining WUE entails comparing the amount of water supplied to the crops with the crop yield that was ultimately obtained. High WUE denotes effective water use, while low WUE denotes possible water loss or ineffective irrigation techniques. The evaluation of WUE aids in identifying areas for development and encourages water-saving practices.

Moreover, we aim to obtain data on crop yield, which is a good indicator of how productive “Regadio” systems are. It shows how much and how well agricultural output is produced per unit of land. Higher agricultural yields are a sign of efficient irrigation techniques, sufficient water supplies, and appropriate crop management techniques. Crop yield analysis aids in identifying elements that favorably or unfavorably affect productivity and guides plans for advancement.

Another important indicator is water productivity, which is the amount of economic production (such as crop value) produced for each unit of water utilized. It provides a thorough analysis of the relationship between water use and agricultural productivity by fusing financial incentives with water usage efficiency. In terms of monetary gains, higher water productivity denotes effective distribution and use of water. Evaluating water productivity aids in the creation of plans for resource efficiency, water waste reduction, and income generation per unit of water utilized.

Future work will also entail a comparison of national and international standards. Portugal’s Cova da Beira region operates within the parameters of national water use efficiency guidelines. The national criteria for WUE in Portugal are typically developed based on the availability of water in the region, crop water needs, and best management practices. These requirements offer a standard against which the “Regadio” systems in the area can be assessed. It is possible to determine whether the irrigation systems in the Cova da Beira region are operating above, at, or below the nation’s average efficiency levels by comparing their WUE to national norms. Depending on the situation and the conditions in “Med-Wet” partner countries, international standards for water use efficiency can change. Guidelines and standards for WUE in agriculture are provided by groups like the Food and Agriculture Organization (FAO) and other research institutions. These criteria take into account things like crop varieties, climate, and irrigation techniques. It is feasible to evaluate the performance of the region against worldwide standards and pinpoint areas for improvement by comparing the WUE of “Regadio” systems in the Cova da Beira region, as well as those in other partner countries’ regions, to international benchmarks. By comparing the water consumption efficiency of the systems in the areas with regional, national, and international standards, the effectiveness of water management techniques may be assessed. The WUE benchmarks used at national and international levels differ based on the type of crop, the climate, and irrigation methods. It is possible to identify areas for improvement and implement improvement actions by comparing the various water consumption efficiency situations to these standards.

For example, the National Irrigation Plan (Plano Nacional de Regadio), which sets strategies and goals for effective water use in irrigation, is a key resource in Portugal. Modernizing irrigation systems, encouraging water-saving technology, and enhancing general water management practices are the main goals of the plan. In addition, the Plano Nacional de Recursos Hdricos (National Water Resources Plan) establishes objectives and priorities for managing water resources, including agricultural water use. The strategy places a strong emphasis on effective irrigation techniques, water loss reduction, and sustainable water use. In order to enable an insightful comparison, these benchmarks and regulatory frameworks are insightful in each of the partner regions involved in the 'Med-Wet' project.

4 Water Management Practices

4.1 Review of Existing Water Management Practices Associated with the "Regadio" System, Including Water Allocation Policies, Reservoir Management, and Irrigation Scheduling

In order to encourage sustainable farming methods and effective water usage, the Cova da Beira region of Portugal's Regadio system currently employs a number of water management strategies. Although specific methods may differ, these are some typical elements of water management connected to the local Regadio system.

4.1.1 Water Allocation and Rights

The distribution of water resources among various users, particularly agricultural operations, is governed by water allocation regulations and rights. Water management authorities in the Cova da Beira region are normally in charge of allocating water, and they do so in accordance with predetermined standards, including crop water needs, water availability, and water use priorities.

4.1.2 Reservoir Management

Reservoirs are crucial for controlling and storing water supplies for irrigation. Reservoir management entails using techniques to maximize water storage and release while taking into consideration variables such as seasonal fluctuations, precipitation patterns, and water demand from various sectors. Management strategies seek to limit water losses from evaporation and seepage while ensuring water supply during dry spells.

4.1.3 Irrigation Scheduling

Optimizing water utilization and reducing water loss requires effective irrigation scheduling. The Cova do Beira region's farmers often adhere to irrigation schedules that take crop water needs, soil moisture levels, and weather variables into account. A variety of instruments and data sources, such as soil moisture sensors, meteorological data, and evapotranspiration models, may be used to determine when and how much water should be applied to crops.

4.1.4 Modern Irrigation Techniques

The Cova da Beira region's "Regadio" system encourages the employment of contemporary irrigation methods to increase water efficiency. To reduce water loss due to evaporation and guarantee targeted irrigation of crop root zones, drip irrigation, sprinkler irrigation, and localized irrigation systems are frequently used. These methods aid in maximizing water consumption, decreasing water waste, and enhancing crop output.

4.1.5 Water Conservation Practices

To reduce water consumption and improve overall effectiveness, various water conservation methods are advocated, including mulching, which aids in lowering soil surface evaporation, and the use of cover crops, which enhances soil moisture retention. Additionally, farmers may be urged to embrace precision farming techniques that make use of cutting-edge technology to maximize the application of fertilizer and water.

4.1.6 Stakeholder Engagement and Education

Stakeholder participation and education play a vital role in ensuring the efficiency of the "Regadio" system's water management procedures. Farmers, agricultural institutions, and water management authorities work together to share knowledge, spread best practices, and encourage water-efficient methods. Farmers are taught about sustainable water management, irrigation methods, and the value of water conservation through training programs and workshops. However, a lot of additional support can be implemented in order to overcome some inefficiencies in the system.

It is crucial to keep in mind that specific water management techniques may differ within the Cova da Beira region, depending on regional circumstances, water availability, and agricultural practices. The techniques listed above give a broad overview of the methods frequently used within the "Regadio" system in the area.

4.2 Examination of Technology Adoption, Data Monitoring, and Automation in Water Management Practices

The Portuguese region of Cova da Beira has seen a significant increase in the use of technology, data monitoring, and automation in water management techniques. Much has been accomplished in this instance with the assistance of regional institutions, which set up cooperative, shared plans to enhance monitoring systems in order to increase productivity and efficiency. This is a synopsis of the current state of these aspects.

4.2.1 Technology Adoption

In the Cova da Beira region, the use of technology in water management measures has increased. In order to maximize agricultural operations and increase water efficiency, farmers and water management authorities have been adopting more and more contemporary irrigation techniques and equipment. Several crops, including the production of cherries, peaches, corn and almonds, are using technologies like drip irrigation, sprinkler systems, and localized irrigation approaches to increase targeted water delivery to crops and decrease water waste. Additionally, precision farming techniques are becoming more popular to optimize water and fertilizer use based on specific crop and soil conditions. These techniques make use of cutting-edge technologies like Global Positioning System (GPS), remote sensing, and data analytics.

4.2.2 Data Monitoring

To implement efficient water management strategies, data monitoring is essential. Methods such as data collection and analysis of water availability, soil moisture content, weather patterns, and agricultural water needs are becoming more and more important in the Cova da Beira area. These data are crucial for making informed decisions on irrigation scheduling, water resource allocation, and water use optimization. Data monitoring strategies include the use of weather stations, soil moisture sensors, and remote sensing technologies. Collected data are helpful in determining water needs, directing irrigation operations, and improving the overall efficiency of water management strategies.

4.2.2.1 Automation

In the Cova da Beira region, automation in water management procedures is gaining traction. Automated systems aid in streamlining procedures, increasing accuracy, and boosting productivity. Precision water supply is made possible by automated irrigation systems, including sensor-based controllers and controlled valves. These systems can be configured to modify irrigation schedules based on crop water requirements, weather conditions, and soil moisture levels. Water distribution networks can be made more efficient and operate more cost-effectively by automating components like canal gates and remote control systems.

4.2.2.2 Decision Support Systems

In the Cova da Beira region, decision support systems (DSS) are being utilized more frequently to aid in water management procedures. These systems make recommendations and improve decision-making processes by using techniques such as modeling, simulation, and data analytics. DSS can help in scheduling irrigation, forecasting crop water needs, analyzing potential water allocation plans, and streamlining water distribution. DSS gives stakeholders the tools they need to make educated decisions and enhance water management procedures by combining diverse data sources and analysis tools.

Overall, the use of technology, data monitoring, and automation in water management procedures is growing in the Portuguese region of Cova da Beira. The application of these innovations helps increase productivity, optimize water consumption, and encourage sustainable agricultural practices. More effective and efficient water management can be made possible in the area with further efforts to encourage technology adoption, data monitoring, and automation.

5 Best Practices and Case Studies

5.1 Presentation of Successful Case Studies and Best Practices Related to "Regadio" Irrigation Systems and Water Management from Similar Regions or Countries

A list of recommended techniques for managing irrigation systems and water is provided below. These techniques may be used to improve the "Regadio" systems in the Cova da Beira region of Portugal. These examples of sustainable water management are drawn from similar nations or regions.

5.1.1 Spain—Almeria Region

The intensive horticultural industry in the Almeria region of Spain has effectively adopted water management measures. The area makes use of cutting-edge technologies for soil moisture monitoring, water recycling, and drip irrigation. These procedures have contributed to increased crop output, reduced water loss, and improved efficiency in water usage.

5.1.2 Israel—Negev Desert

Known for its water management prowess, Israel has used cutting-edge technologies, like precision irrigation, desalination, and wastewater recycling, to address the problems associated with water scarcity in desert areas. These methods have made it possible to cultivate crops in the Negev desert, resulting in more efficient water use and greater crop yields.

5.1.3 Australia—Murray-Darling Basin

The Murray-Darling Basin in Australia is a highly productive agricultural region that has experienced challenges related to water constraints. To increase water efficiency and productivity, the area has introduced water allocation regulations, water trading networks, and enhanced irrigation techniques, like sprinkler and drip irrigation. These actions have led to more effective water management and environmentally friendly agricultural methods.

5.1.4 California, USA—Central Valley

California's Central Valley is known for its extensive agricultural production. The region has implemented water management practices, such as adopting precision irrigation technologies, using soil moisture sensors, and implementing water pricing mechanisms to encourage efficient water use. These initiatives have contributed to enhanced water use efficiency and increased agricultural productivity.

5.1.5 Netherlands

The Netherlands is well known for its cutting-edge greenhouse farming methods, which include closed-loop irrigation systems, water recirculation, and climate control innovations. In greenhouse agriculture, these techniques have dramatically decreased water use and improved water use efficiency, resulting in increased crop yields and resource preservation.

These case studies and best practices offer illustrations of efficient water management techniques that could provide information and possible plans for improving the "Regadio" systems in the Cova da Beira region. It is crucial to modify and customize these methods to the unique climatic, environmental, and socioeconomic circumstances of the area while taking into account regional laws and resources.

6 Challenges and Opportunities

6.1 Identification and Analysis of the Challenges and Barriers Faced in Improving the Efficiency and Productivity of "Regadio" Systems and Water Management

There may be a number of difficulties and obstacles in the way of improving "Regadio Cova da Beira" systems and water management. Here are a few typical examples.

6.1.1 Water Scarcity

Like many other agricultural regions, the Cova da Beira region may experience problems with water scarcity as a result of variables like scarce water supplies, climatic change, and rising water demand. A lack of water might make it difficult to increase the effectiveness and productivity of irrigation systems.

6.1.2 Aging Infrastructure

The irrigation networks, pipelines, and "Regadio" systems in the Cova da Beira region may suffer from deteriorated infrastructure. Water losses, ineffective water distribution, and difficulties in implementing cutting-edge irrigation systems can all be caused by aging infrastructure.

6.1.3 Limited Financial Resources

It costs money to implement changes to irrigation systems and water management. The adoption of cutting-edge irrigation technologies, the implementation of infrastructural improvements, and the employment of effective water management techniques can all be hampered by a lack of funding.

6.1.4 Technical Adoption

The lack of information, awareness, and training may prevent the widespread use of cutting-edge technologies and precise irrigation methods. For farmers and water management authorities to comprehend and use new technology effectively, they need to be educated and guided.

6.1.5 Governance and Policy

The development of "Regadio" systems and water management techniques might be hampered by ineffective water allocation policies, insufficient laws, and confusing governance structures. Policy frameworks must encourage efficient behavior, support sustainable water usage, and promote stakeholder cooperation.

6.1.6 Limited Access to Information and Data

For efficient water management, precise data on water resources, soil moisture, and crop water needs must be available and easily accessible. The implementation of data-driven water management strategies might be hampered by a lack of access to data and information.

6.1.7 Stakeholder Engagement

Enhancing production and efficiency in "Regadio" systems requires collaboration and coordination between a variety of stakeholders, including farmers, water management agencies, academic institutions, and local communities. Implementing efficient water management strategies may be hampered by a lack of stakeholder participation and cooperation.

6.1.8 Institutional Capacity

Enhancing the capability of water management organizations, giving farmers instruction and technical assistance, and fostering knowledge sharing are crucial for promoting gains in production and efficiency. Increasing institutional capability can assist in removing obstacles and promoting the adoption of best practices.

A multifaceted strategy, including infrastructure investment, legislative reforms, knowledge dissemination, capacity building, and stakeholder engagement, is needed to address these obstacles and problems. For the Cova da Beira region to overcome these obstacles and advance sustainable water management techniques, it is crucial for relevant agencies and stakeholders to collaborate.

6.2 Exploration of Opportunities, Such as Adopting Precision Irrigation Techniques, Promoting Water-Saving Practices, and Leveraging Modern Technologies Within the "Regadio" System

Numerous advantages can be attained by investigating and implementing precise irrigation methods, encouraging water-saving habits, and utilizing contemporary technologies within the "Regadio Cova da Beira" system. Several benefits are as follows.

6.2.1 Increased Water Use Efficiency

Drip irrigation and sprinkler systems are examples of precision irrigation methods that provide accurate water application to the plant's root zone. This focused strategy reduces water losses from evaporation or runoff, leading to increased water use efficiency. Farmers can increase crop production per unit of water utilized by maximizing their water use.

6.2.2 Enhanced Crop Productivity

Utilizing contemporary irrigation techniques and technologies can give crops the ideal amount of water for each stage of their growth. Improved crop health, decreased water stress, and increased crop yield are all benefits of a reliable and effective water supply. Additionally, it may result in higher-quality food with a higher market value.

6.2.3 Resource Conservation

The "Regadio Cova da Beira" systems help conserve water resources by using water-saving techniques. Farmers can contribute to the sustainable management of the region's water resources by minimizing water waste and maximizing water consumption, particularly during times of water scarcity or drought. Future agricultural operations and the ecosystem gain from this conservation.

6.2.4 Improved Nutrient Management

Improved nutrient management is made possible by precise irrigation systems, which provide water and nutrients to plants. Farmers can minimize fertilizer runoff and leaching by managing water application to ensure that nutrients are effectively absorbed by the crops. This encourages efficient fertilizer management, lessens the negative effects on the environment, and improves crop nutrition.

6.2.5 Cost Savings

Farmers can achieve cost savings by using water more effectively and using targeted irrigation techniques. Farmers can cut their water bills and operating costs by using less water. Additionally, optimal agricultural yields brought on by better water management may result in higher income, boosting profitability.

6.2.6 Environmental Sustainability

Using contemporary technologies and water-saving techniques is in line with environmental sustainability objectives. Farmers help preserve local ecosystems, maintain water quality, and lessen the ecological impact of agricultural activities by conserving water use and lowering water pollution through effective irrigation management.

6.2.7 Resilience to Climate Change

Using precise irrigation methods and contemporary technology can help increase the "Regadio Cova da Beira" systems' resistance to the effects of climate change. Effective water management techniques help farmers adapt to changing conditions, maximize water use, and reduce risks related to water scarcity or extreme weather events as climatic patterns become more unpredictable.

The "Regadio Cova da Beira" system can gain from increased productivity, resource conservation, cost savings, environmental sustainability, and climate change resistance by looking into opportunities and embracing innovations in irrigation technologies and water management methods. These advantages help the region's agriculture remain viable and sustainable in the long run.

7 Policy Recommendations

7.1 Proposal of Policy Recommendations to Enhance the Efficiency and Productivity of "Regadio" Irrigation Systems and Water Management, Taking into Account Local and Regional Contexts

Here are some policy suggestions that take into account local and regional settings in order to increase the effectiveness and productivity of irrigation systems and water management in "Regadio Cova da Beira":

- Create comprehensive water management plans and put them into action: these plans should include strategies for effective water allocation, infrastructure improvements, and the promotion of water-saving technologies. These strategies should take into account the region's stakeholder interests, crop water needs, and local water availability.
- Encourage farmers to adopt modern irrigation techniques: these methods may include drip irrigation, the use of precise sprinklers, and soil moisture monitoring. Incentives, educational opportunities, and technical support should be made available to encourage the adoption of these technologies. Work together with regional agricultural groups, research institutes, and irrigation specialists to give direction and share expertise.
- Promote data monitoring and decision support systems: to acquire precise data on soil moisture levels, weather patterns, and crop water needs, promote the use of data monitoring technologies, such as remote sensing and real-time monitoring systems. Create decision-supporting systems, which give farmers timely information for planning and optimizing irrigation.
- Implement water pricing mechanisms: this will encourage effective water use and compensate farmers who use water-saving techniques. Graduated water tariffs can stimulate farmers to invest in effective irrigation technologies and urge them to manage water resources wisely.
- Improve coordination and cooperation between stakeholders: these stakeholders include farmers, water management agencies, research organizations, and the local people. Create forums for discussion, knowledge exchange, and participatory decision-making to collaboratively solve the difficulties in water management.
- Boost irrigation and infrastructure networks: spend money on repairing and updating irrigation infrastructure, such as distribution systems, pipes, and canals. By upgrading infrastructure, water losses will be reduced and effective water distribution to farmers will be made possible. To supplement the water supply, look at options for on-farm water storage and rainfall collection.
- Offer financial support and capacity building: provide financial assistance in the form of grants, subsidies, and support programs to farmers who adopt water-efficient methods and technology. To enhance the capacity of farmers and water management professionals in effective irrigation techniques and water management strategies, conduct training programs, workshops, and demonstrations.
- Strengthen research and development: invest in projects that seek to enhance the Cova da Beira region's agricultural methods, crop choices, and water use efficiency. Encourage the conduct of studies, trials, and pilot projects by agricultural and research institutes to determine the efficacy of novel techniques and methods.
- Increase awareness and education: run awareness campaigns to inform farmers, decision-makers, and the general public about the significance of effective water management, the advantages of implementing water-saving techniques, and the function of "Regadio" systems in sustainable agriculture. Encourage educational initiatives that include themes of water management in the agricultural curriculum.

- Track progress: establish monitoring and evaluation methods to evaluate the efficacy of adopted policies and measures. Regularly review crop yield, water use efficiency, and other important indicators to assess success, spot potential weak points, and alter policies as necessary.

By taking into account local and regional contexts, these policy ideas seek to increase the effectiveness and productivity of "Regadio Cova da Beira" irrigation systems and water management. It is crucial to adapt these suggestions to the region's particular needs, resources, and difficulties while also including stakeholders and fostering ongoing collaboration.

7.2 Emphasis on Sustainable Practices, Water Conservation Measures, and Stakeholder Engagement to Ensure the Long-Term Viability of "Regadio" Systems

Here are several sustainable practices, water conservation measures, and stakeholder engagement techniques that can be used to guarantee the long-term viability of the "Regadio Cova da Beira" systems:

- Put crop rotation and diversification into practice: encourage farmers to use crop rotation and expand their range of agricultural pursuits. Crop rotation enhances soil health, lowers the threat of pests and diseases, and maximizes water utilization. The cultivation of less water-intensive crops, such as legumes or drought-tolerant types, can be a part of diversification, which lowers overall water demand.
- Use precise irrigation methods: encourage the use of precise irrigation methods that supply water to the plant's root zone, such as drip irrigation and microsprinklers. These methods promote effective water use by reducing water loss due to evaporation or runoff.
- Use cover crops and mulching: promote the use of cover crops and mulching in agricultural fields. Mulching aids in controlling soil temperature, weed development, and moisture retention. Cover crops boost water penetration, prevent soil erosion, and enhance the structure and fertility of the soil.
- Encourage the implementation of water recycling and reuse systems: this entails either treating and reusing city wastewater for irrigation or treating and reusing farm wastewater. To guarantee water quality and reduce any potential dangers, appropriate filtration and treatment procedures should be put in place.
- Encourage agroforestry and windbreaks: build windbreaks and incorporate agroforestry techniques into agricultural landscapes. In agroforestry systems, trees or shrubs are grown alongside agricultural crops in order to provide shade, lower wind speeds, and reduce evapotranspiration. Windbreaks protect crops from wind damage, help conserve soil, and lessen water loss due to wind evaporation.
- Inform farmers about effective irrigation techniques: conduct workshops and training sessions to inform farmers about the value of effective irrigation tech-

niques. Give advice on how to check soil moisture, schedule irrigation properly, and maintain their irrigation systems. Sharing successful local strategies and case studies might motivate farmers to implement water-saving practices.

- Promote innovation in water-efficient technologies: encourage the development of water-efficient technologies and assist in their integration into the "Regadio" systems. To maximize water use, this involves boosting the use of irrigation automation systems, remote sensing technology, and soil moisture sensors. The most recent developments in water conservation can be found and put into practice with the assistance of research institutions and technology vendors.
- Promote stakeholder collaboration and engagement: involve farmers, water management organizations, neighborhood groups, academic institutions, and other interested parties in decision-making processes. Encourage cooperation using participatory methods, such as multistakeholder forums, water user associations, or agricultural cooperatives. This participation guarantees the consideration of various viewpoints and cultivates a sense of ownership and responsibility for sustainable water management.
- Promote water governance and policy coordination: at the local, regional, and national levels, strengthen water governance frameworks and policy coordination. Make sure that laws and policies encourage the use of sustainable water management techniques in farming. Create channels of communication and coordination between the various governmental, agricultural, and water management groups.
- Monitor and evaluate the performance of sustainable practices and water conservation measures by establishing monitoring and evaluation systems: regularly keep an eye on environmental indicators, crop productivity, and water use efficiency. These data will aid in pinpointing problem areas, monitoring development, and directing the next decision-making procedures.

The "Regadio Cova da Beira" systems can maintain long-term viability, resilience, and sustainability in the face of water scarcity and changing environmental conditions if these sustainable practices, water conservation measures, and stakeholder engagement initiatives are put into effect.

8 Conclusions

The present study allowed us to conduct a descriptive analysis to assess the performance of a region within the "Med-Wet" consortium—the Portuguese Cova da Beira "Regadio." We have identified a set of tools and measures to highlight the behavior of the irrigation systems, their drivers, and roadblocks.

In addition, we have gathered data from the Associação de Regantes da Cova da Beira, which allowed us to run a set of statistical estimations to evaluate the performance of the irrigation system in the region.

When estimating panel data and quantile regression, we conclude that the impacts of the size of the units of exploration on the performance of the irrigation system are significant, being positive as the size increases. However, in the panel estimation, the influence of the region's economic returns indicates a strong and positive impact on the system's performance. This effect is not seen in the quantile regression, where the environmental impact is negative.

8.1 Reinforcement of the Importance of Improving the Efficiency and Productivity of "Regadio" Systems for Sustainable Agriculture in the Cova da Beira Region

It is crucial for sustainable agriculture in the Cova da Beira region to increase the effectiveness and production of "Regadio Cova da Beira" systems. Here are several main explanations on why.

8.1.1 Water Resource Management

The Cova da Beira region faces water scarcity and limited water resources, just like many other places. Water losses can be reduced, and water can be used more effectively by improving the efficiency of irrigation systems. This allows for better management of available water resources, ensuring their long-term viability and meeting the water requirements of various industries, including agriculture.

8.1.2 Sustainable Water Use

The agricultural sector consumes a significant amount of water, and ineffective irrigation techniques can result in wasteful overuse. By optimizing water delivery, reducing water losses due to evaporation and runoff, and aligning water application with crop water requirements, improvements in the "Regadio" system aid in the achievement of sustainable water use. It supports the preservation of water resources and encourages good water management practices.

8.1.3 Environmental Protection

Sustainable agriculture strives to reduce the harmful effects that agricultural activities have on the environment. Ineffective irrigation techniques can cause water pollution by allowing fertilizers and pesticides to permeate into the water. The amount of water and agrochemicals required for irrigation can be decreased by increasing efficiency, which lowers the risk of environmental pollution and maintains the quality of soil, water bodies, and ecosystems.

8.1.4 Economic Viability

The increased productivity and efficiency of the Regadio system have made the agriculture of the Cova da Beira region more economically feasible and have led to a growing crop diversity. Farmers can increase output levels and profitability by optimizing water consumption and crop yields. This promotes rural economies, maintains farmer livelihoods, and ensures the long-term viability of agricultural activities in the area.

8.1.5 Climate Change Resilience

Agricultural production faces more uncertainties and difficulties because of climate change. "Regadio" systems can become more resistant to the effects of climate change, such as altered precipitation patterns, droughts, and heat waves, by increasing their efficiency. The sensitivity of agriculture to hazards associated with climate change is decreased by efficient irrigation practices that aid farmers in adapting to water scarcity and variability.

8.1.6 Food Security and Nutrition

Increasing the effectiveness and productivity of the "Regadio Cova da Beira" systems is crucial for ensuring food security and nutrition in the area. Farmers can contribute to increasing agricultural productivity and food supply by maximizing crop yields with limited water resources. This increases local populations' access to a variety of wholesome food options and contributes to meeting the rising demand for food.

8.1.7 Goals for Sustainable Development (SDGs)

Improving the effectiveness and productivity of the "Regadio" systems aligns with several Sustainable Development Goals: Goal 2 (Zero Hunger), Goal 6 (Clean Water and Sanitation), Goal 12 (Responsible Consumption and Production), and Goal 13 (Climate Action). Efficiency-focused sustainable agricultural methods support the accomplishment of these global development objectives and encourage a more resilient and sustainable future.

For sustainable agriculture in the Cova da Beira region, it is crucial to increase the effectiveness and productivity of the "Regadio Cova da Beira" systems. It promotes the management of water resources, environmental preservation, economic viability, climatic adaptability, food security, and the accomplishment of sustainable development goals. The area can preserve the long-term resilience and viability of its agricultural industry while protecting natural resources for future generations by adopting sustainable irrigation practices.

8.2 *Call to Action for Policymakers, Farmers, and Relevant Stakeholders to Implement the Proposed Recommendations to Optimize "Regadio" Irrigation Systems and Water Management Practices*

The current report provides a series of recommendations for improving the irrigation systems and water management procedures at "Regadio Cova da Beira." The stakeholders can guarantee the long-term survival, sustainability, and resilience of the agricultural sector through cooperation. Therefore, the following call to action is made.

8.2.1 For Policymakers

(a) Adopt sustainable water management practices: recognize the significance of sustainable water management in agriculture and give top priority to creating and implementing laws that encourage smart water use, the employment of technology, and the observance of conservation practices within the "Regadio" systems.
(b) Establish supportive policies and incentives: develop and implement policies that promote the adoption of water-efficient practices and technology. Give farmers financial incentives, subsidies, and grants to help them adopt sustainable irrigation techniques.
(c) Foster collaboration and stakeholder engagement: encourage cooperation between farmers, water management organizations, academic institutions, and other pertinent parties. To ensure that policies and programs are carried out successfully, forums for knowledge exchange, capacity building, and participatory decision-making should be established.

8.2.2 For Farmers

(a) Embrace efficiency and conservation practices: adopt water-saving technologies, such as precision irrigation methods, soil moisture monitoring, and crop diversity in recognizing the significance of effective water usage. Use farming techniques that minimize environmental effects while maximizing output and water use.
(b) Seek knowledge and training: follow the most recent developments in conservation techniques and irrigation technology. Participate in training courses, seminars, and information-sharing forums to get an understanding of effective irrigation planning, crop water needs, and optimum management techniques.
(c) Collaborate and share experiences: engage in discussions on effective irrigation practices with other farmers, agricultural groups, and the community at large. We can mutually improve water management and increase agricultural productivity by exchanging information and experiences.

8.2.3 For Relevant Stakeholders

(a) Provide support and expertise: encourage the adoption of water-saving technologies and sustainable practices among farmers by providing them with support, financial aid, and technical advice. Work together with academic institutions, agricultural extension agencies, and technology suppliers to spread information and offer advice on effective irrigation methods.
(b) Foster awareness and education: promote effective irrigation methods and the advantages of sustainable water management through awareness campaigns, seminars, and field exhibits. Provide farmers with the information and resources they need to make educated decisions about their irrigation strategies.
(c) Monitor progress and share findings: create systems for monitoring and evaluating the performance of adopted policies and procedures. Share research results and lessons gained with decision-makers, farmers, and other stakeholders to help guide choices and encourage ongoing development.

Capital Investment Planning of Infrastructure Assets in the Water Sector

Pedro Aguiar Costa, Jaime Gabriel Silva, Ligia B. Ramos, Nuno Marques de Almeida, and Ana S. Camanho

Abstract All companies have limited available capital to make investments in infrastructures. Consequently, it is imperative that investments are well funded to ensure proper asset functioning and high service standards at a manageable risk level and at the lowest possible cost. These dimensions are quantified through multiple financial and operational indicators. This chapter proposes a more pragmatic approach to the calculation of one of these indicators—the infrastructure value index (IVI), which is being tested in the Portuguese water sector. A revised formulation for the IVI is proposed and illustrated with an empirical case study consisting of water infrastructures managed by the Portuguese company Águas do Douro e Paiva (AdDP).

Keywords Asset management · Water systems · Water infrastructure · Investment planning

P. A. Costa · A. S. Camanho (✉)
Faculdade de Engenharia, Universidade do Porto & INESC TEC, Porto, Portugal
e-mail: acamanho@fe.up.pt

J. G. Silva · L. B. Ramos
Águas do Douro e Paiva (AdDP), Porto, Portugal
e-mail: g.silva@ADP.PT; ligiaramos@adp.pt

N. M. de Almeida
CERIS, Instituto Superior Técnico, Universidade de Lisboa, Lisboa, Portugal
e-mail: nunomarquesalmeida@tecnico.ulisboa.pt

J. C. Correia Leitão et al. (eds.), *Sustainable Water Management and Irrigation Systems*, Natural Resource Management and Policy 62,
https://doi.org/10.1007/978-3-031-93587-9_13

1 Introduction

Asset management is a holistic concept that spans several areas. It is focused on the use of assets to generate value and achieve the goals of the organization. The concept of value should be defined by each firm according to its mission.

Therefore, asset management must consider the nature of the organization, the operational context, the financial limitations, and the organization's needs to define the goals to be achieved and the means required to maximize the value obtained from the assets.

There is a vast number of accepted definitions for asset management. One that is considered to be quite comprehensive is the definition from the Department of Water and Environment Affairs of South Africa, which states that infrastructure asset management (IAM) is an integrated process of decision-making, planning, and control over the acquisition, use, safeguarding and disposal of assets to maximize their service delivery potential and benefits and to minimize their related risks and costs over their entire life (Bhagwan, 2012).

Organizations with well-developed asset management procedures may gain a competitive advantage as this enables the achievement of better financial performance, more assertive decision-making, and improvements in the quality of services. Overall, asset management can enhance organizations' efficiency and sustainability.

Asset management is essential in the water supply and sanitation sector, given its intensive capital nature. High investments have to be continuously made by utilities of the sector, often involving irreversible decisions. Consequently, investments must be adequately substantiated so that the proper functioning of assets and high service levels can be achieved at a manageable risk level and with the lowest possible cost. This involves designing strategies in the context of informed decision-making, which requires integrating asset management approaches into planning and management processes to generate benefits for both service users and stakeholders.

To achieve the best execution of these strategies, a trade-off between the dimensions "risk," "cost," and "performance" becomes essential. These dimensions can be analyzed through multiple financial and operational indicators. This chapter contributes to the literature on asset management by discussing and testing the use of an indicator that reflects the risk and cost dimensions. This indicator, called the infrastructure value index (IVI), is already in use within the water sector in Portugal, at the request of the national regulator, although initially as a test of its quantification and only for pipelines. However, utilities can develop and test its calculation for other types of assets.

The IVI has been explored in various research studies and industry projects in the water sector. It has been recognized as an effective tool that supports long-term investment planning. It is calculated as the ratio of the current value of an infrastructure to its replacement cost. The current value of an infrastructure is its market value in the present condition. The replacement cost is the current price that must be paid for a current alternative to the infrastructure under study.

Conceptually, the IVI is a powerful tool because it translates pertinent information in a simple way. It allows decision-makers to correctly compare the condition of the various assets and estimate the impact that future investments will have on the condition of the assets and the overall system.

The IVI indicator was first introduced by Alegre (2008) and was subject to improvements by Alegre et al. (2014). Although organizations have already started using this indicator, its calculation depends on subjective information as its components can be interpreted in different ways. More recently, Vieira et al. (2020a) presented a methodology to consider the uncertainty associated with the useful lives of assets, when forecasting the IVI in future years as a function of planned capital investments. Regardless of whether a stochastic approach is used in the forecast, the calculation of the IVI for future years makes the indicator valuable in empirical contexts and allows it to be embedded proactively in companies' capital investment planning processes.

Although the IVI is mostly used in the water supply and sanitation sector, this paper aims to present a tool that can be extended to other industries, making this indicator truly impactful in different asset management contexts. For example, Salvado et al. (2020) have adapted and tested this methodology for managing large building assets' portfolios.

This chapter aims to build upon previous efforts and test a new approach for long-term investment planning focused on annual capital expenditures and asset deactivation. Relative to earlier studies, the novel contribution of this chapter is the development of a methodology where the useful life of the assets is adjusted to their age and current condition. This is an enhancement in comparison with the common practice of using only the state of condition to determine the current value of the asset. Another contribution is a more objective approach to the long-term IVI calculation proposed by Vieira et al. (2020b). This improvement is centered on simplifying the formula originally presented, reducing the number of subjective variables and mitigating the associated error. The methodology developed was applied in a Portuguese water supply company. A dashboard was also constructed to enable the visualization of the information obtained. The new methodology was developed, in cooperation with company collaborators, to ensure the inclusion of enterprise knowledge and proper validation of the results obtained.

2 Review of Capital Investment Planning for Infrastructure/Asset Management

This section describes the approach proposed for the calculation of the IVI in an asset management system and how companies and the Portuguese regulator are currently using it.

2.1 Traditional IVI Formulation

As stated in the Introduction, the IVI is the quotient between the infrastructure current value and the infrastructure replacement cost, as shown in Expression 1.

$$\text{Infrastructure value index}\left(\%\right) = \frac{\text{Infrastructure current value}}{\text{Infrastructure replacement cost}} \tag{1}$$

The infrastructure current value (ICV_t) estimates how much the company would profit from decommissioning and selling the assets that comprise the infrastructure after disposal costs have been accounted for using Expression 2. Note that this value, in reality, can be negative if the asset is so depreciated that what the company earns with the sale of the assets does not cover the disposal costs (Howard et al., 2015).

$$ICV_t = \sum_{i=1}^{N} ACV_{it} \tag{2}$$

In Expression 2, N is the number of assets that compose the infrastructure and ACV_{it} is asset i ($i = 1,\ldots,N$) current value in year t, expressed in monetary terms (e.g., in euros).

The Australian Infrastructure Financial Management Manual (Howard et al., 2015), published by the Institute of Public Works Engineering Australasia (IPWEA), defines depreciation as a function of residual useful life to reflect the pattern of consumption of future economic benefits. The residual useful life is the difference between the total technical useful life and the age of the asset in question. The total technical useful life used in this analysis is a national theoretical average.

The depreciation of an asset can then be calculated by the product of its annual depreciation, given by the ratio between its replacement cost and its total technical useful life, and its age. The asset current value is then given by the difference between its replacement cost and its depreciation.

$$ACV_{it} = rc_{it} - \frac{rc_{it} \times a_{it}}{ul_i} \tag{3}$$

In Expression 3, rc_{it} is the asset i replacement cost in year t, expressed in monetary terms (e.g., in euros); a_{it} is the asset i age in year t; and ul_i is the total technical useful life of asset i.

Depreciation can be calculated on an accounting or economic basis. The choice between using an economic or an accounting approach will depend on how the variables rc_{it} and ul_i are obtained.

The accounting method is static and uses as a reference the beginning of the asset's life. This method divides the acquisition cost by the total technical useful life and obtains a static depreciation associated with each year of the asset's service.

The economic approach is dynamic. That is, it is adjusted every year according to the change in the asset's economic value in that period. The interesting thing in this type of analysis is to know the reality of an infrastructure, taking into account possible obsolescence and its state of condition. Despite the fact that economic depreciation is possibly a better approach, it should be noted that it entails additional difficulty and is prone to errors, requiring a more significant effort from the asset management team for a careful and reasoned analysis.

The Portuguese Water and Waste Services Regulation Authority (ERSAR) suggests using an economic approach since it provides a more accurate estimate of reality. However, in the absence of adequate information to use this method, it may be better to use an accounting approach.

Replacement costs can be calculated in two ways. The first is to use, for each asset, the acquisition cost as the replacement cost. Despite the limitations of this approach, it can be used in the absence of better alternatives. The second, and the most appropriate according to the United States Environmental Protection Agency (USEPA) (2005), is through the modern equivalent asset (MEA) methodology. This alternative assigns to the replacement cost what the company would currently have to pay for an asset that meets the project's specifications. Due to technological advances and currency inflation, which is critical to account for, this value is different from the initial acquisition value. Hence, according to Howard et al. (2015), it is a more correct alternative than the alternative of using the acquisition cost.

The concept of the current value of an asset proposed by IPWEA, which underlies the specification of Expression 3, is also adopted in the definition of the current value presented and elaborated upon in the ERSAR guide (Covas, 2018). In this guide, the current value of an asset is given by its residual useful life multiplied by the estimated annual depreciation.

The mathematical manipulation illustrated in Expression 4 shows how the approach presented by IPWEA results in the asset current value proposed by ERSAR.

$$\begin{aligned} ACV_{it} &= rc_{it} - \frac{rc_{it} \times a_{it}}{ul_i} \\ &= rc_{it} - \frac{rc_{it} \times \left(uL_i - rl_{it}\right)}{ul_i} \\ &= rc_{it} - rc_{it} + \frac{rc_{it} \times rl_{it}}{ul_i} \\ &= \frac{rc_{it} \times rl_{it}}{ul_i} \end{aligned} \tag{4}$$

In Expression 4, rl_{it} is the residual useful life of asset i in year t.

Using the final expression in Expression 4, it is possible to replace ACV_{it}, used in Expression 2, and arrive at the ERSAR proposal for the estimation of ICV_t, as shown in Expression 5.

$$ICV_t = \sum_{i=1}^{N}\left(\frac{rc_{it} \times rl_{it}}{ul_i}\right) \quad (5)$$

The suggested formulation for the IVI in Eq. 1, at a given period (year t), can be calculated as shown in Expression 6:

$$IVI_t\,(\%) = \frac{ICV_t}{\sum_{i=1}^{N} rc_{it}} \quad (6)$$

The value of the IVI varies between 0 and 1. A value of 0 corresponds to total degradation, and 1 indicates the full preservation of the infrastructure. In the water sector, according to Alegre (2008) and Covas et al. (2018), the expected value of stable infrastructures ranges from 0.4 to 0.6. Values below 0.4 signal that the infrastructures have deteriorated and require significant investments. Values above 0.6 may indicate overinvestment in rehabilitation/replacement or a recent and nonstabilized infrastructure.

From a computational perspective, the estimation of the IVI is straightforward. However, it can be challenging to estimate the values of assets' residual useful life, total technical useful life, and replacement cost accurately. The estimation of these values implies the collaboration of several entities within the organization, including the maintenance, operations, and financial departments.

2.2 *Long-Term IVI*

All the aspects discussed so far present a point estimate for the IVI. However, in the context of asset management, it is also relevant to estimate the future value of the indicator. Predicting how the infrastructure condition (IVI) varies with the planned capital investments is necessary to make it a more helpful indicator for investment decisions.

Once the IVI for the current year has been obtained, its estimation for future years requires comprehensive infrastructure knowledge. Over the years, renovation, replacement, and expansion operations will be carried out on the infrastructure, which should be accounted for. To calculate the IVI on a long-term basis, Vieira et al. (2020b) proposed an approach where the expected CapEx on renovation, replacement, and expansion operations influences the condition of the assets. It is prudent to state that assets or infrastructures subject to high capital investments may see their current condition increase.

The model presented by Vieira et al. (2020b) proposes the use of Expression 7 for the estimation of the replacement cost in the future year t and Expression 8 for the residual useful life of the various assets of each infrastructure.

$$rc_{it} = rc_{i(t-1)} \times \left(1 - \frac{1}{ul_i}\right) + \text{CapExRen}_t \times \alpha_{ren} + \text{CapExRep}_t \times \alpha_{rep} + \text{CapExExp}_t \times \alpha_{\exp} \quad (7)$$

In Expression 7, CapExRen_t, CapExRep_t, and CapExExp_t are the capital expenditure values, respectively, in the investment of renewals, replacements, and expansions, at year t, and α_{ren}, α_{rep}, α_{exp} are the coefficients that reflect the impact of a given intervention on the residual useful life.

$$rl_{it} = \beta \times rl_{i(t-1)} + \text{CapExRen}_t \times \alpha_{ren} + \text{CapExRep}_t \times \alpha_{rep} + \text{CapExExp}_t \times \alpha_{\exp} \quad (8)$$

In Expression 8, β is the coefficient that translates the level of deactivation of existing assets in replacement interventions, where the replaced assets are deactivated and removed from the existing assets.

In the event of a renewal or replacement, the coefficients α_{ren} and α_{rep} have a positive value that typically ranges between 0.5 and 2. When the intervention contributes fully to the increase of the residual life and is made with assets with the same characteristics as those to be decommissioned, these coefficients have a value equal to one. If the interventions include technological advances and materials with better properties, the coefficients will be higher than one. If the interventions are for lower-quality assets, the values of the coefficients will be less than one.

The expansion interventions represented by α_{exp} are easier to determine. If there are expansion interventions that fully contribute to the increase of the useful life of its assets, then this coefficient is equal to 1.

β is the abatement coefficient, ranging from 0 to 1, and represents the degree of asset deactivation as a function of the planned replacement associated with each capital investment. A beta value close to 0 means that the asset under study will be almost totally dismantled, whereas a β value close to 1 means that the investment will deactivate only a few assets. The coefficient becomes more challenging to obtain when it is far from the limits (near 0.5).

All these coefficients are obtained subjectively since they require engineering judgment and an evaluation of the impact that the interventions will have on the useful life of the infrastructures.

This methodology enables complementing a financial tool with engineering considerations, providing a quantitative tool to estimate the efficiency of the planned asset management interventions.

3 Enhancing the Infrastructure Value Index: A Methodological Advancement

This study discusses possible enhancements to the formulation described in the previous section to make the estimation of the IVI indicator more pragmatic and aligned with business reality. This discussion culminates with the proposal of a new method for calculating the IVI and extending it to future years.

3.1 A New Approach to Estimate the IVI at Time t

The formulation proposed in this chapter for calculating the IVI is based on Expression 6. It differs from what is available in the literature in the way the variable ICV_t is calculated. The variables subject to changes in the ICV_t formulation shown in expression 5 are the assets' estimated residual useful life and the total technical useful life. These variables are adjusted to express a more accurate infrastructure current value:

$$ICV_t = \sum_{i=1}^{N}\left(\frac{rc_{it} \times arl_{it}}{aul_i}\right) \tag{9}$$

In Expression 9, arl_{it} is the asset i adjusted residual useful life and aul_i is the total adjusted useful life of asset i in year t.

Establishing a residual useful life value by considering a linear depreciation of the asset as a function of age and total technical useful life is easy but leads to inaccurate information. A better approach to estimating the residual useful life of an asset requires accounting for its state of condition. An asset at the end of its theoretical life that has been subject to good maintenance actions and capital investment can still be in good condition, consequently increasing the residual useful life. To standardize the classification of the assets' condition (c), the approach proposed by the United States Environmental Protection Agency (EPA) (USEPA, 2005) and adapted by Alegre (2008) is used in this study, as shown in Table 1.

The state of condition varies between 1 and 5 and refers to the degradation of the inherent characteristics of each asset at the date of evaluation. Thus, more degraded assets in need of major rehabilitation interventions will have a lower state of condition, translated by a higher value variable c.

At an international level, there are examples of institutions and entities that use only the state of condition (c) to determine the residual useful life of an asset. As pointed out in Alegre (2008) the US EPA uses a linear relationship between the state of condition and residual useful life.

Another example is that of a simple methodology used by Carollo Engineers Inc. in California (Carollo, 2015), where residual life is given by the following expression (10):

Table 1 Conservation status classification as proposed by the US EPA (Alegre, 2008)

State of condition (c)	Description	Required rehabilitation rate
1	In perfect condition	0%
2	It has minor anomalies	5%
3	Presents anomalies requiring curative maintenance	10–20%
4	Requires renewal	20–40%
5	Practically unusable asset	>50%

$$\text{Residual useful life} = (1 - \text{required rehabilitation rate}) \times \text{Original useful life} \quad (10)$$

While these approaches may be better than considering a linear depreciation of the asset as a function of age and total technical useful life, using only the infrastructure condition to determine useful lives seems limiting. It is expected that two infrastructures with very different ages and similar states of condition will have slightly different residual useful lives. Therefore, Carollo Engineers Inc. uses a formulation (Eq. 11, whose graphical representation is presented in Fig. 1) for the residual useful life that accounts for the condition and age of the infrastructure (H2Opt, 2020). The graphical representation was made for a total technical useful life of 60 years to present the concept.

$$arl_{it} = ul_i - \left(\frac{ul_i}{5} \times c\right) \times \left(\frac{a_{it}}{ul_i}\right)^{1/c+1} \text{[years]} \quad (11)$$

In Expression 11, c is the condition coefficient ranging from 1 to 5 (USEPA classification shown in Table 1).

The graphical representation of Eq. 11, shown in Fig. 1, can also be used to determine arl_{it}. With the condition state of the infrastructure established, the line with the respective color is chosen, and the arl_{it} is determined as a function of the age of the infrastructure.

The proposed methodology is based on the linear variation of the total theoretical useful life with the infrastructure age. It tries to adjust this linear variation, as a function of the condition state and total theoretical useful life, for each condition

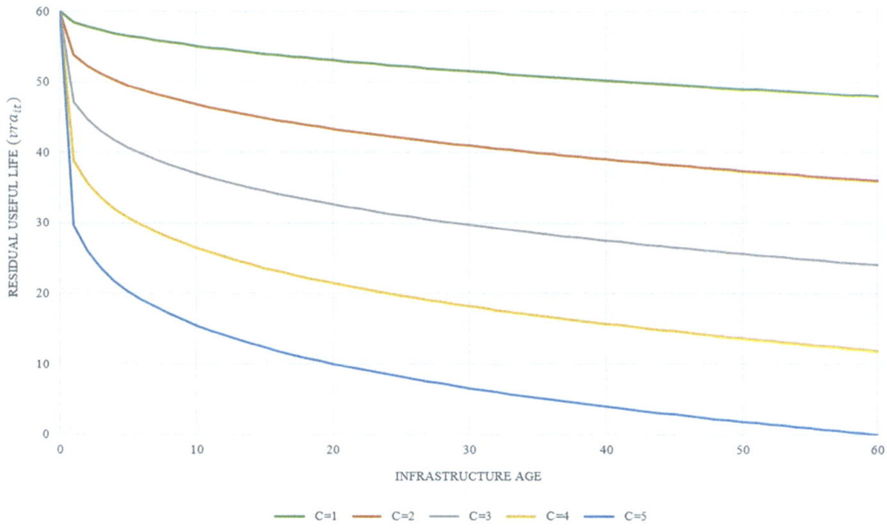

Fig. 1 Adjusted residual useful life as a function of infrastructure condition

state, as a function of the age, considering an approximation, in power, that is inversely proportional to the condition state.
With the determination of the adjusted residual useful life, the total theoretical useful life is no longer used as the adjusted total useful life provides a better estimation on the actual condition of the asset. This adjusted total useful life is given by the sum of the infrastructure age and the arl_{it}.

3.2 Improvements to Long-Term IVI Calculation

This section proposes the extension of the estimation of the IVI indicator to future periods, enhancing the approach first presented by Vieira et al. (2020b). A limitation of the Vieira et al. (2020b) formulation is that it draws on several variables that are estimated without resorting to information of the actual condition observed in the real-world system, such as β, α_{rep}, α_{exp}, and α_{ren}. Another issue is that in some cases, the IVI can be greater than 1. It should first be emphasized that the IVI, by its nature, is an indicator that only makes sense when applied to a wide range of assets—a portfolio of assets, such as a water supply system or a part of it (subsystem). It would lose its interpretative logic if it were applied to an isolated asset or even to an infrastructure such as a pipeline. But in order to provide an example in which the IVI becomes greater than 1, the case of an infrastructure (a set of assets) can be considered, assuming the criticism regarding the interpretative difficulty of the indicator in this situation. In these circumstances, the IVI could be higher than 1 if the infrastructure is completely replaced without having been fully depreciated.

Expressions 12 and 13 are proposed for the calculation of the variables rl_{it} and rc_{it} in order to avoid these pitfalls.

$$rc_{it} = rc_{i(t-1)} \times (1-\beta) + CapEx_t \tag{12}$$

$$rl_{it} = ul_i \times \frac{rc_{i(t-1)} \times (1-\beta) \times \left(rl_{i(t-1)} - 1/ul_i \right) + CapEx_t}{rc_{it}} \tag{13}$$

With the variables rl_{it} and rc_{it} defined and substituting them into Eq. 6, an estimate for future IVIs can be obtained.

The value of β in this formulation is a percentage rather than a coefficient. Thus, in Eq. 13, β is the percentage of decommissioning inherent to the various capital investment interventions in renewal, replacement, or expansion. In the current formulation, the classification of interventions is simplified from three categories (renewals, replacements, and expansions) to two (renewals or replacements, and expansions), since, in practice, renewal activities typically involve asset replacement and decommissioning. What separates the replacements or renewals from the expansions is the β value. When there is a replacement or renewal, the β value corresponds to the percentage of the asset deactivated, and in expansions, the β is automatically 0.

In addition, another adjustment has been introduced. Whenever an asset or a set of assets is renewed, it is known as the percentage of assets that have been deactivated. In practice, it is as if it is possible to automatically update the total replacement value of the assets and no longer use the α coefficients. For example, when replacing one-third of an asset, its new replacement cost is two-thirds of its previous replacement cost, plus the CapEx planned for that asset.

With this new approach, an IVI greater than 1 cannot occur as the replacement values are revised based on the percentage of the asset replaced and the value of that replacement.

4 Case Study

This section describes an empirical application of the new approach proposed for the IVI calculation to the water supply system of company Águas do Douro e Paiva (AdDP).

AdDP was created in 1995. It is a utility belonging to a Portuguese state group (Águas de Portugal (AdP)), with the social purpose of exploring and managing the multimunicipal water supply system for the southern area of the Greater Porto region for 30 years. Later, the company underwent some changes, including aggregation with other AdP Group companies and subsequent unbundling, as well as some expansions of its area of operation. Currently, the AdDP system integrates 20 municipalities and covers an area of 2715 km^2 with 1.7 million inhabitants. The concession period was also extended to 2036.

AdDP comprises a wide range of infrastructures, such as water intakes, pipelines, pumping stations, water treatment plants (WTP), and reservoirs. Due to the large size of the company and the fact that water supply is a capital-intensive sector, the best methodologies to identify the infrastructure condition must be implemented to make informed decisions on where to invest the available capital, hence the interest in undertaking this study. The IVI indicator is used by AdDP as one of its decision support tools. The organization has been determining IVI values for all different types of infrastructures since 2017.

Until now, AdDP has used the IVI to determine infrastructure conditions in the year under analysis. The formulation used is the one presented in Expression 6.

To determine the replacement cost of the infrastructure, AdDP uses ERSAR Technical Guide 23 (Covas et al., 2018). In this document, ERSAR proposes replacement values for the different types of infrastructure that make up a water supply system, according to specific characteristics. The guide divides the infrastructure into two groups—equipment (electro-mechanical) and civil construction—and proposes replacement costs for each one.

The useful lives of the assets that AdDP considers are 60 years for civil construction and 25 years for equipment. Regarding residual useful lives, AdDP subtracts the infrastructure age from the useful theoretical life. The company uses reports that discriminate when the civil construction and equipment groups were built or rehabilitated.

The methodology used so far does not fully reflect the real situation of the infrastructures because the infrastructure assets have different ages and distinct total technical useful lives. Furthermore, the costs presented by ERSAR (Covas, 2018) for each group are based on the representative averages of the national panorama, having been developed based on budgets of a broad set of public water supply and urban sanitation contracts carried out in recent years, so they may not represent accurately the specific situation of a given system. Finally, the estimation of residual useful lives based solely on age, without taking good maintenance policies into account, can be misleading. Although using the indicator in these conditions has allowed the company to gain experience in handling it, it is not a robust estimate of the IVI. Therefore, it is necessary to use a methodology that gets as close as possible to the reality of the infrastructure condition.

More recently, in 2020, a study was carried out for AdDP by the consulting firm H2Opt to introduce improvements in the methodology used in the IVI formulation. Thus, an equation was formulated for calculating the residual useful life as a function of the asset age and condition. This study represents a relevant improvement since it accounts for maintenance work without disregarding the inevitable deterioration of assets over time. Despite providing a more reliable result, the methodology calculates the IVI by aggregating the assets into equipment and construction, forecasting equal costs, ages, and states of condition for these groups.

Also, AdDP is aware that the company's calculations of IVI are not yet considering, in the IVI formulation, its extension to future years in order to calculate the influence of investments on the infrastructure condition. This is crucial if the IVI is to be used as an indicator that helps investment planning.

The methodology developed, and presented in this paper, presents a proposal for calculating the IVI, which attempts to respond to these difficulties.

The proposed methodology is applied to two infrastructures: the Jovim pumping station and the Lever Upstream Intake pumping station. However, replicating this approach for the other infrastructures of the company would be advisable. Regarding the IVI forecast for future dates, the window chosen for the analysis is 5 years for Jovim PS and 10 years for Lever Upstream Intake. This time window was chosen because it is the time window for which the CapEx investments were recently forecasted in AdDP.

The Lever water supply system is divided into two subsystems: the North Lever and the South Lever. The Jovim pumping stations are an integral part of the North Lever subsystem. The water comes directly from the Lever Upstream water intake, the final pumping station of the Lever WTP, or through the Lagoa reservoir.

The older Jovim pumping station (PS1) was integrated into the multimunicipal system in 1996/1997, and a new pumping station (PS2) was built in 1998, enabling the old lift to be also rehabilitated in 1998. Both pumping stations have four electric pump groups and raise the water to Ramalde, each one through its respective pipeline.

PS1 has a lifting capacity of 1347 l/s and reaches a manometric height of 57.1 mwc, while PS2 has a lifting capacity of 2718 l/s and 56.2 mwc.

The Lever Complex Upstream Abstraction dates from the 1980s. It consists of two deep wells where the water is collected by radial drains installed on two levels. After naturally seeping into the bed of the Douro River, the water is sent by gravity to a collecting well about 20 m deep, from where it will be elevated. A platform equipped with eight pumps is installed over the collecting well. Six of these pumps raise the water to Jovim and two to the Lever Water Treatment Plant for subsequent treatment.

The pumps responsible for raising the water to the Jovim reservoir can rise to 600 l/s. As regards pumping to the WTP, each pump can lift a variable flow rate of up to 1500 l/s.

4.1 Collection of Asset-Related Data

For the calculation of the IVI, only assets that are directly related to the water supply service are accounted for. Therefore, assets such as office equipment, fire extinguishers, and vehicles are excluded from the analysis.

The asset-related data must be clear, complete, and intuitive. The assets that set the infrastructure must have information about what they are, their function, condition, and installation dates and costs. It is also important to break the assets into their components. It is requested that condition (c) be properly defined and with strict judgment since it is directly related to the asset's adjusted residual useful life.

This relationship is evidenced in Expression 11. Determining the condition is the most subjective part of the calculation and, therefore, should be as thoroughly and identically calculated as possible for all assets in the company. The parameters required for estimating this variable are as follows:

- Replacement, renewal, and expansion works that have been made during the asset's lifetime
- The frequency of maintenance routine work
- The need for inspection/supervision and operational intervention
- The availability of the functional unit and the existence of malfunctions that jeopardize its performance
- The existence of equipment redundancy to ensure uninterrupted operation in case of breakdown or preventive maintenance of one of them

Each of these parameters requires collecting information that may be needed to define assets' condition. As each case has its particularities, this type of approach may not be applicable to all situations, and each company should use the method considered the most appropriate to the system under assessment as well as the data available.

4.2 Asset Breakdown Structure

The infrastructures are broken down to a level where it is comfortable to assign values to the variables mentioned in Expression 6. The proposed decomposition is based on five levels:

- *Infrastructure* (first level): in a water supply network, there are pumping stations, pipelines, reservoirs, and abstractions.
- *Group* (second level): it is where the infrastructure assets are separated into civil construction, electrical installation, and equipment.
- *Subgroup* (third level): this indicates the function the assets within the group relate to (such as hydraulic circuit or lifting).
- *Assets* (fourth level): the components are organized into an item or set of items that have value to the organization (such as a pumping group or electrical boardroom).
- *Components* (fifth level): these are the individual parts that make an asset (such as screws or nuts).

The decomposition proposed is shown in Table 2. The table only shows information for PS2 and up to the fourth level to make it easier to visualize.

The breakdown of the Lever Upstream Abstraction and Jovim PS1 is made in the same way (Appendix A).

Table 2 Breakdown in the first four levels of Jovim PS2

First level	Second level	Third level	Fourth level
PS2 Jovim	Equipment	Production	Generator set
		General support systems	Traveling crane
		Hydraulic circuit	Hydraulic circuit
		Instrumentation, automation, control, and measurement	Instrumentation, automation, control, and measurement equipment
		Lifting	Pumping group no. 1
			Pumping group no. 2
			Pumping group no. 3
			Pumping group no. 4
			Ventilators
		Water hammer protection	Compressed air system
			Air vessel RSP 3—Remetal, 35 m3 (PS2)
		Sampling	Sampling system
	Electrical facilities	Distribution	Electrical boardroom
			Transformer substation
		Safety systems	Security system
	Civil works	Civil works	Building
			Pipe system

Decomposing infrastructures, in most cases, can be a complex and time-consuming task. Information is not always complete and up to date, the best way of division may not be intuitive, and an iterative approach may be required. A critical approach and excellent articulation between the various departments of the organization are imperative. The better this division is worked on and cared for, the better the understanding of the infrastructure will be.

AdDP infrastructures are subdivided into five levels according to the information in the database. The first level is already previously defined, which is the infrastructure where the study will be carried out—Jovim PS and Lever Upstream Abstraction. The assets that make up the PS and the Abstraction are then grouped into three categories—civil works, equipment, and electrical facilities—according to the terminology used by the company and proposed in Vieira et al. (2020b). Selecting the most appropriate decomposition and identifying the relevant assets of the water system is a time-consuming process, which requires detailed knowledge of the infrastructure and a critical analysis of the database.

4.3 Infrastructure Replacement Cost

Regarding the infrastructure's replacement costs, it is essential to consider how detailed the technical team can specify the information. It is clear that assigning a replacement cost to the infrastructure is challenging and may not be done with much precision. Therefore, disaggregating the infrastructures into several levels and calculating replacement costs for a given level seems to be the viable alternative. This case study is conducted at the asset level, and therefore it is ideal to use data and calculate replacement costs at this level. Nevertheless, data at a more aggregated level can be used when the information at the asset level is missing.

Companies' catalogs are used at this stage, and the technical team is consulted to reach a consensus on costs. This task is made simpler since forecasting replacement costs for single assets is more straightforward than finding plausible values for replacing the three asset categories or even the entire station. PS's asset reports can be used to find the acquisition value of the assets and then account for currency inflation to derive an initial value for the replacement cost. This value is purely theoretical and has to be assessed and confirmed by the technical team as technological advances may mean different prices from those used when acquiring the assets.

With the assets' replacement cost, it is possible to determine the weights that the different groups of assets (in this case, civil works, electrical installation, and equipment) will have in the total cost of the infrastructure replacement. It is essential to realize that these relative weights are only used to have an idea of the distribution since what goes into the IVI formula (Eq. 6) are the costs of each asset. This distribution is critical to understand if the analysis is aligned with the national panorama.

Different infrastructures are going to have different replacement costs and asset category weights. Infrastructures with different objectives will have different assets. It is to be expected that a PS, whose sole purpose is to lift water to other destinations, has a high weight of equipment category due to the pumping groups. In contrast, due to the wells responsible for collecting water, an abstraction has a higher weight of civil works. ERSAR suggests the weights that each group of assets will have in the infrastructure in its Technical Guide 23 (Covas et al. (2018)). The expected distribution weights for PS's is shown in Fig. 2 and for abstractions in Fig. 3.

It is essential in the end to compare the weights obtained, through research and dialogue with the technical team, with the weights suggested in the ERSAR guide. If the weights are very different, it is advisable to find the cause.

In the Jovim PS, civil construction weighs 30% since the station is oversized compared to other stations with the same characteristics. The equipment and electrical installation weigh 49% and 21% respectively, which also diverge from the guide values because it is a facility that requires sizeable hydraulic power and consequently large pumping groups.

Regarding the Lever Upstream Abstraction, the case also diverges from the values presented in the ERSAR 23 Guide, illustrated in Fig. 3. In this abstraction, the estimated weight of civil construction is 53%, which is larger than the 43% foreseen in the guide. To understand the discrepancy, it is worth highlighting the assets that are part of the construction category of the Lever Upstream Abstraction. These assets are two semi-superficial abstractions in the middle of the river, one collecting well, pipes, and the building where the equipment and electrical installations are located.

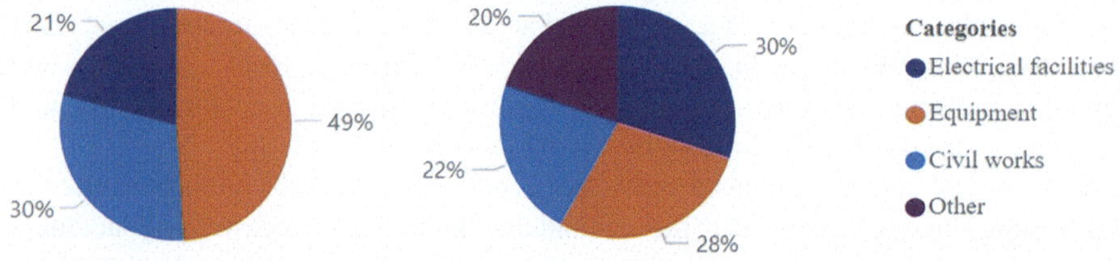

Fig. 2 Distribution of the PS category weights by the ERSAR guide (right) and the actual values in AdDP for the Jovim PS (left)

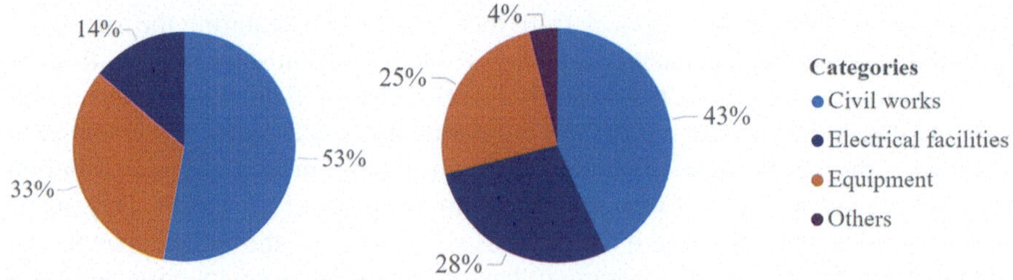

Fig. 3 Distribution of the superficial abstraction category weights by the ERSAR guide (right) and the actual values in AdDP for the Lever Upstream Abstraction (left)

The abstractions and the well were built due to an opportunity that arose when the river dam was built on the Douro River. The drying of the river was an unavoidable stage in the construction of this abstraction. The capital needed to make this infrastructure today, with the river in its natural bed, is much more than it was at the time the dam was built, with the course of the river diverted from its bed.

Not arising from external needs, the draining of the river is not feasible. The abstraction wells are located in the middle of the river, which has a powerful flow, which would make the costs of drying out exorbitant. The alternative would be to build an embankment with sandy material to serve as a working platform and remove it at the end. In any case, the cost associated with constructing the wells is currently very high and consecutively higher than the national panorama.

The equipment category also exhibits a high relative weight. The Lever Upstream Abstraction is responsible for extracting 90% of the water from the Lever North subsystem. As a result, the hydraulic power required is substantial, leading to significantly higher replacement costs compared to those associated with electrical installations.

Taking this characteristic of the abstraction system into account, the forecasted costs for the full reconstruction of the infrastructure are estimated. The weight of each category is then 53% for civil construction, 33% for equipment, and 14% for electrical installations.

The use of the weights for each of the categories (equipment, electrical installations, and civil construction) proposed by ERSAR is an estimate based on average national costs and may therefore differ from the reality of each infrastructure. The ERSAR guide is logically applicable in the case of infrastructures that are close to those average conditions. In the present case, which does not fall into this situation, it is considered appropriate to use the estimate developed.

4.4 Residual Useful Life

The crucial part at this point is determining the condition of the assets that are part of the PS and the Abstraction. Therefore, it is necessary to compare the condition suggested in the company reports with the breakdowns and interventions made over the years and, in case of discrepancies, confront the technical team and arrive at the reason. To determine the theoretical useful life, the average useful life in Portugal suggested by ERSAR (Covas et al., 2018) can be used, as shown in Table 3.

With the condition and the useful theoretical life of the assets defined, it remains to apply Eq. 11 to get the adjusted residual useful life.

Table 3 Assets' technical lifespan

Component type	Asset	Technical lifespan (average in Portugal)
Construction	Buildings and reservoirs	40–50 years
Pipes	Cast iron and steel	40 years
	Concrete	60 years
Equipment	Electric pump groups	20 years
	Valves	15–20 years
	Electronic equipment	15 years
	Control equipment	15 years

5 Results and Discussion

5.1 Current IVI Calculation

The values of the IVI formula, shown in Expression 6, can be found in Appendix B. The results obtained for the Jovim PS in the current year are shown in Table 4.

The overall IVI can be formulated according to Expression 14:

$$IVI_t = \frac{\sum_{i=1}^{N} rc_{it} \times IVI_{it}}{\sum_{i=1}^{N} rc_{it}} \quad (14)$$

In Expression 14, N is the number of categories that compose the infrastructure; rc_{it} is the cost of replacement of the category i ($i = 1,\ldots,N$) in year t, expressed in monetary terms (e.g., in euros); and IVI_{it} is the IVI value of the category i in year t.

The overall IVI can also be obtained directly by substituting the variables of each asset into Eq. 6.

The overall IVI is 0.45, and being in the range [0.4,0.6] means that the condition of the infrastructure is in the acceptable range. A closer analysis shows that although everything seems to conform, at first sight, the IVI of the electrical installations is 0.28, which is considerably lower than the acceptable lower limit. As the relative weight of electrical installations is lower than that of construction and equipment, the overall IVI does not have an accentuated drop. To better understand why this value is so low, it is necessary to analyze the IVI of the assets that compose the electrical facility at the third level and see if there is any specific asset pulling the value down or if it is the group as a whole (see Appendix B). The electrical installation is the same since PS2 was inaugurated. Despite being in good condition as a result of a good maintenance policy, the total technical useful life of electrical installations is relatively short. This information suggests the need for CapEx investments.

Regarding the Lever Upstream Abstraction, the results obtained are shown in Table 5.

Table 4 Jovim PS total and by category IVI

Category	Current value	Replacement cost	IVI
Civil work	810,281.25	1,290,000.00	0.63
Equipment	989,008.81	2,222,944.44	0.44
Electrical facilities	279,655.91	1,004,000.00	0.28
Total	**2,078,945.97**	**4,516,944.44**	**0.46**

Table 5 Lever Upstream Abstraction total and by category IVI

Category	Current value	Replacement cost	IVI
Civil work	2,970,452.74	6,758,637.40	0.44
Equipment	1,806,597.22	5,755,500.00	0.31
Electrical facilities	792,000.00	2,440,000.00	0.32
Total	**5,569,049.96**	**14,954,137.40**	**0.37**

The overall IVI is below the range of acceptable values. A detailed analysis of the categories that make up the infrastructure (see Appendix B) shows that the categories of equipment and electrical facilities are the main causes that lower the overall value of the IVI.

The assets with the highest weight in the equipment category are the pumping groups. Six of the eight pumping groups that make up the infrastructure were installed in 1987, and although they are operating without significant problems, their advanced age must be taken into account. The difficulty of finding spare parts for these groups may occur. Thus, even in acceptable conditions, its residual useful life is low since the age of the asset is a variable of Eq. 11. This consequently translates into a low IVI since the asset's residual life is a variable of Eq. 6. The two remaining pumping groups, although more recent, are beginning to show decreases in performance. All this translates into equipment category IVI of 0.31.

The electrical facilities have assets with already advanced age, and several corrective interventions have occurred in the past. The electrical facilities category presents an IVI of 0.32.

Regarding the civil works category, the IVI value is within the acceptable range, with a value of 0.46. Although the abstraction drains require intervention, these infrastructures have a high total technical useful life. Pipes and buildings are in good condition. Although the construction IVI is reasonable, its weight in the global IVI is low.

5.2 Long-Term IVI Calculation

The planned capital investments for the two infrastructures are presented in Fig. 4. The values of CapEx are reported in million euros.

Regarding the pumping station, the CapEx and the decommissioning percentage inherent to each investment, represented by the β coefficient, are presented in

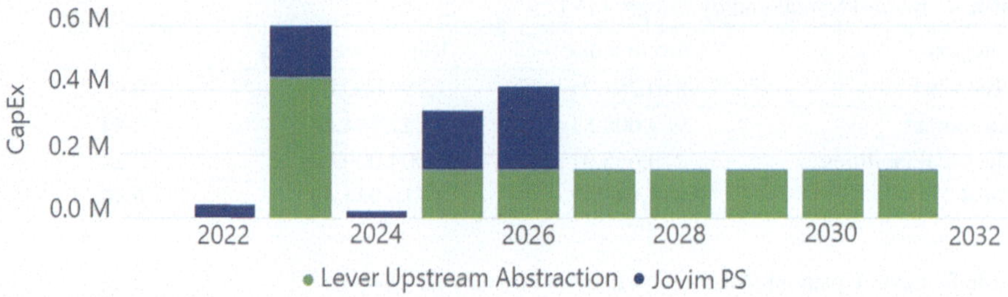

Fig. 4 Lever Upstream Abstraction and Jovim PS planned CapEx

Table 6 Jovim PS's long-term investments

Year	Asset	Beta	CapEx
2022	Electrical boardroom—PS1	0.15	15,000
	Transformer substation	0.07	25,000
2023	Ventilators PS2	1	20,000
	Pumping group no. 1 PS2	0.15	31,250
	Pumping group no. 2 PS2	0.15	31,250
	Pumping group no. 3 PS2	0.15	31,250
	Pumping group no. 4 PS2	0.15	31,250
	Electrical boardroom—PS2	0.15	15,000
2024	Electrical boardroom—PS2	0.15	20,000
2025	Instrumentation, automation, control, and measurement	0.40	60,000
	Pumping group no. 1 PS1	0.15	31,250
	Pumping group no. 2 PS1	0.15	31,250
	Pumping group no. 3 PS1	0.15	31,250
	Pumping group no. 4 PS1	0.15	31,250
2026	Pumping group no. 4 PS1	0.55	150,000
	Instrumentation, automation, control, and measurement	0.40	60.000
	Pumping group no. 2 PS1	0.08	25,000
	Pumping group no. 3 PS1	0.08	25,000

Table 6. All components were easily determined with the help of the technical team and the infrastructure company reports.

The overall IVI value, represented in Fig. 5, varies between 0.45 and 0.4. To reach these values, it is necessary to determine the replacement costs and the residual useful lives along the future years through Eqs. 12 and 13, respectively, where the variables β and CapEx_t are those presented in Table 6. Next, an IVI is estimated, through Eq. 6, for each future year, as a function of the replacement cost and residual useful life. Despite being at the lower limit of the acceptable condition, it is necessary to consider that the electrical installation IVI drops from 0.28 to 0.18, which is not desirable. Although investments are already planned for this group

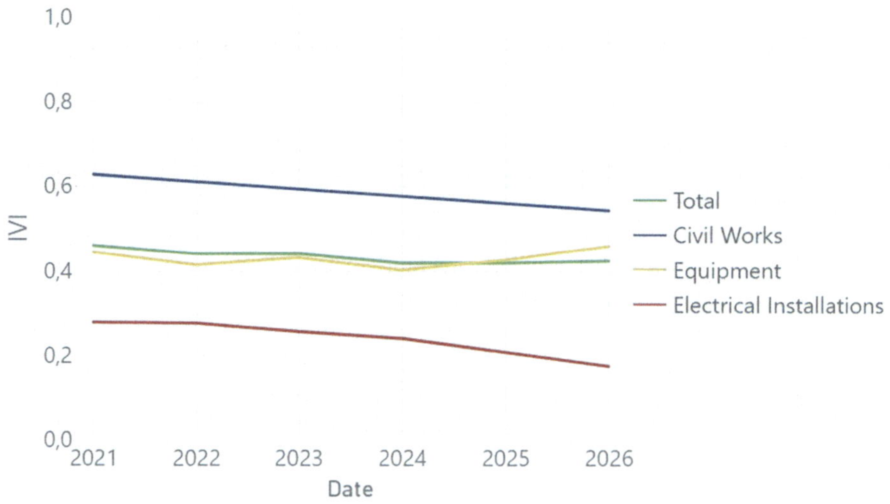

Fig. 5 Total and by category Jovim PS IVI forecast

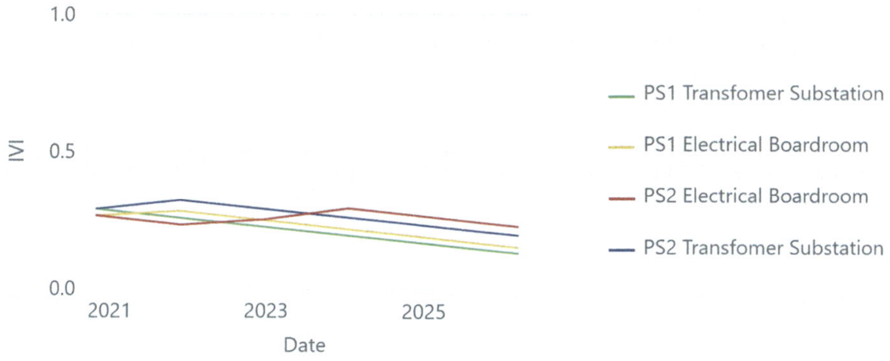

Fig. 6 Jovim PS Electrical installation IVI forecast

during the period, they are insufficient, making it crucial to discuss the need for more CapEx in electrical installations.

To better understand what is happening in the electrical installation, it is necessary to analyze the group in detail. To do this, it is required to break down the information at the asset level, and this information is presented in the chart of Fig. 6.

It is noticeable that there is no particular asset undermining the condition of the group. The poor condition is due to the group's assets being approximately the same age, combined with the fact that the total technical useful life of the electrical installations is less than that of the equipment and construction.

Civil construction will not be subject to investments, the justification being that the building is recent and this category of assets depreciates very slowly. The minimum IVI of 0.54 during the period under analysis corroborates this statement.

The category subject to the most CapEx is equipment. The condition of this group will remain approximately constant over 5 years of analysis, varying only between 0.42 and 0.37.

The β and CapEx for the Lever Upstream Abstraction are shown in Table 7.

The coefficients β for the year 2023 of the Lever Upstream Abstraction present an added difficulty in being obtained. The more easily β is estimated, the more disaggregated the infrastructure is.

By considering the semi-surface abstraction as an asset, there is an aggregation of components with very different total technical useful lives. This point is easily made when weighing up two elemental components that make up a structure like an abstraction well: the concrete and the paint. The lifetime of concrete is around 100 years, and the paint is about 10. If a paint job is planned for the structure and the "paint job" is considered an asset, it is easy to consider β 1 for that investment. However, suppose the asset is the wall composed of concrete and paint components. In that case, estimating the beta becomes more challenging, and it becomes even more so if the asset is a complete abstraction, composed of concrete, paint, drains, reinforcement, etc.

The coefficients obtained for the years 2023–2031 are not error-free. However, any aggregation of components has an associated uncertainty. Any aggregation made will be composed of elements with different total technical useful lives. The teams must choose an aggregation in which they feel comfortable estimating the variables necessary to determine the IVI with maximum accuracy.

The total IVI and by categories can be analyzed in Fig. 7. The total IVI of the abstraction varies between 0.34 and 0.19. Apart from the fact that the IVI at present is already below the lower acceptable limit (0.4), it tends to worsen until 2032.

Regarding civil construction, the investment planned for 2023 allows the IVI of this asset category not to drop much until 2032, decreasing only from 0.46 to 0.41.

These values are within those foreseen for the proper functioning of the facility and therefore do not translate great alarm. However, 0.41 is already close to the lower acceptable limit and should therefore be reviewed and an investment potentially proposed.

The IVI of the equipment shows a drop from 0.31 to 0.19. Even though pumping groups 1–6 have planned rehabilitation for the period under analysis, these are not

Table 7 Lever Upstream Abstraction long-term investments

Year	Asset	Beta	CapEx
2023	Semi-superficial abstraction no. 1	0.15	220,000
	Semi-superficial abstraction no. 2	0.15	220,000
2025	Transformer substation (LPE)	0.75	150,000
2026	Pumping group no. 1	0.4	150,000
2027	Pumping group no. 2	0.4	150,000
2028	Pumping group no. 3	0.4	150,000
2029	Pumping group no. 4	0.4	150,000
2030	Pumping group no. 5	0.4	150,000
2031	Pumping group no. 6	0.4	150,000

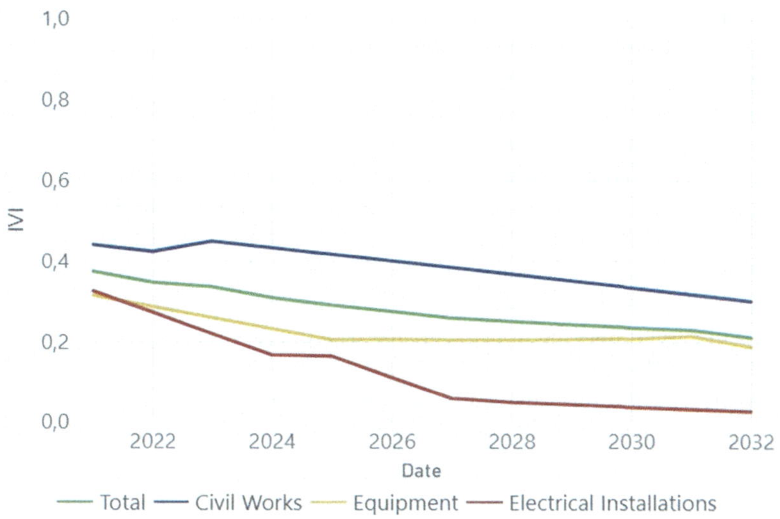

Fig. 7 Total and by category Lever Upstream Abstraction IVI forecast

sufficient investments for the category size, which presents 33% of the weight of the infrastructure.

However, it is in the electrical installations where there are considerable investment shortfalls. The Portuguese average total technical useful life of electrical installations is 15 years. It is expected that the electrical installations, due to their good condition, last longer than 15 years. However, until 2032, only one investment is foreseen in one of the category's assets, which makes the group's remaining assets deteriorate almost wholly. The IVI of 0.32 at present, which is already low, is expected to drop to 0.03 in 2032.

In addition to this indicator helping employees understand in which infrastructure the capital is best used, it also informs which categories need the most investment.

5.3 *Critical Reflections on the IVI*

What makes this indicator impactful is that it requires all teams that deal with assets, such as finance, engineering, and operations, to collaborate. The fact that it requires the contribution of all parties means that the results obtained are equally accepted by all the departments involved, and independent areas of the company can use the information it conveys.

Despite being a powerful indicator, it uses subjective information to obtain it. To determine the β coefficient, subjective information from the technical team is used, which may be misleading. It would be interesting to review the methodology to improve this coefficient's accuracy.

The indicator is valuable but has its shortcomings. The β presented translates the level of deactivation that each investment will bring to the respective assets. It means that investments may have significant weights on the asset structure as a whole but may translate into a low increase in the residual useful life. For example, when comparing painting the exterior of an abstraction in the river with replacing its drains—both interventions having similar costs and weights—the estimated IVI, using the formulation created, is identical in both cases. However, the replacement of the drains leads to a more significant increase in the life of the abstraction than the exterior painting. Thus, there is value in a further study of the indicator to address its limitations. A proposal for future research is to study the feasibility of introducing a variable α that quantifies the improvement—as a function of capital investment—in the residual useful life of an asset, expressed as a percentage in Eqs. 13 and 12.

Despite the limitations, this study represents a satisfactory result for AdDP for the period under analysis. An interesting future work would be to carry out the same survey for other infrastructures.

6 Conclusion

This chapter focuses on an indicator that is already utilized by the water supply industry, though it is still relatively new and has potential for improvement.

According to ERSAR, the indicator reflects the condition of infrastructure and serves as an appropriate tool for setting sustainability targets in infrastructure management (Covas et al., 2018). The indicator is indeed relevant for reliably determining and planning investments as well as for defining targets. This requires extending the point estimate of the indicator and defining an indicator trend as a function of capital investments.

It is essential to bridge the gap between engineering practices in infrastructure management and the financial approach to investment planning. Achieving this requires more rigorous and accurate estimates of equipment valuation, so that the IVI indicator can effectively support strategic investment decision-making.

With a single indicator, the proposed methodology allows to obtain information regarding the present condition of several infrastructures and assets as well as the effect of CapEx investments on the condition of assets. Despite being difficult to obtain, this indicator is easy to understand and analyze, making it a powerful key performance indicator (KPI) in its environment. It allows finance, operations, and engineering teams to be informed of the consequences of decisions made for future investments, which leads to better decision-making.

Obtaining a detailed breakdown of the infrastructure across the four levels (i.e., infrastructure, group, sub-group, assets), accurately assessing the condition of the assets, and appropriately weighing their values are all critical steps in the process.

The case study was carried out on a typical water supply system, made possible through the collaboration of various departments of the company and the availability of credible and organized information. However, it can also be adapted to other contexts and industries as long as proper precautions are taken.

The investment distribution applied is the one proposed by AdDP. The objective of this case study is to verify its impact on the infrastructure's condition using the proposed distribution. However, it would be relevant for future research to confirm if an increase in the global IVI is possible by allocating capital differently.

Appendix A Jovim PS1 and Lever Upstream Abstraction Infrastructures Breakdown (Tables 8 and 9)

Table 8 Breakdown in the first four levels of Lever Upstream Abstraction

First level	Second level	Third level	Fourth level
Lever upstream	Equipment	Production	Generator set
Abstraction		General support systems	Travelling crane
		Instrumentation, automation, control and measurement	Instrumentation, automation, control and measurement equipment
		Hydraulic shock protection	Compressed air system n°1 (2 compressors for air vessels)
			Compressed air system n°2 (compressor for wells)
			AVAC system - (TS) - air vessels
		Lifting	Pumping group no. 1 (direct start)
			Pumping group no. 2 (direct start)
			Pumping group no. 3 (direct start)
			Pumping group no. 4 (direct start)
			Pumping group no. 5 (direct start)
			Pumping group no. 6 (direct start)
			Pumping group no. 7 (variable speed drive)
			Pumping group no. 8 (variable speed drive)
	Electrical facilities	Distribution	Transformer substation (TS)—global
			Transformer substation (TS)—Air vessels
			Power substation
			Electrical boardroom
	Civil works	Civil works	Semi surface abstraction no. 1
			Semi surface abstraction no. 2
			Pipe system
			Building
			Collector well

Table 9 Breakdown in the first four levels of Jovim PS1

First level	Second level	Third level	Fourth level
Jovim PS1	Equipment	General support systems	Travelling crane
		Hydraulic circuit	Hydraulic circuit
		Instrumentation, automation, control and measurement	Instrumentation, automation, control and measurement equipment
		Lifting	Pumping group no. 1
			Pumping group no. 2
			Pumping group no. 3
			Pumping group no. 4
			Ventilators
		Water hammer protection	Compressed air system
			Air vessel 30m3, PN10 - RSP 2 (PS1) I RAC
			Air vessel 30m3, PN10 - RSP 1 (PS1) I RAC
		Sampling	Sampling system
	Electrical facilities	Distribution	Electrical boardroom
			Transformer substation
		Safety systems	Security system
	Civil works	Civil works	Building
			Pipe system

Appendix B Jovim PS and Lever Upstream Abstraction Variables for Current IVI Calculation (Table 10 and 11)

Table 10 Variables for Jovim PS current IVI calculation

Designation	Age	Condition	Theoretical total useful life	Replacement Cost
Civil works PS Jovim	22	1.2	50	645,000.00 €
Pipe system Jovim	22	1.2	40	645,000.00 €
Electrical boardroom - PS1 old	22	2.1	15	250,000.00 €
Electrical boardroom - PS2 new	22	1.9	15	250,000.00 €
Transformer substation PS1 old	22	1.7	15	224,000.00 €
Transformer substation PS2 new	22	1.6	15	280,000.00 €
Emergency generator PS2	21	1.5	20	20,000.00 €
Intrumentação, automação, controlo e medição	9	1	15	100,000.00 €
Pumping group n. 1 PS1	19	1.4	20	180,000.00 €
Pumping group n. 2 PS1	19	1.4	20	180,000.00 €
Pumping group n. 3 PS1	19	1.4	20	180,000.00 €
Pumping group n. 4 PS1	19	1.4	20	180,000.00 €
Pumping group n. 1 PS2	21	1.3	20	285,312.39 €
Pumping group n. 2 PS2	21	2.1	20	285,312.39 €
Pumping group n. 3 PS2	21	1.3	20	285,312.39 €
Pumping group n. 4 PS2	21	2.1	20	285,312.39 €
Ventilators PS1	22	1.7	15	20,000.00 €
Ventilators PS2	22	1.3	15	20,000.00 €
Air vessel 30m3, PN10 - RSP 2 (PS1) I RAC	9	1.7	40	56,388.62 €
Air vessel 30m3, PN10 - RSP 1 (PS1) I RAC	9	1.7	40	56,388.62 €
Air vessel Remetal, 35 m3 - RSP 3 (PS2) I RAC	9	1.6	40	88,917.66 €

Table 11 Variables for Lever Upstream Abstraction current IVI calculation

Designation	Age	Condition	Theoretical total useful life	Replacement cost
Pipe system	34	1	40	100,000.00 €
Civil works	34	1	50	300,000.00 €
Transformer substation (TS) - global	8	3	15	160,000.00 €
Transformer substation (TS) - air vessels	8	3	15	200,000.00 €
Collector well	34	4.5	100	316,453.40 €
Instrumentation, automation e control	14	2	15	210,000.00 €
HVAC system - (TS) - air vessels	14	1	15	55,000.00 €
2 air vessels	34	1	50	230,000.00 €
Travelling crane	34	1	40	100,000.00 €
Emergency generator - L. Montante	14	2	20	50,000.00 €
Power substation	14	3	15	2,000,000.00 €
Electrical boardroom	14	3	15	80,000.00 €
Compressed air system no. 1 L. Montante (2 compressor for air vessels)	14	1	15	6500.00 €
Compressed air system no. 2 L. Montante (compressor for wells)	14	2	15	4000.00 €
Pumping group no. 1 (direct start)	34	3	20	600,000.00 €
Pumping group no. 2 (direct start)	34	2	20	600,000.00 €
Pumping group no. 3 (direct start)	34	2	20	600,000.00 €
Pumping group no. 4 (direct start)	34	2	20	600,000.00 €
Pumping group no. 5 (direct start)	34	2	20	600,000.00 €
Pumping group no. 6 (direct start)	34	2	20	600,000.00 €
Pumping group no. 7 (variable speed drive)	14	2	20	750,000.00 €
Pumping group no. 8 (variable speed drive)	14	2	20	750,000.00 €
Semi surface abstraction no. 1	34	4.5	100	3,021,092.00 €
Semi surface abstraction no. 2	34	4.5	100	3,021,092.00 €

References

Alegre, H. (2008). Water infrastructure asset management. In *Research program, series «thesis and research programmes»*. LNEC.

Alegre, H., Vitorino, D., & Coelho, S. (2014). Infrastructure value index: A powerful modelling tool for combined long-term planning of linear and vertical assets. *Procedia Engineering, 89*, 1428–1436.

Bhagwan, J. (2012). A compendium of best practices in asset management. *Water Intelligence Online, 11*. https://doi.org/10.2166/9781780404622

Carollo. (2015). *Project memorandum 2.4—Condition assessment—Final draft, public works integrated master plan—Water.*
Covas D (2018) *Technical guide no. 23—custos de construção de infraestruturas associadas ao ciclo urbano da água.*
Covas, D., Cabral, M., Pinheiro, A., Marchionni, V., Antunes, S., Lopes, N., Mamouros, L., & Brôco, N. (2018). *Custos de construção de infraestruturas associadas ao ciclo urbano da água—Guia Técnico n° 23.* Entidade Reguladora de Águas e Resíduos.
H2Opt. (2020). *Inspeção e avaliação da condição funcional e do risco dos reservatórios da AdDP.*
Howard J, Dixon J, Comrie J (2015) Australian Infrastructure Financial Management Manual (AIFMM). Institute of Public Works Engineering Australaisa (IPWEA).
Salvado, F., Marques de Almeida, N., & Vale e Azevedo, Á. (2020). Future-proofing and monitoring capital investments needs throughout the life cycle of building projects. *Sustainable Cities and Society, 59*, 102159. https://doi.org/10.1016/j.scs.2020.102159. https://www.sciencedirect.com/science/article/pii/S2210670720301463
USEPA (2005) USEPA advanced asset management workshop. https://19january2017snapshot.epa.gov/sustainable-water-infrastructure/advanced-asset-managementworkshop-materials_.html
Vieira J, Almeida N, Silva JG, Castro C, Lima B, Trindade M (2020a) Using indicators to deal with uncertainty in the capital renewals planning of an industrial water supply system: Testing the infrastructure value index. Engineering assets and public infrastructures in the age of digitalization.
Vieira, J., Cabral, M., Almeida, N., Silva, J. G., & Covas, D. (2020b). Novel methodology for efficiency-based long-term investment planning in water infrastructures. *Structure and Infrastructure Engineering, 16*(12), 1654–1668.

Index

A
Agriculture, 4, 14, 49, 72, 92, 130, 176, 189, 234, 254, 275
Aquaculture, 7, 92–105, 107–110
Aquaponics, 7, 103–110
Asset management, 66, 74, 77, 80, 82–84, 304–309
Automatic tracking, 118, 124, 125

B
Bangladesh, 8, 101, 168–182, 232–247

C
Central Asia, 7, 47–67, 78
Climate change, 1–9, 14–19, 21, 23, 25, 28, 34–41, 47–50, 52, 54, 55, 63–66, 74, 78, 85, 92–96, 110, 130, 162, 168–182, 189–191, 194, 234, 236, 237, 277–278, 294, 299
Crop production, 15, 17, 39, 49, 98–100, 105, 133, 180, 240, 272, 273, 293

D
Dam safety, 7, 50, 57–58, 67
Drip irrigation, 8, 50, 93, 131, 134, 137, 143–144, 149, 156–158, 163, 191, 192, 204, 211, 272, 277, 279, 281, 287, 290, 293, 295, 296

E
Efficient irrigation systems, 93, 253, 267, 285, 299
Energy efficiency, 7, 50–51, 66, 120
Environmental risk, 15

F
Financing mechanisms, 9, 58, 162, 163, 253, 258, 264, 267
Funding programs, 9, 264, 267

G
Governing agencies, 172

H
Health, 2, 4, 7, 8, 15, 23, 36, 38, 48, 72, 74, 81, 88, 93, 94, 107, 108, 162, 170–174, 177–179, 181, 182, 188, 189, 227, 232–247, 293, 296

I
Infrastructures, 5, 7, 9, 36, 38, 41, 50, 51, 53, 55, 58, 60, 63, 64, 66, 73, 74, 76, 78–83, 85, 86, 130, 153, 156, 168–171, 173–178, 181, 182, 237, 243, 254, 258, 262, 263, 267, 273, 278–280, 291, 292, 295, 304–329
Integrated farming system, 92, 96

J. C. Correia Leitão et al. (eds.), *Sustainable Water Management and Irrigation Systems*, Natural Resource Management and Policy 62,
https://doi.org/10.1007/978-3-031-93587-9

Investment planning, 9, 304–327
Irrigation, 4, 14, 48, 75, 93, 130, 179, 190, 232, 252, 272
Irrigation systems, 15, 16, 18, 19, 38, 41, 52, 98, 130–132, 134–136, 152, 155, 156, 158, 159, 163, 170, 172, 179, 190, 191, 193–197, 211, 226, 232, 233, 236, 237

M
Micro-irrigation, 4, 7, 8, 14–41, 130–164, 188–227
Micro-irrigation systems, 8, 41, 134, 135, 227, 253
More crop per drop, 92, 93, 97, 110
Morocco, 5, 8, 9, 130–164, 253–258, 264, 265, 267

R
Rain sensors, 118, 121, 124
Renewable energy, 4, 5, 109, 117–120, 124, 125, 236, 237

S
Self-regulating, Low Energy, Clay based Irrigation (SLECI) technology, 4, 8, 130–164, 191, 195–197, 201, 203–205, 208–221, 223–225, 227, 253–261, 264, 267
Solar panels, 118–121, 124
Sustainability, 3, 5, 7, 8, 14–21, 23–25, 34, 35, 38–41, 49, 63, 74, 80, 82, 84, 86, 88, 96, 97, 99, 104, 110, 117, 119, 124, 130, 131, 154, 159, 163, 168, 170, 175, 179, 188–190, 204, 227, 234, 236, 237, 263, 273, 274, 280, 294, 297, 300, 304
Sustainable development, 7, 38, 40, 47–67, 72, 84, 92, 93, 97, 118, 119, 125, 130, 177, 181, 182, 190, 252, 263, 267, 299

T
Technology, 1, 15, 53, 79, 96, 119, 130, 172, 188, 236, 252, 278
Teesta Barrage Irrigation Project (TBIP), 8, 232–247

U
Urban Bangladesh, 168–182

W
Water economy, 2, 3, 5, 253, 254
Water efficiency, 8, 34, 39, 180–182, 193, 252, 254, 267, 274, 287, 288, 290
Water infrastructures, 3, 9, 52, 57–63, 79, 82, 83, 85, 169, 170, 173, 176, 178, 179, 181, 278
Water management, 1–9, 14–41, 58, 61–63, 65, 72, 74, 78, 83, 84, 87, 92–110, 168–182, 188–227, 234, 254, 259, 260, 267, 272–301
Water productivity, 14, 23, 40, 86, 93, 97–100, 103, 108, 252, 263, 264, 285
Water quality, 3–5, 49, 53, 54, 76, 79, 81, 82, 85, 87, 94–96, 99, 106–107, 110, 118, 168, 172, 175, 176, 181, 223, 234, 236, 237, 294, 296
Water resources, 1–5, 7, 8, 14–19, 21, 23–25, 28, 36, 38–41, 52, 60, 64, 65, 72, 78, 81, 84, 85, 92–98, 100, 107, 110, 119, 130, 152, 168–172, 174, 177–182, 194, 232, 252, 253, 256, 258, 266, 273, 274, 276–279, 281, 286, 292, 293, 295, 298, 299
Water-saving irrigation technology, 130, 154, 155, 252, 267
Water sector, 50–51, 63, 65–67, 79, 80, 85, 304–327
Water systems, 7, 14, 16, 50, 63, 65, 72–88, 168, 170–172, 175, 181, 317
Water utilities, 51, 74, 77–88

The manufacturer's authorised representative in the EU is Springer
Nature Customer Service Centre GmbH, Europaplatz 3, 69115 Heidelberg,
Germany. If you have any concerns regarding our products, please
contact ProductSafety@springernature.com

Printed and bound by CPI Group (UK) Ltd, Croydon, CR0 4YY
01/10/2026
02225971-0012